126 Dynamical Systems and Semisimple Groups

The theory of dynamical systems can be described as the study of the global properties of groups of transformations. The historical roots of the subject lie in celestial and statistical mechanics, for which the group is the time parameter. In some of its recent developments, the theory is concerned with the dynamics of more general, bigger groups than the additive group of real numbers, particularly semisimple Lie groups and their discrete subgroups. Some of the most fundamental discoveries in this area are due to the work of G. A. Margulis and R. Zimmer. This book comprises a systematic, self-contained introduction to the Margulis–Zimmer theory and provides an entry into current research.

Taking as prerequisites only the standard first-year graduate courses in mathematics, the author develops in a detailed and self-contained way the main results on Lie groups, Lie algebras, and semisimple groups, including basic facts normally covered in first courses on manifolds and Lie groups plus topics such as integration of infinitesimal actions of Lie groups. He then derives the basic structure theorems for the real semisimple Lie groups, such as the Cartan and Iwasawa decompositions, and gives an extensive exposition of the general facts and concepts from topological dynamics and ergodic theory, including detailed proofs of the multiplicative ergodic theorem and Moore's ergodicity theorem.

This book should appeal to anyone interested in Lie theory, differential geometry, and dynamical systems.

T0297141

CAMBRIDGE TRACTS IN MATHEMATICS

General Editors

B. BOLLOBAS, F. KIRWAN, P. SARNAK,
C. T. C. WALL

126 Dynamical Systems and Semisimple Groups

RENATO FERES
Washington University in St. Louis

Dynamical Systems and Semisimple Groups: An Introduction

CAMBRIDGE
UNIVERSITY PRESS

CAMBRIDGE UNIVERSITY PRESS
Cambridge, New York, Melbourne, Madrid, Cape Town, Singapore,
São Paulo, Delhi, Dubai, Tokyo

Cambridge University Press
The Edinburgh Building, Cambridge CB2 8RU, UK

Published in the United States of America by Cambridge University Press, New York

www.cambridge.org
Information on this title: www.cambridge.org/9780521142168

First published 1998
This digitally printed version 2010

A catalogue record for this publication is available from the British Library

ISBN 978-0-521-59162-1 Hardback
ISBN 978-0-521-14216-8 Paperback

To my parents

Contents

Preface

An action of a group G on a set M is a homomorphism, $g \to \Phi_g$, from G into the group of invertible transformations of M. Traditionally, actions of $\mathbb{R}$ and $\mathbb{Z}$ have been the main objects of concern of the theory of dynamical systems. For example, if X is a smooth vector field on a compact manifold M, its flow defines an action of $\mathbb{R}$ on M. One thinks of the group as parametrizing "time," so that the orbit

$$G \cdot x := \{ \Phi_g(x) \mid g \in G \}$$

describes the evolution of the system, starting from an initial state represented by $x \in M$.

For actions of a more general group G, the "time evolutions" associated to its various one-parameter subgroups are "interlocked" according to the algebraic structure of G. Thus, suppose that instead of a single smooth vector field on M we have a family of them, $X_1, \ldots, X_m$, whose Lie brackets satisfy

$$[X_i, X_j] = \sum_{k=1}^{m} a_{ij}^k X_k,$$

where the a_{ij}^k are constants. This means that these fields span a finite-dimensional Lie algebra, associated to a Lie group G. By Lie's second fundamental theorem (essentially theorem 3.9.8), these fields integrate to an action of (the universal covering group of) G. If, for example, the constants a_{ij}^k vanish, G is an abelian group, and the flows of $X_1, \ldots, X_m$ may be thought to define "noninteracting" evolutions.

On the opposite extreme to abelian groups are the semisimple groups. A – not very revealing – definition of semisimplicity is that the matrix (c_{ij}), with entries given by

$$c_{ij} = \sum_{k=1}^{m} \sum_{l=1}^{m} a_{il}^k a_{kj}^l,$$

is nonsingular. The various subgroups of a semisimple group are tightly inter-woven, and one might expect the dynamical properties of actions of G to be accordingly constrained.

The actions considered in this book will, for the most part, be assumed to possess a finite invariant measure. Invariance of a measure μ means that for each measurable subset $A \subset M$ we have $\mu(\Phi_g(A)) = \mu(A)$ for each $g \in G$. The existence of a finite invariant measure forces upon the long-term evolution of the system a kind of statistical regularity, given by the ergodic theorems of chapters 8 and 9, most of which are for actions of $\mathbb{R}$ or $\mathbb{Z}$. This is the basic setup of ergodic theory. Actions of $\mathbb{R}$ and $\mathbb{Z}$ very commonly admit (possibly singular) invariant measures – this is always the case if the action is continuous and M is compact. For nonabelian groups, however, this is a strong requirement.

One can obtain much useful information about a group action on M (assuming that the action is differentiable) by studying its linearization along the orbits. A good understanding of this linearization can yield information about the global properties of the system. (For $\mathbb{Z}$-actions, a modest attempt to justify this claim will be made at the end of chapter 9, in a brief discussion on Pesin theory.) One of the main results of the book can be formulated as follows: If the $\mathbb{Z}$-action is part of a differentiable action by a noncompact semisimple Lie group G on an n-dimensional manifold M, preserving a finite measure μ, then it is possible to give a very precise description of the linearization, along almost every orbit (relative to μ), in terms of the representations of G in dimension n. The key result behind this vague assertion is Zimmer's cocycle superrigidity theorem, which will be studied in chapter 10.

The ergodic theory of actions of semisimple Lie groups and their discrete subgroups has grown in the past decade into a large and very active chapter of the general theory of dynamical systems. It is also a subject with deep foundations; most notably, the work by G. A. Margulis concerning rigidity and arithmeticity of lattices in semisimple groups and the work of R. Zimmer, in particular his cocycle superrigidity theorem. The main purpose of this book is to serve as a relatively gentle introduction to the rigidity theorems of Margulis and Zimmer.

Passing from some knowledge of the linearization of the system along orbits to an understanding of its global topological structure is a hard and wide-open problem. R. Zimmer conjectured in [37] that the actions of Lie groups and lattices considered in this book should be, in some sense, "classifiable" on the basis of a few classes of well-understood examples. This classification program gained momentum with the introduction of ideas from hyperbolic dynamics and the theory of Anosov diffeomorphisms, by S. Hurder, A. Katok, and J. Lewis, about six or seven years ago, and continues today with great vitality. In spite of all the recent progress, it is clear that this is not an area of research approaching exhaustion any time soon.

Although I make no attempt to survey the current research activity, the book does contain a few new ideas. The proof of the cocycle superrigidity theorem given in chapter 10 is new, and is due to F. Labourie and myself [11]. The presentation is also different from that in [36] in that it uses throughout a differential-geometric language that some readers may find more natural, or at least more congenial, than the language of cocycles over a group action. I hope that the experts will see some novelty here. In any event, the book was written having in mind primarily the nonexpert, especially the graduate student interested in Lie theory and dynamical systems.

The reader is assumed to have a good working knowledge of measure and integration and the basic theory of differentiable manifolds. Even though first courses on manifolds usually provide some acquaintance with Lie groups and Lie algebras, I have chosen to develop this subject in detail and from the beginning, up to those results that are needed for chapters 8, 9, and 10.

The text, in effect, integrates two courses in one. First, it contains an introduction to part of the modern theory of dynamical systems. This comprises chapters 1, 2, 8, and 9. The dynamics "subcourse" culminates with a detailed proof of the multiplicative ergodic theorem of Oseledec and a hurried discussion of Pesin theory. Taken in isolation, this is necessarily a lopsided account of dynamics. For an even-handed and thorough introduction to dynamical systems, the reader cannot presently do better than to go to [16]. The second "subcourse" is on Lie groups, Lie algebras, and semisimple groups, and comprises chapters 3, 5, and 7. The reader who wishes to pursue this topic further will have no difficulty in continuing on with, say [17], from the point where chapter 7 ends.

There is a large degree of independence among chapters 1 through 9, although almost everything that is developed in them is put to use in chapter 10. The main exceptions are chapter 6, which relies on facts about algebraic actions discussed in chapter 4; section 9.3, which uses some of the language introduced in section 6.1; and section 8.3 on Moore's ergodicity theorem, which uses some of the structure theorems for semisimple Lie groups. Chapter 4 is independent of the first three, although it is also the least self-contained in the book. It consists of a very pedestrian introduction to algebraic geometry and algebraic actions.

The shortest path to arrive at the main results of the book, namely, theorems 10.4.1, 10.6.1, and 10.6.2, is to read chapters 4, 6, and 7, sections 8.2 and 8.3, and the first sections of 10, referring to the earlier chapters for the definition of some occasional unfamiliar term. All the other chapters are there to provide a dynamical systems "context" for the results of chapter 10.

It seemed to me appropriate from the point of view of the "narrative" to place chapter 6, on geometric structures, immediately after a discussion on the classical groups, even though the results from that chapter are not needed until chapter 10. The reader should keep this in mind in case the discussion in chapter 6 seems at first too formal and unmotivated.

The exercises are an integral part of the text. Their main purpose is to expand or illustrate an idea under discussion. Occasionally, an exercise may also be referred to in proofs. They should always be read, even if not always worked on.

The book began as a short series of lectures given at Penn State University in 1996, and was later expanded after a one-semester course I taught at Washington University, in St. Louis, in 1997. I am grateful to Anatoly Katok for inviting me to give the lectures at Penn State and for bringing my notes to the attention of Cambridge University Press. Had I pursued the subject just a little further the impact of his own work would also become apparent in the text. I am also especially indebted to Robert Zimmer, from whom I learned much of this subject firsthand; to François Labourie, my collaborator in [11], the reference on which much of chapter 10 is based; to Scot Adams, for many enlightening conversations and, in particular, for explaining to me the proof of Moore's ergodicity theorem given in chapter 8; and to Mohan Kumar, Mark de Cataldo, and Vladimir Masek, for lending their expertise on algebraic geometry. Finally, I would like to thank Peter Lampe, Michelle Penner, Holly Lowy, Lawrence Roberts, Mark de Cataldo, and Meeyoung Kim for suggesting many improvements and corrections to the original manuscript. Thanks to them, many – but certainly not all – embarrassing mistakes, obscure phrases, wrong signs, and barbarisms of the first draft will not appear in print.

Renato Feres

Washington University, St. Louis
August, 1997

1

Topological Dynamics

The theory of dynamical systems, loosely speaking, studies those properties of group actions that are asymptotic in nature, that is, that become apparent as we "go to infinity" in the group. We call a set X equipped with an action of a group G a *dynamical system* with group G or, alternatively, a *G-space*.

After introducing a few notions that apply to general group actions, we focus our attention on some of the basic properties of topological G-spaces. Other aspects of dynamical systems, relating, for example, to their measurable or smooth properties, will be discussed in later chapters.

1.1 G-Spaces

Let G be a group and X a set. A *G-action* on X is a map $\Phi : G \times X \to X$ that satisfies the following two properties:

1. $\Phi(e, x) = x$ for all $x \in X$, where e is the identity of G.
2. $\Phi(g_2, \Phi(g_1, x)) = \Phi(g_2 g_1, x)$ for all $g_1, g_2 \in G$ and $x \in X$.

For each $g \in G$, let $\Phi_g : X \to X$ be defined by $\Phi_g(x) := \Phi(g, x)$. Then Φ_g is a bijection from X onto itself, with inverse $\Phi_{g^{-1}}$, and the map $g \mapsto \Phi_g$ from G into the group of bijective self-maps of X is a group homomorphism. We often write $g \cdot x$ or $g(x)$, or simply gx, instead of $\Phi(g, x)$. The definition of G-action just given is usually called a *left action* of G. By a *right action* of G on M we mean a map $\Phi : M \times G \to M$ such that property 2 is replaced with

$$\Phi(\Phi(x, g_1), g_2) = \Phi(x, g_1 g_2).$$

For each $x \in X$, we define the *orbit* of x by

$$Gx := \{\Phi_g(x) \mid g \in G\}.$$

1

The orbits of a G-action partition X into disjoint sets; namely, the Gx are the equivalence classes of the relation

$$x \sim y \text{ if and only if there exists } g \in G \text{ such that } x = gy.$$

The *orbit space* is the set of equivalence classes, denoted $G \backslash X$. The action Φ is called *transitive* if the G-space has only one orbit, that is, $X = Gx$ for some x.

Typically, the G-action will leave invariant, or preserve, some structure on X such as a topology, a measurable structure, a smooth manifold structure, or an algebraic variety structure. Of course, these structures are not independent. For example, when studying a smooth group action on a compact Riemannian manifold whose volume form is invariant under the action, one could find it useful at times to focus, say, on the underlying measure-space structure determined by the Borel-measurable sets and the measure obtained by integrating the volume form. The group G, however, will always be regarded here as being, at least, a *topological group*, that is, a Hausdorff space that is also an abstract group for which multiplication and inversion are defined by continuous maps.

A *topological G-space* consists of a topological space X and a continuous action Φ of G on X. In this case each Φ_g, $g \in G$, is a homeomorphism of X. Some of the basic properties of topological G-spaces are discussed in this chapter.

A *smooth G-space* consists of a smooth (C^∞) manifold X and a smooth action Φ of a Lie group G. In this case, each Φ_g is a C^∞ diffeomorphism of X. We discuss smooth actions in chapter 3.

A *measurable G-space* consists of a measurable space $(X, \mathcal{B})$, where $\mathcal{B}$ is a σ-algebra of subsets of X, and a measurable action $\Phi : G \times X \to X$. We will often be interested in actions that preserve a finite measure μ on $(X, \mathcal{B})$. In that case, μ can be normalized so that $\mu(X) = 1$, and μ is then called a *probability measure*. The study of group actions on measurable spaces is the subject of *ergodic theory*, to which we return later, beginning in the next chapter.

An *algebraic G-space* consists of an algebraic variety X defined over some field k (which, in this book, will almost always be $\mathbb{R}$ or $\mathbb{C}$) and an algebraic group G and an algebraic action Φ, both defined over k. Algebraic actions will play an important role in some of the results described in this book. Definitions and general properties concerning them are discussed in chapter 4.

The remainder of the chapter concentrates on the elementary properties of *topological dynamical systems*, that is, topological G-spaces.

Let H be a closed subgroup of G. Then the coset space

$$G/H = \{gH \mid g \in G\}$$

has the quotient topology induced by the natural projection $\pi : G \to G/H$, $\pi(g) = gH$; namely, the open subsets of G/H are $\pi(U) = \{gH \mid g \in U\}$ for

all open sets $U \subset G$. With respect to the quotient topology, π is continuous
and open and G/H is a Hausdorff space.

A *discrete subgroup* of G is a subgroup that is a discrete subset in the topology
of G. If G is connected and locally arcwise connected and H is a closed
subgroup, it can be shown that H is a discrete subgroup if and only if π is a
covering map. If H is discrete and G/H is compact, we say that H is a *uniform
lattice* of G.

The *kernel* of an action Φ, denoted $\mathrm{Ker}(\Phi)$, is the kernel of the homomorphism $g \mapsto \Phi_g$, which is a normal subgroup of G. When $\mathrm{Ker}(\Phi)$ is trivial, the
action is said to be *effective*. If the action is not effective, Φ induces an effective
action of $G/\mathrm{Ker}(\Phi)$ on X. The action is called *locally effective* if $\mathrm{Ker}(\Phi)$ is a
discrete subgroup of G.

For each $x \in X$, the *isotropy group of* x is defined by

$$G_x := \{g \in G \mid gx = x\}.$$

G_x is a subgroup of G and it is immediate that $G_{gx} = gG_x g^{-1}$ for each $g \in G$
and $x \in X$. Moreover, $\mathrm{Ker}(\Phi) = \bigcap_{x \in X} G_x$. If $G_x = \{e\}$ for all $x \in G$, we
say that the G-action is *free*. The action is called *locally free* if G_x is a discrete
subgroup of G for all x in X.

Recall that a topological space X is said to be T_1 if each point $x \in X$ is closed.
It is an easy consequence of the definitions that, whenever X is a T_1 G-space,
each isotropy group G_x as well as the kernel of Φ are closed subgroups of G and
that $G/\mathrm{Ker}(\Phi)$ is a topological group in a natural way. Moreover, the induced
(effective) action of $G/\mathrm{Ker}(\Phi)$ makes X a topological $G/\mathrm{Ker}(\Phi)$-space.

1.2 The Orbit Space

Until we impose any further requirements G will be a locally compact second
countable topological group and X a complete second countable metrizable G-
space. We give $G\backslash X$ the quotient topology induced by the natural projection
that to each $x \in X$ associates its orbit.

It will be apparent from some of the examples described later that the orbit
space $G\backslash X$ can easily fail to have good separation properties, due to the existence of orbits that wander about in X in a complicated way. This is not the
case, however, when Φ is a *proper action*. By definition, Φ is a proper action
if for each $x, y \in X$ there exist neighborhoods U of x and V of y such that

$$\{g \in G \mid V \cap gU \neq \emptyset\}$$

is relatively compact. Clearly, the action is proper whenever G is compact.

Exercise 1.2.1 Show that the orbit space $G \backslash X$ of a proper action is Hausdorff. In particular, each orbit Gx is closed in X. Also show that, for each $x \in X$, the map $\phi_x : G/G_x \to Gx$, $gG_x \mapsto gx$, is a homeomorphism.

A somewhat more complicated situation, but still rather simple from the viewpoint of the general theory of dynamical systems, corresponds to the case in which the σ-algebra $\mathcal{B}$ of Borel sets, that is, the σ-algebra generated by the open sets in $G \backslash X$, is *countably separating*. This means that there is a sequence $B_i \in \mathcal{B}$ such that for each pair of points in X one can find a B_i that contains exactly one of the two points. In this case, the G-action will be called *tame*. Notice that a proper action is tame. In fact, the quotient topology of $G \backslash X$ is second countable, since X is second countable, and Hausdorff, so points can already be separated by open sets.

The next theorem gives a useful characterization of tame actions. It is taken from [36], where a tame action is called *smooth*. The result is due to Glimm and Effros. The orbit Gx of a topological G-space X is *locally closed* if it is open in its closure $\overline{Gx} \subset X$.

Theorem 1.2.2 Suppose that Φ is a continuous action of a locally compact second countable group G on a complete second countable metrizable space X. Then the following are equivalent:

1. All orbits are locally closed.
2. The action is tame.
3. For every $x \in X$, the natural map $G/G_x \to Gx$ is a homeomorphism, where Gx has the relative topology as a subset of X.

Proof. The implication $2 \Rightarrow 1$ is the hardest to prove and will not be discussed here. We refer the reader to [36, 2.1.14] for a proof. We begin with the assertion $1 \Rightarrow 2$. Since the topology of X has a countable basis and the projection $\pi : X \to G \backslash X$ is open, the topology of $G \backslash X$ also has a countable basis. To prove that the Borel-measurable structure is countably separating it suffices to show that $G \backslash X$ is a T_0-space, that is, that we can separate any two points by an open set that contains only one of the points. Let $x, y \in X$. If $\pi(x)$ and $\pi(x)$ are not separated by an open set, $Gy \subset \overline{Gx}$ and $Gx \subset \overline{Gy}$. Therefore Gy is dense in $\overline{Gx}$. But by assumption Gx is open in its closure, so $Gy \cap Gx \neq \emptyset$. This implies that $\pi(x) = \pi(y)$.

We now show that 3 and 1 are equivalent. We may assume without loss of generality that Gx is dense in X. If this is not the case, simply let X denote the closure of that orbit. We begin with $3 \Rightarrow 1$ and assume that $G/G_x \to Gx$ is a homeomorphism. Then Gx with the subspace topology satisfies the Baire category theorem, because G/G_x satisfies it. (G is locally compact, hence a Baire space. It follows that the quotient is also a Baire space.) Now, G is σ-compact, being second countable and locally compact. Therefore, by Baire's theorem, some compact set $A \subset Gx$ contains a nonempty open set, that is, for some nonempty open set $U \subset X$, $U \subset U \cap \overline{Gx} \subset A$. Thus $Gx = GU$, which is open.

For the converse, suppose that Gx is open in X. Notice that $G/G_x \to Gx$ is continuous, so it suffices to prove that it is also open. We call $U \subset G$ a *symmetric set* if $g \in U$ implies $g^{-1} \in U$. Any open neighborhood V of e contains a symmetric neighborhood: $V \cap V^{-1}$. We claim that it suffices to show that for any compact symmetric set $U \subset G$ whose interior is an open neighborhood of $e \in G$, Ux contains a nonempty open set. Namely, let N be any neighborhood of e and choose a compact symmetric U with $U^2 \subset N$. If Ux contains a neighborhood of some ux, $u \in U$, then $u^{-1}Ux$ contains a neighborhood of x, and hence so does Nx. Therefore $G/G_x \to Gx$ is an open map. To show that Ux contains an open set, choose a countable dense set $\{g_i\} \subset G$. Then $Gx = \bigcup_i g_i Ux$, a union of compact sets, so by the Baire category theorem, we have that one $g_i Ux$ contains an open set, and hence so does Ux. □

Exercise 1.2.3 Show that the $\mathbb{R}$-action on $\mathbb{R}^2$ defined by $\Phi_t(x, y) := (e^t x, e^{-t} y)$, where $t \in \mathbb{R}$ and $(x, y) \in \mathbb{R}^2$, is tame but not proper. Verify that the orbits are locally closed, but the orbit space is not Hausdorff.

The following definitions are concerned with different orbit types. An element x in a G-space X is said to be a *fixed point* if $G_x = G$. It is a *periodic point* if G/G_x is compact. A (topological) G-space X is said to be *topologically transitive* if some G-orbit is dense in X. If *all* orbits are dense, the action is called *minimal*. A subset $A \subset X$ is called *G-invariant* if for each $x \in A$ and $g \in G$, $gx \in A$. An equivalent definition of minimal action is that X does not have a proper closed G-invariant set, since the closure of a G-invariant set is a G-invariant set. A point x of a topological G-space X will be called *recurrent* if for each neighborhood U of x and each compact $K \subset G$, there is g in the complement of K such that $gx \in U$. It is immediate from the definitions that periodic points are recurrent. Furthermore, if both the orbit of x and its complement are dense in X, then x is a recurrent point. We leave the verification of this last claim as an exercise to the reader. Notice that the action of G on itself by translations is topologically transitive – in fact, transitive – but not recurrent.

The preceding notions can all be illustrated with actions defined on the n-torus. Let $\mathbb{T}^n = \mathbb{R}^n/\mathbb{Z}^n$ denote the n-dimensional torus, defined as the quotient of the abelian group $\mathbb{R}^n$ by its integer lattice subgroup $\mathbb{Z}^n$. The element $x = v + \mathbb{Z}^n$ will also be denoted $[v]$. $\mathbb{T}^n$ can alternatively be described as the product of n copies of the circle $S^1 = \{z \in \mathbb{C} \,|\, |z| = 1\}$; namely,

$$(a_1, \ldots, a_n) + \mathbb{Z}^n \mapsto (e^{2\pi i a_1}, \ldots, e^{2\pi i a_n})$$

is a diffeomorphism between $\mathbb{T}^n$ and $S^1 \times \cdots \times S^1$.

$\mathbb{R}^n$ acts transitively on $\mathbb{T}^n$ via the smooth action $\Phi : \mathbb{R}^n \times \mathbb{T}^n \to \mathbb{T}^n$ such that

$$\Phi : (u, [v]) \mapsto [u + v].$$

For each $u = (u_1, \ldots, u_n) \in \mathbb{R}^n$ define the translation $\tau_u = \Phi(u, \cdot)$. Then τ_u generates a $\mathbb{Z}$-action on $\mathbb{T}^n$ by $(m, [v]) \mapsto \tau_u^m([v])$. If the components of u are rational numbers, the orbit of each x is periodic and all orbits are finite with same cardinality, as one can easily check.

Real numbers $x_1, \ldots, x_s$ are called *rationally independent* if given integers k_i such that $\sum_{i=1}^{s} k_i x_i = 0$, then $k_i = 0$ for all i.

Proposition 1.2.4 Fix a vector $u = (u_1, \ldots, u_n) \in \mathbb{R}^n$ and consider the $\mathbb{Z}$-action on $\mathbb{T}^n$ generated by τ_u. Then the following statements are equivalent:

1. The action is topologically transitive.
2. The action is minimal.
3. The numbers $1, u_1, \ldots, u_n$ are rationally independent.

Proof. The proof is taken from [16]. It is clear that $2 \Rightarrow 1$. On the other hand, if some $x \in \mathbb{T}^n$ has a dense $\mathbb{Z}$-orbit, then all points have dense orbits since we can get from x to any other point by a translation and all translations commute with the $\mathbb{Z}$-action. Therefore, 1 and 2 are equivalent.

We now show $1 \Rightarrow 3$. Notice that if $1, u_1, \ldots, u_n$ are not rationally independent, we can find integers k_i, not all 0, such that $\sum_{i=1}^{n} k_i u_i = k_0$. Therefore, the function

$$\varphi(v) := \sin\left(2\pi \sum_{i=1}^{n} k_i v_i\right)$$

is a continuous $\mathbb{Z}$-invariant function on $\mathbb{T}^n$, that is, $\varphi \circ \tau_u^m = \varphi$ for all $m \in \mathbb{Z}$. But φ is not a constant function, so there exists $c \in \mathbb{R}$ such that the sets $U = \varphi^{-1}(\{t \in \mathbb{R} \,|\, t > c\})$ and $V = \varphi^{-1}(\{t \in \mathbb{R} \,|\, t < c\})$ are nonempty and disjoint. Furthermore, U and V are invariant sets since φ is $\mathbb{Z}$-invariant. It follows that the action cannot be topologically transitive.

If the action is not topologically transitive, there is a nonempty $\mathbb{Z}$-invariant open set U such that $\bar{U} \neq \mathbb{T}^n$. In fact, if no such U exists, one obtains a dense orbit as follows. Let $U_1, U_2, \ldots$ be a countable base of open sets for the topology of $\mathbb{T}^n$. By assumption, there exists an integer N_1 such that $\tau_u^{N_1}(U_1) \cap U_2 \neq \emptyset$. Let V_1 be a nonempty open set such that $\bar{V}_1 \subset U_1 \cap \tau_u^{-N_1}(U_2)$. There exists an integer N_2 such that $\tau_u^{N_2}(V_1) \cap U_3 \neq \emptyset$. Again, take an open set V_2 such that $\bar{V}_2 \subset V_1 \cap \tau_u^{-N_2}(U_3)$. By induction, we construct a nested sequence of open sets V_n such that $\bar{V}_{n+1} \subset V_n \cap \tau_u^{-N_{n+1}}(U_{n+2})$. The intersection $V = \bigcap_{n=1}^{\infty} \bar{V}_n = \bigcap_{n=1}^{\infty} V_n$ is nonempty since the $\bar{V}_n$ are compact. If $x \in V$, then $\tau_u^{N_{n-1}}(x) \in U_n$ for each $n \in \mathbb{N}$. This shows that the orbit of any $x \in V$ intersects each open set in a basis for the topology of $\mathbb{T}^n$. Therefore, the $\mathbb{Z}$-orbit of x is dense.

The previous claim can now be used to show $3 \Rightarrow 1$. Thus, suppose that the action is not topologically transitive, which implies by the claim that there exists an open nonempty $\mathbb{Z}$-invariant set U whose closure is not $\mathbb{T}^n$. Let χ be the characteristic function of U. In what follows, we think of χ as a function on $\mathbb{R}^n$ that is periodic in each variable. Since U is invariant, we have $\chi \circ \tau_u = \chi$. Take the Fourier expansion

$$\chi(x_1, \ldots, x_n) = \sum_{k \in \mathbb{Z}^n} c_k e^{2\pi i k \cdot x},$$

where $k \cdot x$ denotes the ordinary dot product $k \cdot x = k_1 x_1 + \cdots + k_n x_n$. Then

$$\chi(\tau_u(x)) = \sum_{k \in \mathbb{Z}^n} c_k e^{2\pi i k \cdot u} e^{2\pi i k \cdot x}.$$

Invariance of χ and uniqueness of the Fourier expansion imply $c_k = c_k e^{2\pi i k \cdot u}$ for each $k \in \mathbb{Z}^n$. If $c_k = 0$ for all nonzero $k \in \mathbb{Z}^n$, it would follow that χ is constant almost everywhere with respect to the Lebesgue measure on $\mathbb{R}^n$. Therefore, the measure of either U or its complement would be 0, which is not the case. Therefore, for some nonzero $k \in \mathbb{Z}^n$, $c_k \neq 0$, whence $e^{2\pi i k \cdot u} = 1$. This shows that the numbers $1, u_1, \ldots, u_n$ are rationally dependent. $\quad\square$

Exercise 1.2.5 Show that if X is a locally compact second countable space and $\Phi : G \times X \to X$ is a topological action, then Φ is topologically transitive if and only if any two nonempty open G-invariant sets intersect. (The argument is essentially contained in the preceding proof.) Also show that if X is a Baire space (e.g., locally compact) and the action is topologically transitive, then the set of points with a dense orbit is a dense G_δ-set, that is, it is a countable intersection of open dense sets.

Orbits of different points of a G-space can be very different, as the next example will show. Let $SL(n, \mathbb{Z})$ be the group of n-by-n matrices of determinant

1 with integer entries. Since the linear action of $SL(n, \mathbb{Z})$ on $\mathbb{R}^n$ leaves invariant the integer lattice $\mathbb{Z}^n$, we obtain an action on the torus

$$\Phi : SL(n, \mathbb{Z}) \times \mathbb{T}^n \to \mathbb{T}^n$$

by setting $\Phi(A, [v]) := [Av]$, where Av denotes matrix multiplication of A and $v \in \mathbb{R}^n$, the latter now viewed as a column vector. The next exercise shows that Φ is topologically transitive but not minimal.

Exercise 1.2.6 Show that each point $[u_1, \ldots, u_2] \in \mathbb{T}^n$ with rational components is a periodic point for the above action of $SL(n, \mathbb{Z})$ on $\mathbb{T}^n$. Using the argument employed in the last proposition, show that the action is topologically transitive. The same argument can be used to show that the $\mathbb{Z}$-action on $\mathbb{T}^2$ generated by the single matrix $\left(\begin{smallmatrix} 2 & 1 \\ 1 & 1 \end{smallmatrix}\right)$ is also topologically transitive.

The example introduced in the previous exercise can be generalized as follows. Let G be a topological group and Γ a discrete subgroup of G such that the quotient $X = G/\Gamma$ is compact or, more generally, such that X admits a G-invariant probability measure. (Invariant measures will be defined and discussed in detail in a later chapter.) Let H be a closed noncompact subgroup of G and define an H-action on X by

$$\Phi(h, g\Gamma) := hg\Gamma.$$

It will follow from results of chapter 8 that if $G = SL(n, \mathbb{R})$ (or any other connected, noncompact, simple Lie group) then for any noncompact closed subgroup H of G, the H-action on X is topologically transitive.

1.3 Suspensions

Starting with an action $\Phi : \Gamma \times X \to X$, where Γ is a discrete subgroup of a connected topological group G, it is possible to define in a canonical way a locally free action $\bar{\Phi}$ of G that "looks transversely like" Φ, called the *suspension* of Φ, or the *induced action* from Φ. It is defined as follows. First notice that Γ acts diagonally on the product $G \times X$ by

$$\gamma \cdot (g, x) = (g\gamma^{-1}, \Phi(\gamma, x)).$$

We denote the orbit space by $S = (G \times X)/\Gamma$ and the element of S represented by (g, x) will be written $[g, x]$.

S is the total space of a fiber bundle $p : S \to G/\Gamma$, $(g, x)\Gamma \mapsto g\Gamma$ (see the beginning of chapter 6 for the definition of a fiber bundle), whose fibers are homeomorphic to X. In fact, let $\pi : G \to G/\Gamma$ be the natural projection (a covering map) and, for each $z \in G/\Gamma$, let U be a sufficiently small connected neighborhood of z such that $\pi^{-1}(U)$ is a disjoint union of open sets in G, each homeomorphic to U via π. Choose one component of the preimage, say U_0, and set $\sigma := (\pi|_{U_0})^{-1} : U \to U_0$. Then

$$(z, x) \mapsto [\sigma(z), x]$$

is a homeomorphism between $U \times X$ and $p^{-1}(U)$, showing local triviality. Notice that by making a different choice of connected component of $\pi^{-1}(U)$, say $U_1 = U_0 \gamma$, then the change of trivialization is given by the map from $U \times X$ onto itself that sends (z, x) to $(z, \Phi(\gamma, x))$ for some $\gamma \in \Gamma$ independent of z.

The group G acts on S by

$$\bar{\Phi}(h, [g, x]) := [hg, x].$$

Notice that $[hg, x] = [hg\gamma^{-1}, \Phi(\gamma, x)]$, so the action is indeed well defined and is clearly continuous.

For example, let $G = \mathbb{R}$ and $\Gamma = \mathbb{Z}$, and consider the $\mathbb{Z}$-action on $\mathbb{T}^n$ generated by a diffeomorphism τ of the n-torus. Then $G/\Gamma = \mathbb{R}/\mathbb{Z} = \mathbb{T}^1$, so the suspension of the $\mathbb{Z}$-action is an $\mathbb{R}$-action on $\mathbb{T}^{n+1}$.

Similarly, one obtains an $SL(n, \mathbb{R})$-action on a fiber-bundle with typical fiber $\mathbb{T}^n$ over $SL(n, \mathbb{R})/SL(n, \mathbb{Z})$, by suspending the $SL(n, \mathbb{Z})$-action on $\mathbb{T}^n$ defined earlier.

Exercise 1.3.1 Show that the Γ-action Φ on X is topologically transitive if and only if the suspension $\bar{\Phi}$ is topologically transitive.

1.4 Dynamical Invariants

Two topological G-spaces X and Y are said to be *(topologically) equivalent* if there exists a homeomorphism $F : X \to Y$ that intertwines the respective G-actions. More precisely, $F(gx) = gF(y)$ for each $x \in X$, $y \in Y$, and $g \in G$. Equivalent G-spaces have the same (topological) dynamical properties; for example, the G-action on X is topologically transitive if and only if the action on Y is, and periodic orbits of X correspond under F to periodic orbits of Y.

Any attempt to classify topological G-spaces in terms of their global dynamical properties immediately calls for characteristic "quantities" that have the potential to distinguish inequivalent systems. An analogy can be made with linear maps of vector spaces. Linear maps $T_i : V_i \to V_i$, $i = 1, 2$, are equivalent if there exists a linear isomorphism $F : V_1 \to V_2$ such that $FT_1 = T_2 F$.

Equivalent linear maps must have the same spectrum of eigenvalues, so the spectrum is an "invariant" in this case. (There is no need to define the concept of an invariant formally, since we will be concerned only with some specific examples of them.)

An interesting example of an invariant for G-spaces can be defined as follows. Let $H_q(X_i, \mathbb{R})$ denote the real qth singular homology spaces of G-spaces X_i, $i = 1, 2$. (For the basic definitions in algebraic topology see, for example, [14]. This example, however, will not be needed later.) Then G acts in a natural way on $H_q(X_i, \mathbb{R})$ by vector space isomorphisms. If the actions on X_1 and X_2 are equivalent, via a homeomorphism $F : X_1 \to X_2$, then the linear actions of G on $H_q(X_i, \mathbb{R})$ are linearly equivalent via the map induced by F on the homology spaces, which we denote $H_q(F)$. Therefore, the linear invariants of $H_q(f_i)$ are also invariants of the G-spaces.

We now restrict our attention to $\mathbb{Z}$-actions. Thus, let $f : X \to X$ be a homeomorphism of a topological space X, generating a $\mathbb{Z}$-action on X. Denote by $P_n(f)$ the number of periodic points of f of period n, that is, the cardinality of the set of fixed points of f^n. Define the *exponential growth rate of periodic points* by

$$p(f) := \limsup_{n \to \infty} \frac{1}{n} \log(\max\{P_n(f), 1\}).$$

It is immediate that $p(f)$ is also an example of a topological invariant for $\mathbb{Z}$-actions.

Exercise 1.4.1 Let $f : \mathbb{T}^n \to \mathbb{T}^n$ be a homeomorphism of the n-torus obtained from an integer matrix A of determinant 1. Suppose that none of the eigenvalues of A lie on the unit circle $\{\lambda \in \mathbb{C} \mid |\lambda| = 1\}$. (For example, $A = \begin{pmatrix} 2 & 1 \\ 1 & 1 \end{pmatrix}$.) Show that

$$p(f) = \sum_{|\lambda| > 1} \log |\lambda|,$$

where the sum ranges over the eigenvalues of A of modulus greater than 1. In particular, the number of periodic points of period m grows exponentially with m. (For the details, see [16]. The key point here is to use the Lefschetz fixed point formula, from which one derives

$$P_m(f) = |\det(I - H_1(f)^m)|,$$

where $H_1(f) = A$ is the linear map induced by f on $\mathbb{R}^n = H_1(\mathbb{T}^n)$.)

An important topological invariant of $\mathbb{Z}$-actions is the *topological entropy*. Roughly speaking, it captures the exponential growth rate, as $m \to \infty$, of the number of orbit segments of length m that can be distinguished with a specified, but arbitrarily fine, precision. To quote Katok and Hasselblatt [16], "In a sense, the topological entropy describes in a crude but suggestive way the total exponential complexity of the orbit structure with a single number."

The precise definition is as follows. Let $f : X \to X$ be a homeomorphism of a compact metric space X with metric d. We define an increasing sequence of metrics d_m^f, $m = 1, 2, \ldots$, such that $d_1^f = d$ and for each $x, y \in X$,

$$d_m^f(x, y) := \max_{0 \le i \le n-1} d(f^i(x), f^i(y)).$$

The open ball relative to the metric d_m^f with center at x and radius a will be denoted $B_f(x, a, m)$.

A set $E \subset X$ is said to be (m, ϵ)-spanning if $X \subset \bigcup_{x \in E} B_f(x, \epsilon, m)$. Let $S_d(f, \epsilon, m)$ be the minimum cardinality of an (m, ϵ)-spanning set. Again to quote [16], "One can verbally express the meaning of $S_d(f, \epsilon, m)$ by saying that it is equal to the minimal number of initial conditions whose behavior up to time m approximates the behavior of *any* initial condition up to ϵ."

Define now the exponential growth rate for $S_d(f, \epsilon, m)$ as $m \to \infty$,

$$h_d(f, \epsilon) := \limsup_{m \to \infty} \frac{1}{m} \log S_d(f, \epsilon, m),$$

and note that $h_d(f, \epsilon)$ must increase as ϵ decreases. The topological entropy of f is now defined by

$$h_d(f) := \lim_{\epsilon \to 0} h_d(f, \epsilon).$$

It is an easy computation to obtain that the topological entropy does not depend on the metric d, so we denote it by $h_{\text{top}}(f)$. It is also an easily obtained fact that $h_{\text{top}}(f)$ is an invariant of topological equivalence of $\mathbb{Z}$-actions. We will have more to say about entropy later in the book. For now, we only mention without proof the next proposition.

Call a homeomorphism f of a compact metric space X *expansive* if there exists a constant δ such that if $d(f^n(x), f^n(y)) < \delta$ for all $n \in \mathbb{Z}$ then $x = y$. For example, the linear toral map of the previous exercise is easily shown to be expansive. More generally, Anosov maps (defined in the next section) are expansive.

Proposition 1.4.2 Let f be an expansive homeomorphism of a compact metric space. Then $p(f) \le h_{\text{top}}(f)$.

The (easy) proof can found in [16, 3.2.13]. The proposition implies that the example of exercise 1.4.1 has positive entropy, indicating that the system is very complex (i.e., the number of distinguishable segments of orbits, at a fixed but arbitrarily fine precision, grows exponentially with the length). That example should be compared with isometric transformations, such as translations of $\mathbb{T}^n$, for which the entropy is easily shown to be 0.

1.5 Anosov Diffeomorphisms

Hyperbolic systems, in particular, the so-called Anosov systems, constitute a substantial and central topic in the general theory of (smooth) dynamical systems. Although none of the deeper results in hyperbolic dynamics (such as Ya. Pesin's theory of nonuniformly hyperbolic systems, a little of which will be touched on in chapter 9) will be required for the main results of this book, the subject may help motivate some of the theorems to be discussed later, and will be needed in some of the applications. A limited survey, mostly without proofs, will be given in this book, beginning with the present section. For details and proofs not given here, the reader should consult [16] and the references cited therein.

Let M be a smooth manifold and $f : M \to M$ a smooth diffeomorphism, generating a $\mathbb{Z}$-action on M. The differential map at a point $x \in M$ will be denoted $Df_x : T_x M \to T_x M$. Hyperbolicity will be defined as a condition on the growth of tangent vectors under the iteration of Df, so it is necessary to fix a Riemannian metric on M, although the precise choice will be immaterial. The norm of a vector $v \in T_x M$ with respect to the chosen Riemannian metric will be denoted $\|v\|$.

Let $\Lambda \subset M$ be a compact f-invariant subset. We say that Λ is a *hyperbolic set* for f if there exist λ, $0 < \lambda < 1$, $C > 0$, and a direct sum decomposition of the tangent bundle of M over Λ, $TM|_\Lambda = E^+ \oplus E^-$, such that, for each $x \in \Lambda$, $Df_x E^\pm(x) = E^\pm(f(x))$ and

$$\|Df_x^{\mp m} v\| \leq C\lambda^m \|v\|$$

for each $x \in \Lambda$, $v \in E^\pm(x)$, and $m \in \mathbb{N}$. This can be expressed in words as: The direct sum decomposition over Λ is $\mathbb{Z}$-invariant and vectors in $E^-(x)$ (resp., $E^+(x)$) decrease exponentially in length under forward (resp., backward) iteration of f. It can be shown that the subspaces $E^+(x)$ and $E^-(x)$ must vary continuously on Λ. (They are, in fact, Hölder continuous on Λ.) We call E^- and E^+ the *stable* and *unstable* subbundles of f over Λ, respectively.

If $\Lambda = M$ we say that f is an *Anosov diffeomorphism*. (In particular, M is assumed to be compact.) For example, the diffeomorphism of $\mathbb{T}^n$ defined in exercise 1.4.1 is Anosov.

Even though the fields of stable and unstable subspaces (i.e., the subbundles E^- and E^+ of TM) are only continuous they can be integrated, in the sense that they are the tangent bundles to a family of smooth submanifolds of M, which themselves vary continuously with x. The precise statement is given in the next proposition. The proof of the theorem is rather technical and will not be given here. It should be noticed that the Frobenius theorem (see chapter 3) does not apply here since the subbundles are only continuous.

Theorem 1.5.1 (Local stable manifolds) Let M be a compact smooth manifold and $f : M \to M$ an Anosov diffeomorphism. Let k_s be the dimension of E^-. Then there is an $\epsilon_0 > 0$ such that for $0 < \epsilon < \epsilon_0$ and each $x \in M$ there exists a smooth embedded k_s-dimensional disk $W^s_\epsilon(x)$, which varies continuously with x (in the space of embeddings of the k_s-dimensional unit disk in Euclidean space equipped with the C^1 topology), such that

1. $T_x W^s_\epsilon(x) = E^-(x)$ and $f(W^s_\epsilon(x)) \subset W^s_\epsilon(f(x))$;
2. there are a positive constant K and $0 < \lambda < 1$ such that for each $x \in M$, $y \in W^s_\epsilon(x)$, and $m \in \mathbb{N}$ we have $d(f^m(x), f^m(y)) \leq K\lambda^m d(x, y)$; and
3. $W^s_\epsilon(x) = \{y \in M \mid d(f^m(x), f^m(y)) \leq \epsilon$ for all $n \in \mathbb{N}\}$.

Substituting f^{-1} for f we obtain analogous statements for E^+, in which case the embedded disks are denoted $W^u_\epsilon(x)$.

The embedded disks $W^s_\epsilon(x)$ (resp., $W^u_\epsilon(x)$) of the theorem are called the *local stable* (resp., *local unstable*) *manifolds* of f. It is also possible to define global stable and unstable manifolds as follows:

$$W^s(x) := \bigcup_{n=0}^{\infty} f^{-n} W^s_\epsilon(f^n(x)),$$

where d is the metric function on M induced from the Riemannian metric. The same definition holds for $W^u(x)$, after replacing f with f^{-1}. $W^s(x)$ and $W^u(x)$ are smooth submanifolds of M and satisfy the following properties, all of which are immediate consequences of the theorem:

1. W^s and W^u are f-invariant. This means that for each $x \in M$, $f(W^s(x)) = W^s(f(x))$ and $f(W^u(x)) = W^u(f(x))$.

2. $T_x W^s(x) = E^-(x)$ and $T_x W^u(x) = E^+(x)$ for each $x \in M$.
3. There are a positive constant K and $0 < \lambda < 1$ such that for each $x \in M$, $y \in W^s(x)$, and $m \in \mathbb{N}$,

$$d(f^m(x), f^m(y)) \leq K\lambda^m d(x, y).$$

A similar statement holds for W^u after replacing f by f^{-1}. Moreover, the sets $W^s(x)$ and $W^u(x)$ are uniquely characterized by

$$W^s(x) = \left\{ y \in M \mid \lim_{m \to +\infty} d(f^m(x), f^m(y)) = 0 \right\},$$

$$W^u(x) = \left\{ y \in M \mid \lim_{m \to -\infty} d(f^m(x), f^m(y)) = 0 \right\}.$$

W^s (resp., W^u), defined as the collection of all $W^s(x)$ (resp., $W^u(x)$), is called the *stable foliation* (resp., *unstable foliation*) of f. They are also referred to as the *Anosov foliations* of f. Notice that if $W^s(x)$ and $W^s(y)$ have a nonempty intersection, then they coincide (similarly for W^u).

The local stable and unstable submanifolds of the Anosov diffeomorphism f give M a *local product structure*, defined by the next theorem. We denote

$$B^n = \{(t_1, \ldots, t_n) \in \mathbb{R}^n \mid |t_i| \leq 1, \quad i = 1, \ldots, n\}.$$

Theorem 1.5.2 (Local product structure) Let $f : M \to M$ be an Anosov diffeomorphism of the compact smooth n-manifold M, and let $x \in M$. Denote by k_s (resp., k_u) the dimension of the stable (resp., unstable) manifolds. Then there exist $\epsilon > 0$, a closed set $B(x) \subset M$ containing an open neighborhood of x, and a homeomorphism $\varphi : B(x) \to B^n$ such that $\varphi(x) = 0$ and the following holds: The local stable manifolds $W_\epsilon^s(z) \cap B(x)$, $z \in B(x)$, are mapped by φ onto the sections of the cube B^n by k_s-dimensional planes parallel to the first k_s coordinate axes, and the local unstable manifolds $W_\epsilon^u(z) \cap B(x)$, $z \in B(x)$, are mapped onto the sections by k_u-dimensional planes parallel to the remaining k_u coordinates.

Exercise 1.5.3 Describe the stable and unstable foliations of the example of exercise 1.4.1. Prove that each of the sets $W^s(x)$ and $W^u(x)$ is a dense submanifold of $\mathbb{T}^2$. Verify that the local product structure described in the theorem holds for that example.

From now on, we make the assumption that our Anosov diffeomorphism does not have *wandering points*. A point $x \in M$ is said to be a *wandering* point if it has a neighborhood U such that $f^m(U) \cap U$ is empty for all positive m. A point x is said to be *nonwandering* if for all neighborhoods U of x there is a positive m such that $f^m(U)$ intersects U. Let $NW(f)$ denote the set of nonwandering points of f. Therefore the new assumption is that $NW(f) = M$. This is the case, for example, when f leaves invariant a Riemannian measure on the compact manifold M. (This is a consequence of Poincaré's recurrence theorem, discussed in the next chapter.)

Theorem 1.5.4 Let $f : M \to M$ be an Anosov diffeomorphism of a compact manifold M and assume that $NW(f) = M$. Then the following statements hold:

1. f is topologically transitive.
2. The set of points with periodic orbits is dense in M.
3. Each stable and unstable manifold is dense in M.

Proof. We first show property 2. The assumption $NW(f) = M$ implies that in each nonempty open set $U \subset M$ we can find a point z such that $f^m(z) \in U$ for some $m \in \mathbb{N}$. Let V be a nonempty open set in M. We wish to show that V contains a periodic point. Choose any $x_0 \in V$ and choose $\epsilon > 0$ sufficiently small such that the set $B(x_0)$ defined in theorem 1.5.2 is contained in V. Recall that $B(x_0)$ decomposes into horizontal and vertical plaques, denoted $P^s(z)$ and $P^u(z)$, $z \in B(x_0)$, corresponding to subsets of local stable and unstable manifolds. By theorem 1.5.1, distances in $W_\epsilon^s(z)$ contract uniformly (exponentially) under iterations of f(similarly for distances in $W^u(z)$, under the iteration of f^{-1}). Therefore, we can choose x_1 sufficiently close to x_0 and m sufficiently large, so that $f^m(x_1) \in B(x_0)$, $f^m(P^s(x_1))$ is contained in $P^s(f^m(x_1))$, and $f^{-m}(P^u(z)) \subset P^u(f^{-m}(z))$, for each $z \in f^m(P^s(x_1))$.

By projecting from $f^m(P^s(x_1))$ back into $P^s(x_1)$ along vertical plaques, we obtain a continuous contraction $\Phi : P^s(x_1) \to P^s(x_1)$. Applying Brower's fixed point theorem, we conclude that Φ has a fixed point $x_2 \in P^s(x_1)$. Therefore, the global unstable manifold $W^u(x_2)$ is an invariant set for f^m. We can now repeat the argument using f^{-m}, which contracts $P^u(x_2)$ into itself, to obtain a fixed point for f^{-m} in $B(x_0)$. That point is the periodic point we seek.

We now prove topological transitivity of f. According to exercise 1.2.5, it suffices to show that for any pair of nonempty open sets $U, V \subset M$, there exist $x \in U$ and a positive integer m such that $f^m(x) \in V$. For that purpose, choose periodic points $x_0, \ldots, x_n$ such that $x_0 \in U$, $x_n \in V$, and

$$W^u(x_k) \cap W^s(x_{k+1}) \neq \emptyset$$

for each k, $0 \le k \le n - 1$. Notice that these points exist due to theorem 1.5.2 and the fact that the set of periodic points is dense in M. Let N be a common periodic point for all x_k. By replacing f with f^N and renaming the map we may assume that these points are fixed by f.

We now claim that $W^u(x_0)$ intersects V. This will imply both topological transitivity and property 3. Since $W^u(x_0)$ is an f-invariant set and intersects $W^s(x_1)$ at some point y_1, it must also intersect $W^s(x_1)$ at a point that can be chosen to lie as close to x_1 as we want, since $f^l(y_1)$ approaches x_1 when $l \to \infty$. By the continuity of the local stable and unstable manifolds, we can conclude that $W^u(x_0)$ intersects $W^s(x_2)$, since this is true for $W^u(x_1)$. We can repeat this argument inductively to conclude that $W^u(x_0)$ intersects $W^s(x_n)$ at some point z. Therefore, for some positive integer q, $f^q(z) \in V$. This concludes the proof.   □

It can be shown [16] that if f is an Anosov diffeomorphism of a compact manifold M such that $NW(f) = M$, then $p(f) = h_{\text{top}}(f)$.

2

Ergodic Theory – Part I

Broadly defined, ergodic theory is the study of measurable G-spaces. Even when studying smooth actions of Lie groups on manifolds, it can be very helpful to focus attention on the underlying measure space, and look for properties of the action that are detectable at a measure-theoretic level. After reviewing some basic facts in measure theory, we discuss Poincaré's recurrence and explain with one example its probabilistic significance. We then introduce the notions of ergodicity and of ergodic decomposition. The chapter ends with the definition of the measure-theoretic entropy and a brief survey of some of its properties.

2.1 Review of Measure Theory

Unless stated otherwise, a *measure* on a set X will mean a nonnegative measure, that is, a countably additive function μ defined on a σ-algebra $\mathcal{A}$ of subsets of X, with values in the extended real line $[0, \infty]$. $\mathcal{A}$ is called the algebra of *measurable sets*. When μ is a finite measure, we often normalize it so that $\mu(X) = 1$ and call it a *probability measure*.

A map from a measure space X with σ-algebra $\mathcal{A}$ to another measure space Y with σ-algebra $\mathcal{B}$ is called *measurable* if the preimage of any set in $\mathcal{B}$ is in $\mathcal{A}$. If $T : X \to Y$ is a measurable map and μ is a measure on $(X, \mathcal{A})$, we define $T_* \mu$ as the measure on $(Y, \mathcal{B})$ such that

$$T_* \mu(B) := \mu(T^{-1} B)$$

for each $B \in \mathcal{B}$.

A measurable map T of a measure space $(X, \mathcal{A}, \mu)$ into itself is said to be *measure preserving* if $T_* \mu = \mu$. We also say that μ is an *invariant measure* for T. If $T_* \mu$ and μ are in the same measure class (i.e., if they have the same sets of measure 0) we say that μ is a *quasi-invariant* measure for T. We say that

T is a *measure-preserving transformation* of $(X, \mathcal{A}, \mu)$ if it is bijective and T and T^{-1} are measurable and measure preserving.

Denote by $L^p(X, \mathcal{A}, \mu)$ the space of $\mathcal{A}$-measurable complex-valued functions f on X such that $|f|^p$ is μ-integrable, that is,

$$\|f\|_p = \left(\int_X |f|^p \, d\mu \right)^{\frac{1}{p}} < \infty,$$

and identify functions that differ only on a set of μ-measure 0. Then $L^p(X, \mathcal{A}, \mu)$ is a Banach space with norm $\| \cdot \|_p$. When $p = 2$, it is a Hilbert space with inner product $\langle f, g \rangle = \int_X f \bar{g} \, d\mu$.

Exercise 2.1.1 For $T : X \to X$ and μ as before, show that $T_* \mu$ is characterized by the property

$$\int_X \varphi \, d(T_* \mu) = \int_X \varphi \circ T \, d\mu$$

for all $\varphi \in L^1(X, \mathcal{A}, \mu)$. Conclude that T is measure preserving if and only if $\int_X \varphi \, d\mu = \int_X \varphi \circ T \, d\mu$ for all $\varphi \in L^1(X, \mathcal{A}, \mu)$. (If $\varphi = I_A \in L^1(X, \mathcal{A}, \mu)$ is the characteristic function of a set $A \in \mathcal{A}$ of finite measure, show that the preceding equality is the same as $\mu(A) = \mu(T^{-1} A)$. For the general case, approximate φ by a finite linear combination of characteristic functions.)

Recall that the σ-algebra of *Borel sets* of a topological space is the σ-algebra generated by the open sets. For the most part in this book our measure spaces will be locally compact second countable metrizable spaces, or more generally Borel subsets of such topological spaces. These are called *standard Borel spaces*. A *Borel measure* on a standard Borel space X is a measure defined on the Borel sets. We say that μ is a *regular* Borel measure if $\mu(K) < \infty$ for each compact set $K \in X$. This can be shown to imply that, for all $E \in \mathcal{B}$ such that $\mu(E) < \infty$, $\mu(E)$ is the supremum of $\mu(K)$, where $K \subset E$, K compact, and for all $E \in \mathcal{B}$, $\mu(E)$ is the infimum of $\mu(U)$, where $E \subset U$, U open.

If μ is a Borel measure on a topological space X, we define the *support* of μ to be the complement of the largest open set whose μ-measure is 0.

Let $C_c(X)$ be the linear space of complex-valued, continuous functions on X with compact support. The space of all continuous functions on X will be denoted by $C(X)$. We define for $f \in C(X)$

$$\|f\| := \sup\{|f(x)| \mid x \in X\},$$

which is a finite norm on elements of $C_c(X)$. The completion of $C_c(X)$ with respect to this sup norm is the Banach space $C_0(X)$ of continuous functions on X that vanish at infinity. Its dual space $C_0(X)^*$, consisting of bounded linear functionals on $C_0(X)$, is given the *weak* topology*, defined as follows: A sequence $\Lambda_n \in C_0(X)^*$ converges to $\Lambda \in C_0(Y)^*$ if for all $f \in C_0(X)$ the sequence of numbers $\Lambda_n(f)$ converges to $\Lambda(f)$.

Theorem 2.1.2 (Riesz representation theorem) Let X be a locally compact Hausdorff space and Λ a positive linear functional on $C_c(X)$. Then there exists a regular Borel measure μ on X such that

$$\Lambda(f) = \int_X f \, d\mu$$

for all $f \in C_c(X)$. If $\Lambda(1) = 1$, μ is a probability measure.

We denote by $\mathcal{M}(X)_1$ the space of Borel probability measures on X. By the Riesz representation theorem $\mathcal{M}(X)_1$ can be viewed as the set of bounded positive linear functionals Λ on $C_c(X)$ such that $\Lambda(1) = 1$.

Given two measures μ, ν on a measurable space $(X, \mathcal{A})$, we say that μ is *absolutely continuous* with respect to ν if $\mu(E) = 0$ for all $E \in \mathcal{A}$ such that $\nu(E) = 0$.

Theorem 2.1.3 (Radon–Nikodym) If ν and μ are σ-finite measures on a measurable space $(X, \mathcal{A})$ such that μ is absolutely continuous with respect to ν, then there exists a unique (ν-a.e.) nonnegative measurable function h, ν-integrable over each set $E \in \mathcal{A}$ of finite measure, such that

$$\mu(E) = \int_E h \, d\nu$$

for all $E \in \mathcal{A}$. A function $g : X \to \mathbb{C}$ is μ-integrable if and only if gh is ν-integrable.

The function h obtained in the theorem is usually denoted $\frac{d\mu}{d\nu}$ and is called the *Radon–Nikodym derivative* of μ with respect to ν.

Recall that a space $\mathcal{C}$ is said to be *convex* if for all ν and μ in $\mathcal{C}$ and each a, $0 \leq a \leq 1$, we have that $a\nu + (1 - a)\mu$ also belongs to $\mathcal{C}$.

Theorem 2.1.4 Let X be a compact metrizable space. Then $\mathcal{M}(X)_1$ is a compact convex metrizable space with respect to the weak* topology.

A consequence of the previous theorem is that any sequence of Borel probability measures of a compact Hausdorff space has a convergent subsequence. A more general fact is stated in the next exercise.

Exercise 2.1.5 Let X be a locally compact Hausdorff space and let $\pi : X \to Y$ be a continuous map into a topological space Y such that Y is the union of a countable sequence of compact sets $K_i, i = 1, 2, \ldots$, and such that $\pi^{-1}(K_i)$ is compact for each i. Let μ be a regular Borel probability measure on Y and define the space $\mathcal{M}_\mu(X)_1$ consisting of all Borel probability measures ν on X such that $\pi_* \mu = \nu$. Show that $\mathcal{M}_\mu(X)_1$ is convex and compact with respect to the weak* topology.

Let $(X, \mathcal{A}, \mu)$ and $(Y, \mathcal{B}, \nu)$ be two measure spaces and define the product space $(X \times Y, \mathcal{A} \times \mathcal{B}, \mu \times \nu)$, where $\mathcal{A} \times \mathcal{B}$ is the σ-algebra of subsets of $X \times Y$ generated by elements of the form $A \times B$, $A \in \mathcal{A}$, $B \in \mathcal{B}$, and $\mu \times \nu$ is the product measure.

Theorem 2.1.6 (Fubini) If f is a nonnegative measurable function on $(X \times Y, \mathcal{A} \times \mathcal{B})$, then the functions

$$ F(x) := \int_Y f(x, y)\, d\nu(y), \qquad G(y) := \int_X f(x, y)\, d\mu(x) $$

on X and Y, respectively, are measurable and

$$ \int_{X \times Y} f\, d(\mu \times \nu) = \int_X \left[\int_Y f(x, y)\, d\nu(y) \right] d\mu(x) $$

$$ = \int_Y \left[\int_X f(x, y)\, d\mu(x) \right] d\nu(y). $$

If f is integrable, but not necessarily nonnegative, there exist measurable sets $A_0 \in \mathcal{A}$ and $B_0 \in \mathcal{B}$ of full measure in X and Y respectively, such that F and G are finite and measurable on A_0 and B_0, respectively, and the preceding equalities still hold.

Exercise 2.1.7 Let H be a group of homeomorphisms of a compact metric space X such that the action $\Phi : H \times X \to X$ is continuous. Show that for each $h \in H$, $h_* : \mathcal{M}(X)_1 \to \mathcal{M}(X)_1$ is a homeomorphism and that the action

$$ \Phi_* : H \times \mathcal{M}(X)_1 \to \mathcal{M}(X)_1 $$

defined by $\Phi_*(h, \mu) := h_*\mu$ is also continuous.

If F is a homeomorphism of a compact metric space X, then by the previous exercise F_* is a homeomorphism of the space $\mathcal{M}(X)_1$ of Borel probability measures on X. Since $\mathcal{M}(X)_1$ is convex, we can define a homeomorphism T_n from $\mathcal{M}(X)_1$ onto itself as follows:

$$ T_n(\mu) : = \frac{1}{n} \sum_{i=0}^{n-1} F_*^i \mu $$

for each $\mu \in \mathcal{M}(X)_1$.

Proposition 2.1.8 A homeomorphism of a compact metrizable space must preserve some nonzero Borel probability measure.

Proof. Keeping the notations X, F, and T_n as before, we have by compactness of $\mathcal{M}(X)_1$ that some subsequence $T_{n_i}\mu$ must converge to a μ_0. The measure μ_0 is fixed by F_* since

$$F_*\mu_0 = \lim_{i\to\infty} F_*T_{n_i}\mu = \lim_{i\to\infty} \left(T_{n_i}\mu + \frac{\mu - (F_*)^{n_i}\mu}{n_i} \right) = \lim_{i\to\infty} T_{n_i}\mu = \mu_0.$$

$\square$

The next exercise is a generalization of the proposition.

Exercise 2.1.9 Let X be a locally compact Hausdorff space and let $\pi : X \to Y$ be a continuous map into a σ-compact space Y such that $\pi^{-1}(K)$ is compact for each compact $K \subset Y$. Let $\Phi : \mathbb{R} \times X \to X$ be a continuous action on X and $\phi : \mathbb{R} \times Y \to Y$ a continuous action on Y such that $\pi \circ \Phi_t = \phi_t \circ \pi$ for each $t \in \mathbb{R}$. Let μ be a ϕ-invariant Borel probability measure on Y. (Invariance means that $(\phi_t)_*\mu = \mu$ for each $t \in \mathbb{R}$.) Show that there exists a Φ-invariant Borel probability measure ν on X such that $\pi_*\nu = \mu$. (Hint: Imitate the proof of the proposition so as to obtain a Φ_1-invariant probability measure ν_0 on X that projects onto μ. Show that $\nu = \int_0^1 (\Phi_t)_*\nu_0 \, dt$ is Φ-invariant and projects onto μ.)

Theorem 2.1.10 (Disintegration) Let X and Y be standard Borel spaces, and let p be a measurable map from X to Y. Let ν be a σ-finite Borel measure on X, and suppose that the projection measure $\mu := p_*\nu$ is also σ-finite. Then, there exists a unique (μ-a.e.) family of Borel measures $\{\lambda_y \mid y \in Y\}$ on X such that

1. the function $y \mapsto \lambda_y(A)$ on Y is measurable for each measurable subset $A \subset X$;
2. λ_y is supported on $p^{-1}(y)$ for μ-a.e. $y \in Y$;
3. for each measurable subset $A \subset X$

$$\nu(A) = \int_Y \lambda_y(A) \, d\mu(y);$$

4. every ν-integrable function f on X is λ_y-integrable for μ-a.e. $y \in Y$ and

$$\int_X f(x) \, d\nu(x) = \int_Y \int_{p^{-1}(y)} f(x) \, d\lambda_y(x) \, d\mu(y).$$

The family of measures $\{\lambda_y\}$ given by the theorem is called the *disintegration* of ν.

2.2 Recurrence

We continue to assume that G is a locally compact second countable topological group. Let $(X, \mathcal{A}, \mu)$ be a measure space. A *measure-preserving action* of G on X is an action defined by a measurable map $\Phi : G \times X \to X$, where $G \times X$ has the product measure structure, and the maps $\Phi_g := \Phi(g, \cdot)$, $g \in G$, are measure-preserving transformations of X. Of course, a G-space defined by a continuous G-action is also a measurable G-space for the σ-algebra of Borel sets. We say that X is a *G-space with invariant measure μ*.

We may also consider a G-space with *quasi-invariant measure μ*, in which case the G-action preserves only the measure class of μ. As before, we will often write $\Phi_g(x)$ simply as gx. In this notation, a G-invariant measure μ satisfies $g_* \mu = \mu$.

The translations of $\mathbb{T}^n$ as well as the action of $SL(n, \mathbb{Z})$ on $\mathbb{T}^n$ discussed in chapter 1 are examples of measure-preserving actions. The Lebesgue measure on $\mathbb{T}^n$ is in each case a finite invariant measure.

If a G-space admits a measure μ that is both invariant and finite, then the $\mathbb{Z}$-action generated by each element of G has the following recurrence property.

Theorem 2.2.1 (Poincaré recurrence) Let X be a $\mathbb{Z}$-space, the $\mathbb{Z}$-action being generated by a transformation g that preserves a probability measure μ on X. Then for any set $E \in \mathcal{A}$, μ-almost every point $x \in E$ returns infinitely often to E. More precisely, there is a measurable subset F of E such that $\mu(F) = \mu(E)$ and, for each $x \in F$, a sequence $n_1 < n_2 < \cdots$ such that $g^{n_i} x \in E$ for all i.

Proof. Take F to be

$$F = E \cap \bigcap_{N=0}^{\infty} \bigcup_{n=N}^{\infty} g^{-n} E = E \cap \bigcap_{N=0}^{\infty} E_N,$$

where $E_N := \bigcup_{n=N}^{\infty} g^{-n} E$. A point x belongs to E_N if and only if for some $n \geq N$, $g^n(x) \in E$; therefore $x \in F$ if and only if $x \in E$ and for each N $g^n(x)$ is back into E for some $n \geq N$. In other words, F is precisely the set of recurrent points of E. Notice that $E_{N+1} = g^{-1} E_N$, so

$$\mu(E_{N+1}) = \mu\big(g^{-1} E_N\big) = \mu(E_N).$$

But E_N is a decreasing sequence of sets, so therefore $\mu(\bigcap_{N=0}^{\infty} E_N) = \mu(E_0)$ and $\mu(F) = \mu(E)$, since $E \subset E_0$. $\square$

We illustrate the theorem with an example that clearly shows the connections Poincaré recurrence has with probability theory. It should be pointed out, however, that the actions we are actually concerned with in this book are of a very different nature.

The following discussion is taken from [25]. Suppose that there are two urns; one of them contains n balls, numbered from 1 to n, and the other is empty. There is also a hat containing n slips of paper numbered from 1 to n. Suppose that every second we draw one slip of paper from the hat, read the number on it, replace it in the hat, and move the ball bearing that number from whichever urn it is in to the other urn. The state of the system at any given time i is described by specifying the number $a_i \in \{0, 1, \ldots, n\}$ of balls in the first urn. For simplicity (and so as to have a $\mathbb{Z}$-action) we assume that the experiment began infinitely long ago and continues on to eternity, so that i can be taken to be in $\mathbb{Z}$.

The experiment can be modeled in terms of a *Markov chain*, which corresponds to a $\mathbb{Z}$-action on a probability space. Let $\Omega = \{0, 1, \ldots, n\}^{\mathbb{Z}}$ be the space of all bi-infinite sequences of numbers taken from $\{0, 1, \ldots, n\}$. An element of Ω will be described as a map $\omega : \mathbb{Z} \to \{0, 1, \ldots, n\}$. Define the shift transformation $\sigma : \Omega \to \Omega$ by

$$\sigma(\omega)(i) = \omega(i + 1), i \in \mathbb{Z}.$$

If we give $\{0, 1, \ldots, n\}$ the discrete topology and Ω the product topology, then Ω becomes a compact space. A subbasis for the product topology is given by the collection of sets of the form $\{\omega \in \Omega \mid \omega(i) = k\}$, and an arbitrary open set U can be written as follows: There exists a finite subset $A \subset \mathbb{Z}$ and a collection of subsets $S_i \subset \{0, 1, \ldots, n\}$, $i \in A$, such that

$$U = \{\omega \in \Omega \mid \omega(i) \in S_i, \ i \in A\}.$$

We denote by $\mathcal{B}$ the σ-algebra of Borel sets.

We now construct a probability measure μ on $(\Omega, \mathcal{B})$. The measure of a set such as U just defined should correspond to the probability of finding the number of balls in the first urn to be any number in S_i for each moment of time $i \in A$. The a priori probability p_k of observing k balls in the first urn at time i is independent of i and is

$$p_k = \frac{1}{2^n} \binom{n}{k}.$$

Denote by p the row vector $p = (p_0, p_1, \ldots, p_n)$. For each $k, l \in \{0, 1, \ldots, n\}$, let a_{kl} denote the conditional probability of having l balls in the first urn given that at the preceding moment there were k balls. Drawing a number m from the hat, we find the probability that ball number m is in the first (resp., second) urn is $\frac{k}{n}$ (resp., $\frac{n-k}{n}$). Since the number of balls in each urn can only jump by one unit, we have

$$a_{k,k-1} = \frac{k}{n}, \qquad a_{k,k+1} = \frac{n-k}{n}, \qquad a_{k,l} = 0, \quad |k-l| \neq 1.$$

Denote the matrix of transition probabilities by $A = (a_{k,l})$. Notice that

$$pA = p.$$

(This is a consequence of the identity $\binom{n}{k} = \binom{n-1}{k} + \binom{n-1}{k-1}$.) The probability of observing in the first urn a sequence of numbers $(k_1, \ldots, k_m)$ is given by $p_{k_1} a_{k_1,k_2} \cdots a_{k_{m-1},k_m}$. Therefore, if $B_l(k_1, \ldots, k_m)$ denotes the set of $\omega \in \Omega$ such that $\omega(l + i) = k_i, i = 1, \ldots, m$, we are led to define

$$\mu(B_l(k_1, \ldots, k_m)) = p_{k_1} a_{k_1,k_2} \cdots a_{k_{m-1},k_m}.$$

Using Kolmogorov's consistency theorem [2] it is possible to show that μ indeed extends to a well-defined Borel probability measure on Ω (whose support is the closed subset of all ω such that $\omega(i)$ and $\omega(i + 1)$ differ by 1 or -1). Notice that the equation $p = pA$ is the consistency condition needed in Kolmogorov's theorem so as to have additivity of the measure for the disjoint decomposition

$$\{\omega \in \Omega \mid w(i) = l\} = \bigcup_{k=1}^{n} \{\omega \in \Omega \mid \omega(i) = l \text{ and } \omega(i-1) = k\}.$$

Exercise 2.2.2 Show that μ is invariant under the shift map $\sigma : \Omega \to \Omega$.

We can now interpret the recurrence theorem in light of the previous example. Take, for example, $E = \{\omega \in \Omega \mid \omega(0) = 0\}$, the set of all the performances of the urn experiment in which at time 0 all the balls are found in the second urn. The measure of E is $\mu(E) = p_0 > 0$. According to the theorem,

$$\mu(\{\omega \in E \mid \text{there is } k > 0 \text{ such that } \sigma^k \omega \in E\}) = \mu(E),$$

that is, with probability 1 the first urn is found empty again and again in the future.

The next result gives the mean time of return to E. Let T be a measure-preserving transformation of a probability space $(X, \mathcal{A}, \mu)$. Define for $E \in \mathcal{A}$ the function $k_E : E \to \mathbb{N}$

$$k_E(x) = \min\{n \geq 1 \mid T^n(x) \in E\},$$

the *first return time* to E.

A measurable transformation T of a measure space $(X, \mathcal{A}, \mu)$, with quasi-invariant measure μ, is said to be *ergodic* if the measure of each measurable invariant set is either 0 or 1. It can be shown that σ is a measure-preserving ergodic transformation of the probability space $(\Omega, \mathcal{B}, \mu)$. (Ergodicity here is a consequence of the fact, which is easy to check, that for some positive integer k all the entries of the powers A^k are positive. See, for example, [24] for the ergodicity of Markov chains.)

Theorem 2.2.3 (M. Kac) Let E be a measurable set with positive measure. The mean value over E of the first return time to E for an ergodic measure-preserving transformation T of $(X, \mathcal{A}, \mu)$ is

$$\frac{1}{\mu(E)} \int_E k_E(x) \, d\mu(x) = \frac{1}{\mu(E)}.$$

Proof. Set $E_n = \{x \in E \mid k_E(x) = n\}$. Then the sets $T^i(E_n)$, $n = 1, 2, \ldots$, $0 \leq i \leq n - 1$, are pairwise disjoint, and

$$\bigcup_{n \geq 0} T^n(E) = \bigcup_{n \geq 1} \bigcup_{i=0}^{n-1} T^i(E_n).$$

Therefore

$$\mu\left(\bigcup_{n \geq 0} T^n(E)\right) = \mu\left(\bigcup_{n \geq 1} \bigcup_{i=0}^{n-1} T^i(E_n)\right)$$

$$= \sum_{n=1}^{\infty} \sum_{i=0}^{n-1} \mu(T^i(E_n))$$

$$= \sum_{n=1}^{\infty} n\mu(E_n)$$

$$- \int_E k_E(x) \, d\mu(x).$$

We claim that $S := \bigcup_{n \geq 0} T^n(E)$ has measure 1. Notice that $T^m S \subset S$ for each positive integer m, whence $S \subset T^{-m} S$. Since μ is T^m-invariant,

$$\mu(S) = \mu(T^{-m} T^m S) = \mu(T^m S)$$

for all $m \in \mathbb{Z}$. It follows that the $\mathbb{Z}$-invariant set $\bigcup_{m \in \mathbb{Z}} T^m S$ has the same measure as S. By ergodicity, $\mu(\bigcup_{m \in \mathbb{Z}} T^m S) = 1$. Therefore, $\mu(S) = 1$, so $\int_E k_E(x) \, d\mu(x) = 1$.   □

Therefore, the mean time between two successive moments when the first urn is found empty is $\frac{1}{\mu(E)} = \frac{1}{p_0} = 2^n$ seconds.

Exercise 2.2.4 Suppose n is even. Find the mean return time to the state for which the two urns have the same number of balls.

Markov chains are important in the study of smooth dynamical systems since they are measurable models for a large class of smooth $\mathbb{Z}$-actions that includes Anosov diffeomorphisms. See [7].

2.3 Ergodicity

A measurable map $F : X \to Y$ between G-spaces is called a *G-map* if

$$F(gx) = gF(x)$$

for all $g \in G$ and $x \in X$. If the G-action on $F(X)$ is trivial, we say that F is *G-invariant*. Given a quasi-invariant measure μ on X, we say that a measurable map F between G-spaces X and Y is a *G-map relative to μ* if for each $g \in G$

$$\mu(\{x \in X \mid F(gx) \neq gF(x)\}) = 0.$$

For emphasis, we call a G-map (without reference to any measure) a *strict* G-map. Similarly, we talk about a G-invariant function relative to μ. A measurable set is G-invariant (relative to μ) if its characteristic function is G-invariant (relative to μ.)

It is natural to ask whether a G-map relative to μ agrees μ-a.e. with some strict G-map. If G is countable, this can easily be shown to hold since a countable union of sets of measure zero also has measure zero. The next proposition, taken from [36, B.5], gives the same property for a general G. Whenever reference is made to a measure class on G, it is assumed to be a class defined by a Haar measure (see the beginning of chapter 8). In particular, open subsets of G have positive measure. A set is said to be *conull* if its complement has measure 0.

Proposition 2.3.1 Let X be a standard Borel G-space with quasi-invariant measure μ and Y a standard Borel G-space. Let $F : X \to Y$ be a G-map relative to μ. Then there is a G-invariant conull Borel subset $X_0 \subset X$ and a Borel (strict) G-map $\tilde{F} : X_0 \to Y$ such that $\tilde{F}$ is μ-almost everywhere equal to F.

Proof. Define $X_0 := \{x \in X \mid g \mapsto g^{-1}F(gx) \text{ is a.e. constant in } G\}$. We claim that X_0 is conull. Namely, by definition of a G-map relative to μ, the measurable set $\mathcal{A} := \{(x, g) \in X \times G \mid g^{-1}F(gx) = F(x)\}$ has the property that for each $g \in G$, $\{x \in X \mid (x, g) \in \mathcal{A}\}$ is conull. Let ν be a Haar measure on

G. Then, by Fubini's theorem, $\mathcal{A}$ is conull with respect to the product measure $\mu \times \nu$. By Fubini's theorem again, we conclude that for μ-a.e. $x \in X$, the set $\{g \in G \mid (x, g) \in \mathcal{A}\}$ is ν-conull. The claim now follows.

For each $x \in X_0$, we can define $\tilde{F}(x) \in Y$ by

$$\tilde{F}(x) := g^{-1} F(gx)$$

for a.e. $g \in G$. $\tilde{F}$ is a Borel map since we can write

$$\tilde{F}(x) = \int_G g^{-1} F(gx) \, d\nu(g),$$

where ν is any probability measure in the class of Haar measures. It also follows from the definitions that $\tilde{F}$ and F agree almost everywhere on X_0. From the expression

$$g^{-1} F(ghx) = h[(gh)^{-1} F(ghx)]$$

we deduce that $hx \in X_0$ and $\tilde{F}(hx) = h\tilde{F}(x)$ for each $x \in X_0$ and each $h \in G$. $\qquad\square$

We now turn our attention to the ergodic theory of G-actions. We have already defined what it means for a single measurable transformation to be ergodic. More generally, a measurable G-space X, with a quasi-invariant measure μ, is said to be *ergodic* if every G-invariant measurable set is either null (i.e, it has zero measure) or conull (i.e., its complement has zero measure). We also say that μ is an *ergodic measure* for the G-space X. Therefore, the action is ergodic if X cannot be decomposed as the disjoint union of two G-invariant measurable subsets, both with positive measure.

A nonergodic G-invariant measure can be "disintegrated" into its "ergodic components." The following theorem makes this precise.

Theorem 2.3.2 (Ergodic decomposition) Let X be a compact metrizable space with a continuous action of a locally compact second countable group G, and let μ be a finite Borel measure on X. Then, there exists a measure space $\mathcal{E}$ with measure ν and for each $\alpha \in \mathcal{E}$ a measure space $(X_\alpha, \mathcal{A}_\alpha, \mu_\alpha)$ such that the sets $\{X_\alpha\}_{\alpha \in \mathcal{E}}$ form a partition of X into measurable G-invariant subsets and

1. for any measurable set $A \subset X$, $A \cap X_\alpha$ belongs to $\mathcal{A}_\alpha$ for ν-almost every $\alpha \in \mathcal{E}$ and

$$\mu(A) = \int_{\mathcal{E}} \mu_\alpha(A \cap X_\alpha) \, d\nu(\alpha);$$

2. for ν-almost every $\alpha \in \mathcal{E}$, X_α is an ergodic G-space relative to the measure μ_α.

The foregoing ergodic decomposition can be described as follows. It has already been pointed out before that the space $\mathcal{M}(X)_1$ of all probability measures on X is a compact convex subset of $C(X)^*$, the dual to the space of continuous functions on X with the weak* topology. Let $\mathcal{M}(X, G)_1$ be the closed subset of $\mathcal{M}(X)_1$ consisting of G-invariant probability measures. Then $\mathcal{M}(X, G)_1$ is itself compact and convex. The ergodic G-invariant measures are the *extreme* points of $\mathcal{M}(X, G)_1$, that is, they are those points that cannot be written as a nontrivial convex combination of two G-invariant probability measures. We denote by $\mathcal{E} := \mathcal{E}(X, G)$ the space of extreme points (ergodic measures) in $\mathcal{M}(X, G)_1$. Then it can be shown that for any G-invariant probability measure μ on X there exists a probability measure ν on $\mathcal{E}$ such that

$$\mu = \int_{\mathcal{E}} \alpha \, d\nu(\alpha).$$

(For the details, see [25].)

The next exercise gives one of the simplest examples of an ergodic action.

Exercise 2.3.3 Show that the $\mathbb{Z}$-action on the 1-torus $\mathbb{T}^1 := \mathbb{R}/\mathbb{Z}$ generated by the transformation $[x] \mapsto [x + \alpha]$, where α is an irrational number, is ergodic with respect to the Lebesgue measure on $\mathbb{T}^1$. (Suggestion: Express the characteristic function of a $\mathbb{Z}$-invariant measurable set in terms of its Fourier series and use invariance to deduce that all Fourier coefficients, except possibly the one for the constant term, vanish.) Use the same argument to show that the action of $SL_n(\mathbb{Z})$ on the n-torus $\mathbb{T}^n$ is also ergodic.

The following proposition shows one way in which ergodicity relates to topological properties of the action.

Proposition 2.3.4 Suppose that X is a second countable topological space and that one is given a continuous action of a locally compact second countable topological group G. Suppose that the action is ergodic relative to a quasi-invariant measure μ that is positive on open sets. Then for almost every $x \in X$, the orbit $\{gx \mid g \in G\}$ is dense in X.

Proof. Let $\{U_i\}_{i \in \mathbb{N}}$ be a countable base for the topology of X. The set

$$\bigcap_{i=1}^{\infty} \bigcup_{g \in G} gU_i$$

is a conull set for which every element has a dense orbit. $\square$

Recall that a measurable space is called *countably separated* if there is a countable family of measurable sets that separates points.

Proposition 2.3.5 Suppose that X is an ergodic G-space, Y is a countably separated measurable space, and $f : X \to Y$ is a strict G-invariant measurable function. Then f is almost everywhere constant.

Proof. Let $\{U_i\}$ be a countable family of measurable subsets of Y that separates points. Each $f^{-1}(U_i)$ is a measurable G-invariant subset of X, whence it is either null or conull. Let V be the intersection of all those U_i for which $f^{-1}(U_i)$ is conull. Then $f^{-1}(V)$ is also conull since the intersection of a countable family of conull sets is conull. We claim that V consists of a single point. If that is not the case there would be disjoint sets U_i and U_j in the countable family, each containing a point of V. But $f^{-1}(U_i)$ and $f^{-1}(U_j)$ are disjoint, so they cannot both be conull; say that $f^{-1}(U_j)$ has measure 0. Therefore, V is contained in U_i, which is a contradiction. $\qquad\square$

2.4 Measure-Theoretic Entropy

We define now a notion of entropy for measure-preserving transformations, called the *Kolmogorov–Sinai entropy*, or *measure-theoretic entropy*. The notion was introduced in ergodic theory by Kolmogorov in 1958, who was influenced by the mathematical work of Claude Shannon on information transmission, initiated in 1948.

Roughly speaking, entropy is an average measure of the rate at which a process modeled by a measure-preserving transformation becomes random. The precise definition is as follows. Let T be a measure-preserving transformation of a probability space $(X, \mathcal{A}, \mu)$ and let $\xi = \{A_1, \ldots, A_n\}$ be a finite partition of X into measurable sets (i.e., the union of the A_i is conull in X and $A_i \cap A_j$ is null for each $i, j, i \neq j$). For each positive integer N, one defines a new partition ξ^N consisting of the intersections

$$A_{i_0} \cap T^{-1}\left(A_{i_1}\right) \cap \cdots \cap T^{-(N-1)}\left(A_{i_{N-1}}\right)$$

for all $1 \leq i_0 \leq \cdots \leq i_{N-1} \leq n$. Notice that a point $x \in X$ belongs to the foregoing element of ξ^N if the orbit segment $f^k(x)$, $0 \leq k \leq N - 1$, falls into A_{i_k} at time k. Introduce now the *entropy of the partition ξ^N*

$$H(\xi^N) := - \sum_{A \in \xi^N} \mu(A) \log \mu(A).$$

(If $\mu(A) = 0$, $\mu(A) \log \mu(A)$ is naturally defined to be 0.) It is clear that $H(\xi^N) \geq 0$.

We define the entropy of T with respect to the partition ξ by the limit

$$h(T, \mu, \xi) := \limsup_{N \to \infty} \frac{1}{N} H(\xi^N) \geq 0.$$

It can be shown, in fact, that the limit exists, so it is not necessary to use the lim sup.

Finally, define the measure-theoretic entropy of T relative to μ by

$$h(T, \mu) := \sup\{h(T, \mu, \xi) \mid \xi \text{ is a finite measurable partition of } X\}.$$

For the details, see [16].

It can be shown, for example, that if $T = \sigma$ is the shift map for a Markov chain with matrix $A = (a_{ij})$ and probability vector p (as defined earlier in the discussion of the urn problem) then

$$h(\sigma, \mu) = -\sum_{i,j=1}^{n} p_j a_{ij} \log a_{ij}.$$

A similar but simpler example is provided by *Bernoulli shifts*. Here, the measurable space $(\Omega = \{1, \ldots, n\}^{\mathbb{Z}}, \mathcal{B})$ and the transformation σ are the same ones defined before for the Markov chains, but the measure μ is defined as follows. Let $0 < p_i$, $i = 1, \ldots, n$, be the probabilities of occurrence of events labeled by i in a complete family of mutually exclusive events, so $\sum_{i=1}^{n} p_i = 1$. Define μ by extending to $\mathcal{B}$ the measure

$$\mu(\{\omega \in \Omega \mid \omega(i_1) = j_1, \ldots, \omega(i_k) = j_k\}) = p_{j_1} \cdots p_{j_k}.$$

Exercise 2.4.1 Let μ be the Bernoulli measure on Ω just defined, associated to the probability vector $(p_1, \ldots, p_n)$, and let ξ be the measurable partition of Ω by the sets $A_i = \{\omega \in \Omega \mid \omega(0) = i\}$, $i = 1, \ldots, n$. Show that $h(\sigma, \mu, \xi) = -\sum_{i=1}^{n} p_i \log p_i$. (It can be shown that in this case $h(\sigma, \mu) = h(\sigma, \mu, \xi)$. See [16] for the details.)

Measure-preserving transformations T_1 and T_2 of probability spaces $(X_1, \mathcal{B}_1, \mu_1)$ and $(X_1, \mathcal{B}_1, \mu_1)$, respectively, are said to be *isomorphic* if there exist conull sets $M_i \in \mathcal{B}_i$, $i = 1, 2$, such that M_i is T_i-invariant and there is an invertible measure-preserving transformation $F : X_1 \to X_2$ with $(F \circ T_1)(x) = (T_2 \circ F)(x)$ for all $x \in M_1$. It is not difficult to see that the measure-theoretic entropy is the same for isomorphic measure transformations.

The question of finding conditions for two Bernoulli shifts to be isomorphic is one of the oldest questions in ergodic theory. It should be pointed out that a large class of measure-preserving diffeomorphisms of smooth manifolds can be shown to be (measurably) isomorphic to Bernoulli shifts, so the question is important for smooth systems as well. For example, it can be shown that an Anosov diffeomorphism of a compact manifold M preserving a smooth measure (i.e., a measure coming from a volume form on M) is isomorphic to a Bernoulli shift.

For example, given two Bernoulli shifts T_1 and T_2 defined by the probability vectors $(\frac{1}{4}, \frac{1}{4}, \frac{1}{4}, \frac{1}{4})$ and $(\frac{1}{2}, \frac{1}{8}, \frac{1}{8}, \frac{1}{8}, \frac{1}{8})$, how can one tell whether they are isomorphic? They are indeed isomorphic, by the the following hard theorem.

Theorem 2.4.2 (Ornstein) Two Bernoulli shifts are isomorphic if and only if they have the same entropy.

The next theorem shows how the topological and measure-theoretic entropies are related.

Theorem 2.4.3 (Variational principle) Let $f : X \to X$ be a homeomorphism of a compact metric space. Then the topological entropy $h_{\text{top}}(f)$ is the supremum of $h(f, \mu)$ over all f-invariant Borel probability measures on X.

A proof of Ornstein's theorem can be found, for example, in [9]. For the variational principle, the reader will find a proof in [16].

3

Smooth Actions and Lie Theory

We develop in this chapter the first steps in Lie theory. In the same way that a differential equation – or vector field – on a manifold M can be integrated to a (local) flow on M, so a (local) action of a Lie group G can be viewed as having an infinitesimal generator given by an "action" of the Lie algebra of G. After introducing the basic notions concerning Lie groups and Lie algebras, we discuss the general problem of integrating smooth actions of Lie algebras – or "infinitesimal" actions of G. We also discuss the notion of "completeness" and the question of existence of global actions.

3.1 Vector Fields and Flows

$\mathbb{R}$-actions arise naturally in the context of differential equations, for example, whenever one integrates a (complete) vector field on a smooth manifold. We review here some of the general facts and notations concerning vector fields on manifolds and their flows.

The word *smooth* will continue to be used as a substitute for C^∞. All manifolds in this chapter will be smooth. Maps are called *differentiable* (resp., C^r differentiable, $r \geq 1$) if they are C^1 (resp., C^r). The default hypothesis (and conclusion!) here is that all maps are smooth, although most of the results in the chapter hold under weaker differentiability assumptions.

We continue to denote by $Df_x : T_x M \to T_{f(x)} N$ the differential of a map $f : M \to N$ between manifolds M and N. Thus, if $v \in T_x M$ and $c : (-\epsilon, \epsilon) \to M$ ($\epsilon > 0$) is a differentiable curve in M such that $c(0) = x$ and $c'(0) = v$, then

$$Df_x v = \frac{d}{dt}\bigg|_{t=0} (f \circ c)(t).$$

32

The differential at $x \in M$ of a differentiable function $f : M \to \mathbb{R}$ will be written as follows:

$$Df_x v = (df_x v) \frac{d}{dt},$$

where $\frac{d}{dt}$ is the coordinate vector field on $\mathbb{R}$ and $df_x : T_x M \to \mathbb{R}$ is a linear functional, called the *exterior derivative* of f.

A smooth vector field on M is a smooth section of the tangent bundle TM, that is, a C^∞ map $X : M \to TM$ such that $X_x := X(x) \in T_x M$ for all $x \in M$. X determines a first-order differential operator on smooth functions that assigns to each $h \in C^\infty(M)$ the smooth function Xh such that $(Xh)(x) = dh_x X(x)$, $x \in M$. The differential operator associated to X uniquely determines X as a section of the tangent bundle. The space of smooth vector fields on M will be denoted $\mathfrak{X}(M)$.

The *bracket* of two vector fields is the map

$$[\cdot, \cdot] : \mathfrak{X}(M) \times \mathfrak{X}(M) \to \mathfrak{X}(M)$$

defined by the property $[X, Y]h = X(Yh) - Y(Xh)$ for all $h \in C^\infty(M)$. If $\mathfrak{X}(M)$ is regarded as a vector space over $\mathbb{R}$, the bracket is a skew-symmetric bilinear map that satisfies the *Jacobi identity*

$$[X, [Y, Z]] = [[X, Y], Z] + [Y, [X, Z]]$$

for all $X, Y, Z \in \mathfrak{X}(M)$.

The pair $(\mathfrak{X}(M), [\cdot, \cdot])$ is a *Lie algebra*. By definition, a Lie algebra over a field k consists of a pair $(V, [\cdot, \cdot])$, where V is a vector space over k and $[\cdot, \cdot] : V \times V \to V$ is a skew-symmetric k-bilinear map that satisfies the Jacobi identity. Given two Lie algebras $(V, [\cdot, \cdot]_V)$ and $(W, [\cdot, \cdot]_W)$, a k-linear map $a : V \to W$ such that

$$a([u, v]_V) = [a(u), a(v)]_W$$

for all $u, v \in V$ is called a *homomorphism* of Lie algebras. If a is an isomorphism of vector spaces and a homomorphism of Lie algebras, we say that it is an *isomorphism of Lie algebras*. We will often refer to V as the Lie algebra, since the bracket operation will usually be clear from the context.

If f is a smooth diffeomorphism from a manifold M onto another manifold N and $X \in \mathfrak{X}(M)$, we denote by $f_* X$ the vector field on N, called the *push-forward* of X, such that

$$(f_* X)(y) := (Df)_{f^{-1}(y)} X(f^{-1}(y))$$

for all $y \in N$. More generally, if f is a (not necessarily invertible) smooth map from M into N, X is a vector field on M, and Y is a vector field on N, then X and Y are said to be f-*related* if for each $x \in M$

$$Df_x X(x) = Y(f(x)).$$

Proposition 3.1.1 Let $X_1, X_2 \in \mathfrak{X}(M)$ and $Y_1, Y_2 \in \mathfrak{X}(N)$, and suppose that X_i and Y_i are f-related for $i = 1, 2$. Then $[X_1, X_2]$ and $[Y_1, Y_2]$ are also f-related. If f is a diffeomorphism, then $f_* : \mathfrak{X}(M) \to \mathfrak{X}(N)$ is a Lie algebra isomorphism.

Proof. Fix $h \in C^\infty(N)$. If $X \in \mathfrak{X}(M)$ and $Y \in \mathfrak{X}(N)$ are f-related, then

$$(X(h \circ f))(x) = (Df_x X_x)h = Y_{f(x)}h = ((Yh) \circ f)(x)$$

for all $x \in M$, that is, $X(h \circ f) = Yh \circ f$. Therefore,

$$X_i X_j (h \circ f) = X_i(Y_j h \circ f) = Y_i Y_j h \circ f$$

It follows that

$$(Df_x[X_1, X_2]_x)h = [X_1, X_2]_x(h \circ f) = ([Y_1, Y_2]h)(f(x))$$

for each $x \in M$. The claims follow.   □

A vector field $\mathfrak{X} \in \mathfrak{X}(M)$ determines a differential equation, $\dot{x} = X(x)$, on M. It is proved in courses on ODEs that for any fixed x_0 in M there exists an open interval I containing 0 and a unique smooth curve $c : I \to M$ such that $c(0) = x_0$ and $c'(t) = X(c(t))$ for all $t \in I$. There is, in fact, a maximal interval $I(x_0)$ that contains 0 on which a solution exists and is unique. Moreover, the solution $c(t)$ depends smoothly on the initial point x_0. More precisely, there is an open set $D \subset \mathbb{R} \times M$ and a smooth map $\Phi : D \to M$ such that $\{t \in \mathbb{R} \mid (t, x) \in D\} = I(x)$ is the maximal interval of existence for the initial point $x \in M$ and $c_x(t) = \Phi(t, x)$, $t \in I(x)$, is the unique solution over $I(x)$ with initial point x. Φ is a *local flow*, that is,

$$\Phi(s, \Phi(t, x)) = \Phi(t + s, x) \text{ and } \Phi(0, x) = x$$

if $(t, x), (s, \Phi(t, x)) \in D$. (It is then automatic that $(t + s, x)$ will belong to D.) If X is, moreover, a smooth function of a number of parameters, the local flow Φ will also depend smoothly on those parameters.

A vector field X is called *complete* if its local flow Φ has full domain, that is, $D = \mathbb{R} \times M$. If X is complete, we obtain a global flow on M, that is, the maps $\Phi_t := \Phi(t, \cdot)$ for $t \in \mathbb{R}$ are diffeomorphisms such that Φ_0 is the identity map and $\Phi_t \circ \Phi_s = \Phi_{t+s}$ for all $t, s \in \mathbb{R}$. In other words, we obtain a smooth $\mathbb{R}$-action on M.

Exercise 3.1.2 Suppose that the domain D of a local flow Φ of $X \in \mathfrak{X}(M)$ contains $I \times M$, where I is some open interval containing 0. Show that Φ is (or, rather, can be uniquely extended to) a global flow for X. If M is compact, conclude that any smooth vector field is complete.

Proposition 3.1.3 Let $X, Y \in \mathfrak{X}(M)$ and denote by Φ_t the local flow of X, with domain D. Then

$$\frac{d}{dt}((\Phi_{-t})_* Y)_x = ((\Phi_{-t})_* [X, Y])_x$$

for all $(x, t) \in D$.

Proof. We will use the following easily obtained fact, which the reader may wish to check as an exercise: If Φ_t is the local flow of $X \in \mathfrak{X}(M)$, then for any smooth function f on M, for each $x \in M$, and for all t sufficiently small,

$$f(\Phi_t(x)) = f(x) + t(Xf)(x) + t^2 \Lambda(x, t),$$

where $\Lambda(x, t)$ is a smooth function of (x, t) such that $\Lambda(x, 0) = \frac{1}{2}(X^2 f)(x)$.
Since $\Phi_{t+s} = \Phi_t \circ \Phi_s$, it follows by the chain rule that

$$(\Phi_{t+s})_* Y = (\Phi_t)_* (\Phi_s)_* Y$$

at points where the expression in defined. Therefore, it suffices to prove that $\frac{d}{dt}|_{t=0} ((\Phi_{-t})_* Y)_x = [X, Y]_x$, for all $x \in M$. Let h be an arbitrary smooth function on M. Then, using the preceding remark,

$$\frac{d}{dt}\bigg|_{t=0} ((\Phi_{-t})_* Y)_x h = \frac{d}{dt}\bigg|_{t=0} Y_{\Phi_t(x)}(h \circ \Phi_{-t})$$

$$= \frac{d}{dt}\bigg|_{t=0} Y_{\Phi_t(x)}(h - tXh + t^2 \Lambda(\cdot, -t))$$

$$= \frac{d}{dt}\bigg|_{t=0} (Yh)(\Phi_t(x)) - \frac{d}{dt}\bigg|_{t=0} tY_{\Phi_t(x)}(Xh)$$

$$= X_x(Yh) - Y_x(Xh)$$

$$= [X, Y]_x h.$$

Therefore, $\frac{d}{dt}|_{t=0}((\Phi_{-t})_* Y)_x = [X, Y]_x.$ $\square$

Exercise 3.1.4 Let f be a smooth map from a manifold M into another manifold N, and suppose that $X \in \mathfrak{X}(M)$ and $Y \in \mathfrak{X}(N)$ are f-related vector fields. Denote by Φ^X and Φ^Y the local flows of X and Y, with domains D_X and D_Y. Show that D_Y contains the set $\{(t, f(x)) \mid (t, x) \in D_X\}$ and that

$$\left(f \circ \Phi_t^X\right)(x) = \left(\Phi_t^Y \circ f\right)(x)$$

for all (t, x) in D_X. In particular, conclude that if $X \in \mathfrak{X}(M)$ is such that $f_* X = X$ for all f in a group of diffeomorphisms acting transitively on M, then X is complete.

Exercise 3.1.5 Let $X, Y \in \mathfrak{X}(M)$ be complete vector fields such that $[X, Y] = \lambda Y$, where λ is a constant. Denote by Φ^X and Φ^Y the global flows of X and Y. Show that $(\Phi_t^X)_* Y = e^{-\lambda t} Y$ and

$$\Phi_t^X \circ \Phi_s^Y = \Phi_{e^{-\lambda t}s}^Y \circ \Phi_t^X.$$

In particular, if $[X, Y] = 0$ the flows of X and Y commute.

Given homeomorphisms f, g of M, we define their *commutator* as

$$(f, g) := g^{-1} \circ f^{-1} \circ g \circ f.$$

If Φ and Ψ are the respective local flows of $X, Y \in \mathfrak{X}(M)$, then it can be shown by a simple but tedious computation that $X + Y = \frac{d}{dt}|_{t=0}(\Phi_t \circ \Psi_t)$ and $[X, Y] = \frac{d}{dt}|_{t=0^+}(\Phi_{\sqrt{t}}, \Psi_{\sqrt{t}})$. More precisely, for each smooth function f on M and each $x \in M$, it can be shown that

$$f((\Phi_\epsilon \circ \Psi_\epsilon)(x)) = f(x) + \epsilon((X + Y)f)(x) + o_1(x, \epsilon),$$

$$f((\Phi_{\sqrt{\epsilon}}, \Psi_{\sqrt{\epsilon}})(x)) = f(x) + \epsilon([X, Y]f)(x) + o_2(x, \epsilon),$$

where $o_i(x, \epsilon)$, $\epsilon \geq 0$, are smooth functions of x and $\lim_{\epsilon \to 0^+} o_i(x, \epsilon)/\epsilon = 0$. If (t, x) lies in the domain of the local flow Φ^X of X, we denote

$$\exp(tX)(x) := \Phi_t^X(x).$$

Notice that we have indeed $\Phi_t^X(x) = \Phi_1^{tX}(x)$ and

$$\exp((t + s)X)(x) = (\exp(tX) \circ \exp(sX))(x),$$

whenever the maps are defined.

Theorem 3.1.6 If $X, Y \in \mathfrak{X}(M)$, then

$$\exp(t(X + Y))(x) = \lim_{m \to \infty} \left(\exp\left(\frac{t}{m}X\right) \circ \exp\left(\frac{t}{m}Y\right)\right)^m (x)$$

$$\exp(t[X, Y])(x) = \lim_{m \to \infty} \left(\exp\left(\frac{\sqrt{t}}{m}X\right), \exp\left(\frac{\sqrt{t}}{m}Y\right)\right)^{m^2}(x),$$

for all $x \in M$ and all t sufficiently small.

Proof. The two identities can be proven in similar ways. We show the second identity and leave the first one as an exercise to the reader. Since the problem is local, we may assume that $M = \mathbb{R}^n$ and that X and Y have compact support. In particular, the flows can be assumed to be global and to have compact support, that is, the set of points where $\exp(tX)(x) \neq x$ or $\exp(tY)(x) \neq x$ for some $t \in \mathbb{R}$ is relatively compact.

Write $A_m := (\exp(\frac{\sqrt{t}}{m} X), \exp(\frac{\sqrt{t}}{m} Y))$ and $B_m := \exp(\frac{t}{m^2}[X, Y])$. Notice that A_m and B_m are diffeomorphisms of $\mathbb{R}^n$ and there is a compact set in $\mathbb{R}^n$ outside of which A_m and B_m are the identity map. Let $\| \cdot \|$ denote the Euclidean norm on $\mathbb{R}^n$. By approximating the commutator of two flows by its first-degree Taylor polynomial, as described before, we obtain

$$\| B_m x - A_m x \| \le c(m) \frac{1}{m^2}$$

for all $x \in \mathbb{R}^n$, where $c(m)$ goes to 0 as m goes to $+\infty$. By the homomorphism property of the flow of $[X, Y]$, there is a constant $K > 0$ such that

$$\left\| \left(DB_m^k \right)_x \right\| \le K$$

for all $k \le m^2$ and all m. (If Θ_t denotes the flow of $[X, Y]$, we can write $B_m^k = \Theta_{tk/m^2}$, so that $K = \max\{ \| (D\Theta_s)_x \| \mid 0 \le s \le 1\}$.)

Let f be an arbitrary smooth function on $\mathbb{R}^n$ with compact support, let $x, y \in \mathbb{R}^n$ be arbitrary points, and let $\gamma : [0, 1] \to \mathbb{R}^n$ denote the straight line segment from x to y. Then, writing

$$\| df \| := \sup\{ \| df_x v \| / \| v \| \mid x, v \in \mathbb{R}^n, v \neq 0\},$$

we have

$$\left| \left(f \circ B_m^k \right)(y) - \left(f \circ B_m^k \right)(x) \right| = \left| \int_0^1 \frac{d}{dt} \left(f \circ B_m^k \circ \gamma \right)(t) \, dt \right|$$
$$\le \int_0^1 K \| df \| \| \gamma'(t) \| \, dt$$
$$= K \| df \| \| y - x \|.$$

Observe that $f(B_m^{m^2} z) - f(A_m^{m^2} z)$ can be written as a telescoping sum as follows:

$$\sum_{k=0}^{m^2-1} \left[f\left(B_m^k \circ B_m \circ A_m^{m^2-k-1} z \right) - f\left(B_m^k \circ A_m \circ A_m^{m^2-k-1} z \right) \right].$$

Since $A_m^{m^2-k-1}$ is a diffeomorphism, we can bound the absolute value of the sum by the expression

$$\sum_{k=0}^{m^2-1} \sup_{z \in \mathbb{R}^n} \left| f\left(B_m^k \circ B_m z\right) - f\left(B_m^k \circ A_m z\right) \right| \le m^2 K \|df\| \|B_m z - A_m z\|$$

$$\le m^2 K \|df\| c(m) \frac{1}{m^2}$$

$$= K \|df\| c(m),$$

which goes to 0 as m goes to $+\infty$. Therefore,

$$\exp(t[X, Y])(x) - \left(\exp\left(\frac{\sqrt{t}}{m} X\right), \exp\left(\frac{\sqrt{t}}{m} Y\right) \right)^{m^2}(x)$$

goes to 0 as m goes to $+\infty$, which is the assertion of the theorem. $\square$

The following notation will be used later, in the context of left-invariant vector fields on Lie groups. If f is a diffeomorphism of M and $Y \in \mathfrak{X}(M)$, we define

$$\mathrm{Ad}(f)Y := f_* Y.$$

If $X, Y \in \mathfrak{X}(M)$ are complete vector fields, it is immediate that

$$\exp(t \, \mathrm{Ad}(\exp X)Y) = \exp X \circ \exp(tY) \circ \exp(-X).$$

3.2 Foliations

The orbits of a locally free smooth action partition the manifold into a family of submanifolds forming a *foliation* of M. We now discuss some of the basic notions regarding foliated spaces.

Let M be a smooth n-dimensional manifold and let E be a smooth subbundle of the tangent bundle TM. More precisely, for each $y \in M$, we have an m-dimensional subspace $E(y)$ of $T_y M$ and, for each $x \in M$, there are smooth vector fields $X_1, \ldots, X_m$ defined on some neighborhood U of x such that $X_1(y), \ldots, X_m(y)$ form a basis of $E(y)$ for each $y \in U$.

We say that a vector field X *belongs* to E if $X(x) \in E(x)$ for each $x \in M$. The subbundle E is called *involutive* if whenever X, Y are differentiable vector fields that belong to E, the bracket $[X, Y]$ also belongs to E.

A submanifold L of M is an *integral manifold* of a subbundle E of TM if for each $x \in L$, $T_x L = E(x)$. A coordinate chart $(x_1, \ldots, x_n, U)$ on M is said to be *adapted* to E if the coordinate vector fields $\frac{\partial}{\partial x_i}$, $i = 1, \ldots, m$, belong to $E|_U$.

Having a coordinate chart adapted to E clearly implies that the m-dimensional slices of U obtained by fixing the values of the remaining $n - m$ coordinate functions are integral manifolds of $E|_U$, and that $E|_U$ is involutive.

Theorem 3.2.1 (Frobenius) Given an m-dimensional smooth subbundle E of TM, E is involutive if and only if for each $x \in M$ there is a coordinate chart adapted to E defined on a neighborhood of x.

Proof. We assume that E is involutive and show the existence of the adapted coordinate system. The other direction has already been noted.

With an appropriate choice of smooth coordinate chart $(z_1, \ldots, z_n, W)$ centered at x, we may assume that E is a smooth subbundle of the tangent bundle of $\mathbb{R}^n$ and x is the origin of $\mathbb{R}^n$. We write $\mathbb{R}^n = \mathbb{R}^m \times \mathbb{R}^{n-m}$ and denote by Π the projection onto the first factor, which we refer to as the horizontal subspace of the product. We can choose the coordinates so that E is tangent to the horizontal subspace at the origin. Fix a neighborhood of the origin in $\mathbb{R}^n$ of the form $V \times U$, where V (resp., U) is a neighborhood of the origin in $\mathbb{R}^m$ (resp., $\mathbb{R}^{n-m}$). By choosing these neighborhoods sufficiently small, we may assume that the restriction of $(D\Pi)_x$ to $E(x)$ is a linear isomorphism onto $T_{\Pi(x)}\mathbb{R}^m$ for all $x \in V \times U$.

Define vector fields $Z_1, \ldots, Z_m$ on $V \times U$ that belong to E and project onto the coordinate vector fields $\frac{\partial}{\partial z_1}, \ldots, \frac{\partial}{\partial z_m}$ on V (i.e., Z_i is Π-related to $\frac{\partial}{\partial z_i}$, for $i = 1, \ldots, m$). Notice that $[Z_i, Z_j]$ also belongs to E since E is involutive, and is Π-related to the zero vector field $0 = [\frac{\partial}{\partial z_i}, \frac{\partial}{\partial z_j}]$. Therefore $Z_1, \ldots, Z_m$ are commuting vector fields that span the subbundle E on $V \times U$.

Denote by Φ^i_t the local flow associated to Z_i near the origin. By making V and U sufficiently small, we may assume that there is an $\epsilon > 0$ such that the domain of Φ^i, where $\Phi^i(x, t) := \Phi^i_t(x)$, contains $V \times U \times (-\epsilon, \epsilon)$. We define a smooth map $\Phi : (-\epsilon, \epsilon)^m \times U \to \mathbb{R}^n$ given by

$$\Phi(t_1, \ldots, t_m, y) = \left(\Phi^1_{t_1} \circ \cdots \circ \Phi^m_{t_m} \right)((0, y)).$$

(Notice that Φ fixes the origin.) A simple computation shows that $(D\Phi)_0$ is the identity map of $\mathbb{R}^n$. Therefore, by the inverse function theorem, Φ has a smooth inverse on some neighborhood of the origin. The coordinate system we seek can now be obtained by restricting the original coordinates z_i to a sufficiently small open subset of W and composing $\varphi = (z_1, \ldots, z_n)$ on the left with the inverse of Φ. $\square$

In the preceding proof of the Frobenius theorem we have produced about each point in M a coordinate chart (φ, W) with the following property: If

$\varphi = (x_1, \ldots, x_n)$, then $\frac{\partial}{\partial x_1}, \ldots, \frac{\partial}{\partial x_m}$ belong to the m-dimensional subbundle E. A coordinate chart having this property will be called a *Frobenius chart*. Given an involutive E, we can fix a coordinate cover of M consisting of Frobenius charts $\{(\varphi_\alpha, V_\alpha) \mid \alpha \in \mathcal{A}\}$, $\varphi_\alpha = (x_1^\alpha, \ldots, x_n^\alpha)$, such that the index set $\mathcal{A}$ is at most countable, $\{V_\alpha \mid \alpha \in \mathcal{A}\}$ is a locally finite open cover of M, each x_i^α ranges over the interval $(-2, 2)$, and if W_α is the open subset of V_α defined by the inequalities $-1 < x_i^\alpha < 1, 1 \le i \le n$, then $\{W_\alpha \mid \alpha \in \mathcal{A}\}$ is an open cover of M. A coordinate cover $\{W_\alpha \mid \alpha \in \mathcal{A}\}$ satisfying all of these properties will be called a *regular Frobenius cover* for E.

If $a = (a_{m+1}, \ldots, a_n)$ belongs to the open cube $(-1, 1)^{n-m}$, the equations $x_i^\alpha = a_i, m + 1 \le i \le n$, specify an integral manifold $P^\alpha(a)$ to $E|_{W_\alpha}$, called a *plaque* of E in W_α. Similarly, there is a plaque $\check{P}^\alpha(a)$ of E in V_α and the closure of $P^\alpha(a)$ is contained in $\check{P}^\alpha(a)$. Also notice that the closure of W_α is given in the coordinates of V_α by the closed cube $[-1, 1]^n$.

Lemma 3.2.2 If E is involutive and (φ, W) is one of the $(\varphi_\alpha, W_\alpha)$ or $(\varphi_\alpha, V_\alpha)$ just defined, then every connected integral manifold for $E|_W$ lies in some plaque of W. Consequently, if N_1 and N_2 are arbitrary integral manifolds of E, then $N_1 \cap N_2$ is an integral manifold of E.

Proof. We write $\varphi = (x_1, \ldots, x_n)$ and recall that $x_{m+1}, \ldots, x_n$ are constant along plaques. The derivative of x_{m+i} along any vector in E is, therefore, 0. Let $p \in W$ and let $P(a)$ be the plaque through p. If $N \subset W$ is also a connected integral manifold through p and if $q \in N$, there is a differentiable curve $c : [0, 1] \to N$ with $c(0) = p$ and $c(1) = q$. But $c'(t) \in E(c(t))$, so $\frac{d}{dt}x_{m+i}(c(t)) = 0$ for each i. It follows that $x_{m+i}(c(t))$ is constant; hence it is equal to a_{m+i} since $x_{m+i}(p) = a_{m+i}$. Therefore, the curve lies entirely in the plaque $P(a)$.

Let N_1 and N_2 be arbitrary integral manifolds to E and assume that $N_1 \cap N_2$ is nonempty. Let $p \in N_1 \cap N_2$ and choose W about p as before. Then, the component of p in $N_1 \cap W$ lies in a plaque $P(a)$, and since N_1 has the same dimension as the plaque, $N_1 \cap W$ is an open subset of $P(a)$. Similarly, the component of p in $N_2 \cap W$ is an open subset of $P(a)$. Therefore, the component of p in $N_1 \cap N_2 \cap W$ is an open subset of $P(a)$, and hence is an integral manifold to E. Since $p \in N_1 \cap N_2$ is arbitrary, $N_1 \cap N_2$ is an integral manifold of E. $\square$

We claim that each plaque $P^\alpha(a)$ of E in W_α meets at most finitely many plaques of E in other charts of the regular Frobenius cover. In fact, since the cover $\{V_\alpha \mid \alpha \in \mathcal{A}\}$ is locally finite, each point x in the closure $\overline{P^\alpha(a)}$ has a connected neighborhood U_x in $\check{P}^\alpha(a)$ that meets only finitely many V_{α_i} and lies entirely in any one that it meets. By the previous lemma, U_x lies entirely in a plaque $\check{P}^{\alpha_i}(a_i)$ of each of these charts. Since $\overline{P^\alpha(a)}$ is compact, the claim follows.

If E is involutive, we say that $x, y \in M$ are *E-equivalent*, and write $x \sim y$, if there are connected integral manifolds $N_1, N_2, \ldots, N_r$ to E such that $x \in N_1$, $y \in N_r$, and $N_i \cap N_{i+1}$ is nonempty for $1 \le i < r$. It is immediate that $\sim$ is an equivalence relation on M. Each equivalence class L is the union of integral manifolds to E. It is therefore natural to topologize L by declaring open sets to be the unions of integral manifolds to E. This indeed defines a topology. Notice that the previous lemma asserts, in particular, that a finite intersection of open sets is open.

The topology of L is second countable and Hausdorff, which can be seen as follows. First notice that we can take the connected integral manifolds N_i that define $\sim$ to be plaques of a regular Frobenius cover of M, so that any two plaques contained in L can be joined by a finite chain of plaques $P_1, P_2, \ldots, P_r$, with $P_i \cap P_{i+1}$ nonempty. Moreover, we know that each plaque intersects only finitely many other plaques. It follows that, fixing a plaque $P_0 \subset L$ and a positive integer m, we can find a finite set $\mathcal{P}_r$ of plaques in L such that any plaque in L that can be joined to P_0 by a chain of r or fewer plaques can be joined by plaques in the set $\mathcal{P}_r$. Therefore, we only need a countable set, $\bigcup_{r \in \mathbb{N}} \mathcal{P}_r$, of plaques to cover L. Since each plaque is a second countable Hausdorff topological space, L also is second countable Hausdorff. It is also clear that L is path connected and locally Euclidean.

Therefore, L is a connected m-dimensional topological manifold. The Frobenius theorem provides a smooth atlas for L and shows that L is immersed in M. L is obviously a maximal connected integral manifold to E. We have thus established the following result.

Proposition 3.2.3 Let E be a smooth m-dimensional involutive subbundle of TM. Then the equivalence classes of $\sim$ are the maximal connected integral manifolds to E. Each class is thus an immersed k-dimensional smooth submanifold of M.

The decomposition of M into E-equivalence classes is called a *foliation* $\mathcal{E}$ of M. Each E-equivalence class L is called a *leaf* of $\mathcal{E}$. If the dimension of M is n and the leaves have dimension m, we say that $\mathcal{E}$ has *codimension* $n - m$.

3.3 Lie Groups and Lie Algebras

A *Lie group* G is a smooth manifold provided with a group structure such that the group operations of multiplication and inversion are smooth mappings (from $G \times G$ to G and from G to G, resp.). A *Lie subgroup* of a Lie group G is a subgroup that is also a smooth submanifold of G. A *Lie group homomorphism* between Lie groups H and G is a group homomorphism $\varphi : H \to G$ that is also a smooth map. If φ is, moreover, a diffeomorphism, then it is called an *isomorphism* and the groups H and G are said to be isomorphic.

Given g_0 in a Lie group G, we define the diffeomorphisms L_{g_0} and R_{g_0} from G into itself, called the *left translation* and *right translation* by g_0, respectively, by letting $L_{g_0}g := g_0 g$ and $R_{g_0}g := g g_0$. Also define

$$C_{g_0} := L_{g_0} \circ (R_{g_0})^{-1} = (R_{g_0})^{-1} \circ L_{g_0},$$

which sends g to $g_0 g g_0^{-1}$. One easily checks that the maps L, R^{-1}, C from $G \times G$ into G defined by $L(g_1, g_2) := L_{g_1}g_2$, $R^{-1}(g_1, g_2) := R_{g_1^{-1}}g_2$, and $C(g_1, g_2) := C_{g_1}(g_2)$ correspond to smooth actions of G on itself.

Since C_{g_0} fixes the identity e, its differential $\mathrm{Ad}(g_0) := (DC_{g_0})_e$ is a linear map of $\mathfrak{g} := T_e G$. For example, if $G = GL(n, \mathbb{R})$ is the group of all invertible real matrices of order n, the tangent space $\mathfrak{g}$ is easily seen to be the space of all real matrices of order n, and we have for $X \in \mathfrak{g}$ that $\mathrm{Ad}(g_0)X = g_0 X g_0^{-1}$.

It follows by the chain rule applied to $C_{gh} = C_g \circ C_h$, $g, h \in G$, (and $C_e = \mathrm{id}_G$) that

$$\mathrm{Ad} : G \to GL(\mathfrak{g})$$

is a homomorphism of G into the group $GL(\mathfrak{g})$ of linear isomorphisms of $\mathfrak{g}$, called the *adjoint representation* of G.

A *left-invariant* (resp., *right-invariant*) vector field X on a Lie group G is a vector field such that $(L_g)_* X = X$ (resp., $(R_g)_* X = X$) for all $g \in G$. Therefore, a left-invariant vector field is completely determined by its value at e since $X(g) = (DL_g)_e X(e)$ for all $g \in G$, and we obtain a linear isomorphism between $\mathfrak{g}$ and the finite-dimensional subspace of $\mathfrak{X}(G)$ of left-invariant vector fields. It is immediate that if X, Y are left-invariant vector fields, the bracket $[X, Y] \in \mathfrak{X}(G)$ also is left-invariant. Similar statements hold for right-invariant fields.

The *Lie algebra* of a Lie group G is the finite-dimensional subalgebra of $\mathfrak{X}(G)$ consisting of left-invariant vector fields.

We refer to $\mathfrak{g} = T_e G$ also as the Lie algebra of G and denote by the same symbol $[\cdot, \cdot]$ the Lie bracket on $\mathfrak{g}$ obtained by the correspondence between elements of $\mathfrak{g}$ and left-invariant vector fields. More precisely, for left-invariant fields X, Y with $X(e) = u$, $Y(e) = v$, we set $[u, v] := [X, Y](e)$.

The flow $t \mapsto \Phi_t$ of a left-invariant vector field X commutes with left translations, that is, $L_g \circ \Phi_t = \Phi_t \circ L_g$ for all $g \in G$. As a consequence, any flow line is a left translate of the curve $c(t) = \Phi_t(e)$ through the identity element.

Exercise 3.3.1 Show that a right-invariant or left-invariant vector field on a Lie group G is complete. (Notice that the invariance property implies that the local flow Φ associated to the vector field may be assumed to contain $G \times (-\epsilon, \epsilon)$ for some $\epsilon > 0$. We can now define $\Phi(g, t) = \Phi_t(g)$ for $t \geq \epsilon$ as $(\Phi_{\frac{t}{n}})^n(g)$ for a sufficiently big n, which depends only on t.)

Exercise 3.3.2 Let $\Psi : H \to G$ be a homomorphism of Lie groups. Show that $\psi :=$ $D\Psi_e : \mathfrak{h} \to \mathfrak{g}$ satisfies

$$\psi([A, B]) = [\psi(A), \psi(B)]$$

for all $A, B \in \mathfrak{h}$.

3.4 Infinitesimal and Local Actions

Let G be a connected Lie group and let M be a smooth manifold. A *local G-action on M* is a map $\phi : \mathcal{D} \to M$ that satisfies the following properties:

1. $\mathcal{D}$ is an open subset of $G \times M$ such that $\{g \in G \mid (g, x) \in \mathcal{D}\}$ is a connected open neighborhood of $e \in G$ for each $x \in M$.
2. For each $x \in M$, $\phi(e, x) = x$
3. If $(h, x), (g, \phi(h, x)), (gh, x) \in \mathcal{D}$, then $\phi(gh, x) = \phi(g, \phi(h, x))$.

Notice that ϕ is a G-action on M if $\mathcal{D} = G \times M$.

We set $\phi_g(x) := \phi^x(g) := \phi(g, x)$. The domain of ϕ^x is, by definition, the set

$$\mathcal{D}^x = \{g \in G \mid (g, x) \in \mathcal{D}\}.$$

For example, the local flow of a smooth vector field on M is a local $\mathbb{R}$-action on M, and if the vector field is complete we have an $\mathbb{R}$-action. The domain $\mathcal{D}^x$ may be taken to be the maximal interval of existence of the solution, with initial value x, of the ODE defined by the vector field.

The Lie algebra $\mathfrak{g}$ of a Lie group G was defined before as the space of left-invariant vector fields on G. It will be convenient to also consider the Lie algebra of right-invariant vector fields, which we denote by $\tilde{\mathfrak{g}}$.

Let M be a smooth manifold. We define a $\mathfrak{g}$-*action* on M to be a Lie algebra homomorphism

$$\theta : \tilde{\mathfrak{g}} \to \mathfrak{X}(M).$$

A $\mathfrak{g}$-action on M will also be called an *infinitesimal G-action on M*. The $\mathfrak{g}$-action θ is called *complete* if $\theta(X)$ is a complete vector field on M for each $X \in \mathfrak{g}$. (It would seem more natural to define a $\mathfrak{g}$-action to be a homomorphism of $\mathfrak{g}$ rather than of $\tilde{\mathfrak{g}}$. We have preferred the given definition since it will make some of the statements to be discussed later assume simpler forms.)

Given a local G-action $\phi : \mathcal{D} \to M$, define a map $\check{\phi}$ from the Lie algebra $\tilde{\mathfrak{g}}$ into $\mathfrak{X}(M)$ by

$$\check{\phi}(X)(x) := (D\phi^x)_e X(e).$$

It is clear that $\check{\phi} : \tilde{\mathfrak{g}} \to \mathfrak{X}(M)$ is a linear map.

Proposition 3.4.1 The map $\check{\phi}$ is a $\mathfrak{g}$-action on M. Furthermore,

$$\check{\phi}(X)(\phi(g, x)) = (D\phi^x)_g X(g)$$

for all $X \in \tilde{g}$ and $(g, x) \in \mathcal{D}$.

Proof. Let $X \in \tilde{\mathfrak{g}}$ be a right-invariant vector field on G, and let f be a smooth function on M. We first show that $\check{\phi}(X)f$ is a smooth function, from which it will follow that $\check{\phi}(X)$ is a smooth vector field. Let $h = f \circ \phi$, so h is a smooth function on some neighborhood of (e, x). The vector field Z on $G \times M$ given by $Z(l, y) = (X(l), 0)$ is smooth, so Zh is a smooth function on some neighborhood of (e, x). Therefore, $(Zh)(e, y)$ is a smooth function of $y \in M$ and we have

$$(Zh)(e, y) = dh_{(e,y)} Z(e, y)$$
$$= df_y (D\phi)_{(e,y)} Z(e, y)$$
$$= df_y (D\phi^y)_e X(e)$$
$$= (\check{\phi}(X)f)(y),$$

proving the claim.

Let $(g, x) \in \mathcal{D}$ and let $y = \phi(g, x)$. Then $\mathcal{D}^y \cap (R_{g^{-1}} \mathcal{D}^x)$ is a neighborhood of e and for all h in it, the points (g, x), (hg, x), $(h, \phi(g, x))$ are in $\mathcal{D}$. Therefore, $\phi(hg, x) = \phi(h, \phi(g, x))$, which can also be written as $(\phi^x \circ R_g)(h) = \phi^y(h)$. Taking the differential at e on each side and using the right invariance of X we obtain $(D\phi^x)_g X(g) = \check{\phi}(X)(y)$. Therefore, $\check{\phi}(X)$ and X are ϕ^x-related and the proposition follows. $\square$

We call the Lie algebra homomorphism $\check{\phi}$ the *infinitesimal generator* of the local G-action ϕ. It will be seen later that the converse of the previous proposition, namely, that an infinitesimal G-action on M is the infinitesimal generator of some local G-action on M, also holds. This fact is known as Lie's second fundamental theorem.

3.5 The Exponential Map

Let G be a Lie group with Lie algebra $\mathfrak{g}$. The *exponential map* of $\mathfrak{g}$ is the map $\exp : \mathfrak{g} \to G$ defined by $\exp(v) := c(1)$, where $c(t)$ is the flow line through e of the right-invariant vector field associated to v.

Exercise 3.5.1 Let G be an arbitrary Lie group. Show that any Lie group homomorphism $\varphi : \mathbb{R} \to G$ is of the form $\varphi(t) = \exp(tA)$ for some $A = \varphi'(0) \in \mathfrak{g}$. Also show that if $\Psi : H \to G$ is a Lie group homomorphism, then $\psi := D\Psi_e : \mathfrak{h} \to \mathfrak{g}$ satisfies $\exp_G(\psi(A)) = \Psi(\exp_H(A))$ for all $A \in \mathfrak{h}$. (Hint: For the first claim, use $\varphi(t+s) = \varphi(t)\varphi(s)$ to show that for each $s \in \mathbb{R}$, the curve $t \mapsto c_t(\varphi(s)) := \varphi(s)\varphi(t)$ satisfies the same differential equation that defines the flow of the right-invariant vector field X such that $X(e) = A$.)

Since the differential $(D\exp)_0$ is the identity map on $\mathfrak{g}$, the inverse function theorem implies that the exponential map is a diffeomorphism from a neighborhood U of 0 in $\mathfrak{g}$ onto a neighborhood W of e in G. By restricting $\exp$ to a smaller neighborhood V of 0 we can guarantee that $\exp(V)\exp(V)$ is still contained in W. It follows that there is a unique smooth map $C : V \times V \to U$ such that

$$\exp A \exp B = \exp C(A, B).$$

C has a Taylor series expansion near the origin of $\mathfrak{g} \oplus \mathfrak{g}$ given by

$$C(u) = C^{(0)} + C^{(1)}(u) + \frac{1}{2}C^{(2)}(u, u) + \cdots,$$

where the kth term $C^{(k)}(u, \ldots, u)$, $u \in \mathfrak{g} \oplus \mathfrak{g}$, is the differential of order k of C at $(0, 0)$. $C^{(k)}$ is a homogeneous polynomial of degree k taking values in $\mathfrak{g}$, in the following sense: Given a basis $\{e_i \mid i = 1, \ldots, n\}$ of $\mathfrak{g}$ with dual basis $\{e_i^* \mid i = 1, \ldots, n\}$, each function $e_j^*(C^{(k)}(u, \ldots, u))$, where $u = (\sum_{i=1}^{n} a_i e_i, \sum_{i=1}^{n} b_i e_i)$, is a homogeneous polynomial of degree k in the variables $a_i, b_i, i = 1, \ldots, n$.

Clearly, $C_0 = 0$. Moreover, since $C(x, 0) = C(0, x) = x$ it follows that $C^{(1)}(x, y) = x + y$ and $C^{(2)}(x, 0) = C^{(2)}(0, x) = 0$. Therefore, the quadratic term $C^{(2)}(x, y)$ must be linear in x and y separately. On the other hand, $C(ax, bx) = (a + b)x$, since $\exp$ is a group homomorphism when restricted to any one-dimensional direction. Substituting $C(x, x) = 2x$ into the

Taylor expansion and comparing terms of order 2 we obtain $C^{(2)}(x, x) = 0$. This implies that $C^{(2)}(x, y)$, $x, y \in \mathfrak{g}$, is an alternating bilinear form. As the proposition below shows, $C^{(2)}$ is exactly the Lie bracket of $\mathfrak{g}$.

Proposition 3.5.2 For all x and y in some fixed neighborhood of $0 \in \mathfrak{g}$ we have

$$\exp x \exp y = \exp\left(x + y + \frac{1}{2}[x, y] + R(x, y)\right),$$

where the remainder term is of third order in $(x, y) \in \mathfrak{g} \oplus \mathfrak{g}$.

Proof. The flows of the left-invariant vector fields X, Y associated to x and y are, respectively, $\Phi_t(g) = g\exp(tx)$ and $\Psi_t(g) = g\exp(ty)$. Using the limit formula for the commutator of vector fields given earlier, we obtain for any smooth function f on G

$$([X, Y]f)(e) = \lim_{t \to 0^+} \frac{1}{t}(f(\exp(\sqrt{t}x)\exp(\sqrt{t}y)$$
$$\times \exp(-\sqrt{t}x)\exp(-\sqrt{t}y)) - f(e)).$$

It follows by the discussion preceding the proposition and by a simple computation that

$$([X, Y]f)(e) = \lim_{t \to 0^+} \frac{1}{t}(f(\exp(tC^{(2)}(x, y) + \text{higher-order terms in } t)) - f(e))$$
$$= C^{(2)}(x, y)f.$$

Therefore, $C^{(2)}(x, y) = [x, y]$. $\qquad\qquad\qquad\qquad\qquad\qquad\qquad\square$

Exercise 3.5.3 Let $X_1, \ldots, X_n$ be a basis for the Lie algebra $\mathfrak{g}$ of G. Show that the map

$$\varphi : (t_1, \ldots, t_n) \mapsto \exp(t_1 X_1)\cdots\exp(t_n X_n)$$

is a smooth diffeomorphism from some neighborhood of $0 \in \mathbb{R}^n$ onto some neighborhood of $e \in G$.

3.6 Lie Subgroups and Homogeneous Spaces

For each Lie subgroup H of a Lie group G, the subspace $\mathfrak{h} := T_e H$ clearly is a Lie subalgebra of $\mathfrak{g} = T_e G$. The converse is given by the next proposition.

Proposition 3.6.1 Let $\mathfrak{g}$ be the Lie algebra of a Lie group G, and let $\mathfrak{h}$ be a Lie subalgebra of $\mathfrak{g}$. Then $\mathfrak{h}$ is the Lie algebra of a unique connected (not necessarily closed) Lie subgroup H of G.

Proof. Since the Lie subalgebra $\mathfrak{h}$ can be viewed as a left-invariant involutive subbundle of TG, we get a foliation $\mathcal{H}$ in G whose leaves are the maximal connected integral manifolds of $\mathcal{H}$. Denote by H the leaf of $\mathcal{H}$ containing e. Notice that left translation by $g \in G$ maps any leaf of $\mathcal{H}$ diffeomorphically onto another leaf. In particular, if $h \in H$, then $hH = H$, since hH is another leaf of $\mathcal{H}$ that intersects H at h (since $he = h$) so the two leaves must coincide. Similarly, if $h \in H$ we have $h^{-1}H = H$, since the two leaves must contain $e = h^{-1}h$. Therefore, H is an abstract subgroup of G and a smooth submanifold. The proposition is now a consequence of the next lemma. $\square$

Lemma 3.6.2 Let G be a Lie group, X a differentiable manifold, and f a one-to-one immersion of X into G such that

$$f(x)f(y) \in f(X) \text{ and } f(x)^{-1} \in f(X)$$

for all $x, y \in M$. Introduce a multiplication on X by $xy := f^{-1}(f(x)f(y))$ and suppose that this map is continuous. Then X is a Lie group under this multiplication.

Proof. It is clear that $f(X)$ is a subgroup of G. Thus, the map from $X \times X$ into G given by $(x, y) \mapsto f(x)f(y)$ is differentiable. Define a map u from $X \times X$ into X by $u(x, y) := f^{-1}(f(x)f(y))$. We wish to show that u is also differentiable. Since u is assumed to be continuous, it suffices to show that there is a neighborhood U of $u(x, y)$ such that for every differentiable function φ defined on U, $\varphi \circ u$ is differentiable at (x, y). Since f is an immersion, we can find neighborhoods U of $u(x, y)$ and V of $f(u(x, y))$, and coordinates $(z_1, \ldots, z_n)$ on V such that $f(U)$ is given by those points of V such that $z_{p+1} = \cdots = z_n = 0$ and $z_1, \ldots, z_p$ are coordinates on U. Thus, any function φ on U can be extended to a function $\bar{\varphi}$ on V by setting $\bar{\varphi}(z_1, \ldots, z_n) := \varphi(z_1, \ldots, z_p)$. But $\varphi \circ u = \bar{\varphi} \circ u$, which is differentiable. $\square$

Exercise 3.6.3 Let H be a closed subgroup of a Lie group G and let $\mathfrak{g}$ be the Lie algebra of G. Let X_i be a sequence of elements of $\mathfrak{g}$ converging to $X \in \mathfrak{g}$ and t_i a sequence of nonzero real numbers converging to 0. Show that if $\exp(t_i X_i) \in H$ for all i, then $\exp(tX) \in H$ for all $t \in \mathbb{R}$.

Theorem 3.6.4 Let G be a Lie group and H a closed subgroup. Then H is a Lie subgroup of G. The coset space G/H carries a unique differentiable structure such that

1. the projection map $\pi : G \to G/H$ is differentiable;

2. for each $x \in G/H$ we can find a neighborhood W and a differentiable map $\sigma : W \to G$ such that $\pi \circ \sigma$ is the identity map on W;

3. left translation by any element of G is a diffeomorphism of G/H.

Proof. Let $\mathfrak{h}$ be the set of all $X \in \mathfrak{g}$ such that $\exp(tX) \in H$ for all $t \in \mathbb{R}$. Using the previous exercise and the identity

$$\exp(A + B) = \lim_{m \to \infty} (\exp(A/m) \exp(B/m))^m,$$

which follows from theorem 3.1.6, we conclude that $\mathfrak{h}$ is a Lie subalgebra of $\mathfrak{g}$.

By the previous lemma, in order to show that H is a Lie group it suffices to show that it is a submanifold of G. In fact, it suffices to show that every $g \in G$ has a neighborhood $\bar{U}$ such that $\bar{U} \cap H$ is a submanifold of $\bar{U}$. This is trivial if g does not lie in H, since we can pick a neighborhood of g that does not intersect H, due to the assumption that H is closed. If $g \in H$, then such a neighborhood will exist if we can find one for e; namely, left translate the neighborhood of e by g so as to obtain such a neighborhood for g.

Choose a complementary subspace $\mathfrak{h}'$ of $\mathfrak{h}$. Thus, $\mathfrak{g} = \mathfrak{h}' + \mathfrak{h}$ is a direct sum of linear spaces. We claim that there is a neighborhood V' of 0 in $\mathfrak{h}'$ such that $0 \neq X' \in V'$ implies $\exp X'$ does not lie in H. In fact, if this were not true, the previous exercise would give an element X of h' such that $\exp(tX) \in H$ for all $t \in \mathbb{R}$, but that would mean that X is also in $\mathfrak{h}$, a contradiction.

Let $\bar{W}$ be a neighborhood of 0 in $\mathfrak{g}$ on which the map $\xi : \mathfrak{g} \to G$ defined by $\xi(X + X') = \exp(X') \exp(X)$ is a diffeomorphism, where $X \in \mathfrak{h}$ and $X' \in \mathfrak{h}'$. (We can find such a $\bar{W}$ as a consequence of the inverse function theorem.) Choose $\bar{W}$ to be of the form $W_1 \times W_2$ with $W_2 \subset V'$ and let $U = \xi(\bar{W})$. If $x \in U$, then $x = \exp X' \exp X$ for $X \in \mathfrak{h}$ and $X' \in V'$. If $x \in U \cap H$, then as $\exp X \in H$, we have $\exp X' \in H$, since H is a group. Therefore, $X' = 0$. The conclusion is that $U \cap H = U \cap \exp \mathfrak{h}$ and H is indeed a submanifold of G.

Assertion 3 is true for any differentiable structure on G/H satisfying 1 and 2. To see this, fix $g \in G$ and let $x \in G/H$, W, and σ be as in 2. Denoting by $\bar{L}_g$ the left translation on G/H by g, we have $\pi \circ L_g = \bar{L}_g \circ \pi$, from which we get $\bar{L}_g = \pi \circ L_g \circ \sigma$, which is differentiable. Since $\bar{L}_g^{-1}$ is the same as left translation by g^{-1}, $\bar{L}_g$ is indeed a diffeomorphism.

We will, in fact, construct a coordinate chart $\varphi : W \to \mathbb{R}^n$ (where n is dim G − dim H) satisfying properties 1 and 2, defined on a neighborhood W of the identity coset eG (that is, defined so that $\varphi \circ \pi$ and $\sigma \circ \varphi^{-1}$ are smooth and $\pi \circ \sigma$ is the identity map). Having done so, we can then define a coordinate chart on a neighborhood of an arbitrary point $x = gH$, also satisfying 1 and 2, by composing the coordinate map φ with left translations. More precisely, define $W_g := \bar{L}_g(W)$, $\varphi_g := \varphi \circ \bar{L}_g^{-1} : W_g \to \mathbb{R}^n$, and $\sigma_g := L_g \circ \sigma \circ \bar{L}_g^{-1}$. Notice that $\varphi_g \circ \pi$ and $\sigma_g \circ \varphi_g^{-1}$ are smooth and $\pi \circ \sigma_g$ is the identity map. It is now easy to check that $\mathcal{A} = \{\varphi_g \mid g \in G\}$ is an atlas for G/H, defining a differentiable structure for which 1 and 2 hold.

We now construct a coordinate system about eH, as claimed in the previous paragraph. Take for W the neighborhood U of eH constructed in the first part of the proof. Recall that $U = \xi(\bar{W})$ and that each $g \in \bar{W}$ can be uniquely written as $g = (\exp X')(\exp X)$, with $X' \in \mathfrak{h}'$ and $X \in \mathfrak{h}$. Therefore, we obtain a well-defined map ψ from W into $\mathfrak{h}'$, which sends gH to X'. Choose a coordinate system on $\mathfrak{h}'$, thus inducing via ψ a coordinate system on W. The map $\exp \circ \psi : W \to \bar{W}$ is clearly differentiable and satisfies 2. The map ξ from $\mathfrak{g}$ into G, sending $X + X'$ into $(\exp X')(\exp X)$, is a diffeomorphism of some neighborhood of 0 onto $\bar{W}$. Thus the map from $\bar{W}$ into $\mathfrak{h}'$ that sends $g = (\exp X')(\exp X)$ into X' is differentiable, which proves the existence of the differentiable chart.

To prove uniqueness of the differentiable structure on W defined by the chart just constructed, suppose that W has two differentiable structures with two coordinate maps φ_1, φ_2 and maps σ_1, σ_2 from W into G as in 2. Then $\varphi_1 \circ \varphi_2^{-1} = (\varphi_1 \circ \pi) \circ (\sigma_2 \circ \varphi_2^{-1})$, which is a product of differentiable maps. Therefore, the two structures are diffeomorphic. $\square$

3.7 Continuous Homomorphisms Are Smooth

Theorem 3.7.1 Every continuous homomorphism of a Lie group H into a Lie group G is smooth.

Proof. We have already seen that $\exp$ is a smooth diffeomorphism from some neighborhood V of 0 in the Lie algebra $\mathfrak{g}$ of G onto some neighborhood U of $e \in G$ and that for each $X \in \mathfrak{g}$, the map $t \mapsto \exp(tX)$ defines a homomorphism from $\mathbb{R}$ into G. We first show that every continuous homomorphism σ from $\mathbb{R}$ into G is differentiable. In fact, since $\sigma(t + s) = \sigma(t)\sigma(s)$, it clearly suffices to show that σ is differentiable at 0.

Choose $\epsilon > 0$ small enough so that $\sigma(t) \in U$ if $|t| < \epsilon$. Since $\exp$ is locally a smooth diffeomorphism, there is a continuous map $t \mapsto X(t)$ from $(-\epsilon, \epsilon)$ into $V \subset \mathfrak{g}$ such that $X(0) = 0$ and $\sigma(t) = \exp X(t)$. Fix t_0 in $(-\epsilon, \epsilon)$. Using the homomorphism property of $\exp$ we get $\sigma(\frac{m}{n}t_0) = \exp(\frac{m}{n}X(t_0))$ for all $m, n \in \mathbb{Z}$ such that $0 < |m| \le n$, whence $X(\frac{m}{n}t_0) = \frac{m}{n}X(t_0)$ for the same m, n. By continuity, $X(\lambda t_0) = \lambda X(t_0)$ for all λ, $|\lambda| < 1$. Written in a different way, $X(t) = t\frac{X(t_0)}{t_0}$, $-t_0 < t < t_0$. This shows that $X(t)$ is smooth (in fact, linear) near 0, so σ is also smooth near 0. Once σ is known to be smooth, it is clear that it must be of the form $\sigma(t) = \exp(t X_0)$ for some $X_0 \in \mathfrak{g}$ and all $t \in \mathbb{R}$, since the infinitesimal generator of the flow $\Phi_t(g) = g\sigma(t)$ (which commutes with left translations) is a left-invariant vector field, hence an element of $\mathfrak{g}$.

To prove that a continuous homomorphism $\rho : H \to G$ is smooth, it is enough to show that ρ is smooth at $e \in H$. Let $X_1, \ldots, X_k$ be a basis of the Lie algebra $\mathfrak{h}$ of H. The map $t \mapsto \rho(\exp(t X_i))$ is a continuous homomorphism from $\mathbb{R}$ into G, so it is actually smooth and there is $Y_i \in \mathfrak{g}$ such that $\rho(\exp(t X_i)) = \exp(t Y_i)$ for all $t \in \mathbb{R}$. Since ρ is a homomorphism,

$$\rho(\exp(t_1 X_1) \cdots \exp(t_k X_k)) = \exp(t_1 Y_1) \cdots \exp(t_k Y_k).$$

Moreover, by exercise 3.5.3,

$$\varphi : (t_1, \ldots, t_k) \mapsto \exp(t_1 X_1) \cdots \exp(t_k X_k)$$

is a smooth diffeomorphism from some neighborhood of 0 in $\mathbb{R}^k$ onto some neighborhood of e in H. Therefore, since $\rho \circ \varphi$ is smooth we conclude that ρ is smooth. $\qquad\square$

It will be seen in a later chapter that the conclusion of the previous result still holds if the homomorphism is only measurable.

Lemma 3.7.2 Let G be a connected Lie group and V any open neighborhood of $e \in G$. Then

$$G = \bigcup_{n=1}^{\infty} V^n,$$

where V^n consists of all n-fold products of elements of V.

Proof. Let $U = V \cap V^{-1}$ and observe that U is an open neighborhood of e such that $U = U^{-1}$. Let $H = \bigcup_{n=1}^{\infty} U^n \subset G$. Then H is an abstract subgroup of G and is an open subset of G. Thus each coset gH of G/H is open in G. But G is the disjoint union of cosets of H, so there can be only one coset since G is connected. Therefore $G = H$. $\qquad\square$

Exercise 3.7.3 If F_1 and F_2 are smooth homomorphisms from a connected Lie group G into a Lie group H such that $(DF_1)_e = (DF_2)_e$, show that $F_1 = F_2$.

3.8 Representations of Compact Groups

Let G be a Lie group and let V be a finite-dimensional vector space over $\mathbb{R}$ or $\mathbb{C}$. A *linear representation* of G on V is a continuous, hence smooth, group homomorphism $\rho : G \to GL(V)$. Therefore, a representation ρ yields a G-action on V by linear transformations.

We say that the representation ρ is *irreducible* if there are no proper G-invariant linear subspaces of V. The representation is said to be *completely reducible* if any G-invariant subspace $W \subset V$ has a G-invariant complement, that is, there is a G-invariant $W' \subset V$ such that $V = W \oplus W'$. We will show next that finite-dimensional representations of compact groups are completely reducible. This result will be used later to show that finite-dimensional representations of semisimple Lie groups are completely reducible.

Lemma 3.8.1 Let G be a connected compact Lie group of dimension m. Then there exists a nonvanishing, both right- and left-invariant, m-form ν on G such that $\nu(G) = 1$.

Proof. Let ν_0 be a nonzero m-form on $T_e G$. By left translating ν_0 over G we obtain a left-invariant m-form ν on G. By multiplying ν by $\nu(G)^{-1}$ we may assume without loss of generality that $\nu(G) = 1$. Since left and right translations commute and the space of left-invariant m-forms on G is one-dimensional, there must exist for each $g \in G$ a real number $\lambda(g) \neq 0$ such that $R_{g^{-1}}^* \nu = \lambda(g)\nu$. One easily sees that $\lambda : G \to \mathbb{R} - \{0\}$ is a smooth homomorphism. But G is compact, so λ takes values in the compact subgroup $\{+1, -1\}$ of $\mathbb{R} - \{0\}$. On the other hand, G is connected, so λ is identically equal to 1. Therefore, ν is also right-invariant. $\square$

Proposition 3.8.2 Any finite-dimensional representation of a compact Lie group G is completely reducible.

Proof. Let $\rho : G \to GL(V)$ be a finite-dimensional representation. G acts on the space of bilinear forms on V by

$$(g_* k)(u, v) := k(\rho(g)^{-1} u, \rho(g)^{-1} v),$$

where $u, v \in V$, $g \in G$, and k is any bilinear form. Choose a positive-definite Hermitian inner product $k_0 = \langle \cdot, \cdot \rangle_0$ on V and average k_0 over G. This can be done as follows. First average k_0 over the connected component of G using integration with respect to the bi-invariant form ν given by the lemma and then average the resulting inner product over the (finitely many) connected components. More precisely, let μ denote the bi-invariant probability measure on G^0 induced by integrating ν. Define

$$\langle u, v \rangle := \int_{G^0} (g_* k_0)(u, v) \, d\mu(g).$$

(For more on invariant integration on groups, see chapter 8.) The resulting inner product is G^0-invariant. We can now average it over the finite group G/G^0 to obtain the desired G-invariant Hermitian form.

If W is a G-invariant subspace of V, the orthogonal complement $W^\perp$ with respect to a G-invariant inner product is a G-invariant complementary subspace. □

Proposition 3.8.3 (Schur's lemma) Let G be an abelian group and let $\rho : G \to GL(V)$ be an irreducible representation, where V is a nonzero, finite dimensional vector space over $\mathbb{C}$. Then V is one-dimensional.

Proof. Pick any $g_0 \in G$ and let v be an eigenvector of $\rho(g_0)$ associated with an eigenvalue $\lambda \in \mathbb{C}$. Since G is abelian, each nonzero vector in the subspace spanned by $\{\rho(g)v \mid g \in G\}$ is an eigenvector of $\rho(g_0)$ associated with λ. By irreducibility, this subspace coincides with V. Therefore, $\rho(g) = \lambda(g)I$ is a scalar matrix for each $\lambda(g) \in \mathbb{C}$. By irreducibility again, dim $V = 1$. □

Corollary 3.8.4 Let G be a compact abelian Lie group and V a finite-dimensional vector space over $\mathbb{C}$. Then any linear representation of G on V decomposes as a direct sum of one-dimensional subrepresentations.

3.9 Integrating Infinitesimal Actions

Let $\mathfrak{g}$ be the Lie algebra of a connected m-dimensional Lie group G, and let $\theta : \tilde{\mathfrak{g}} \to \mathfrak{X}(M)$ be an infinitesimal action of G on a manifold M. Recall from section 3.4 that θ is, by definition, a Lie algebra homomorphism from the Lie algebra $\tilde{\mathfrak{g}}$ of right-invariant vector fields on G into $\mathfrak{X}(M)$. (The Lie algebra $\mathfrak{g}$ comprises the left-invariant vector fields.) In this section we show that there is a local action of G on M for which θ is the infinitesimal generator. We also discuss the issue of *completeness* of infinitesimal actions. The key idea will be to define an involutive subbundle E of $T(G \times M)$ using θ, apply to E the Frobenius theorem so as to obtain a foliation $\mathcal{E}$ on $G \times M$, and show that the foliation thus obtained is, in a sense, the graph of the G-action we seek.

To each $X \in \tilde{\mathfrak{g}}$ we define a vector field $\theta^*(X)$ on $G \times M$ by

$$\theta^*(X)(g, x) := (X(g), \theta(X)(x))$$

for all $(g, x) \in G \times M$.

$E := \theta^*(\mathfrak{g})$ is an m-dimensional subbundle of $T(G \times M)$, spanned at each point (g, x) by the vectors $\theta^*(X)(g, x)$, $X \in \tilde{\mathfrak{g}}$. Furthermore, E is involutive since, for $X, Y \in \tilde{\mathfrak{g}}$,

$$[\theta^*(X), \theta^*(Y)] = [(X, \theta(X)), (Y, \theta(Y))]$$
$$= ([X, Y], [\theta(X), \theta(Y)])$$
$$= ([X, Y], \theta([X, Y]))$$
$$= \theta^*([X, Y]).$$

Also notice that E is right-invariant, in the following sense. For each $h \in G$, let $\bar{R}_h$ denote the diffeomorphism of $G \times M$ such that $\bar{R}_h(g, x) = (gh, x)$. Right invariance of $X \in \tilde{\mathfrak{g}}$ then implies that $(\bar{R}_h)_* \theta^*(X) = \theta^*(X)$, so E is also invariant under $\bar{R}_h$.

Let $\mathcal{E}$ be the foliation of $G \times M$ that integrates E. The leaf of $\mathcal{E}$ containing (e, x) will be denoted by L_x. The natural projections from $G \times M$ onto G and M will be denoted, respectively, Π_G and Π_M. Notice that if L is the leaf of $\mathcal{E}$ that contains $(g, x) \in G \times M$, then $\theta^*(X)(g, x)$ is the only vector in $E(g, x)$ such that

$$(D\Pi_G)_{(g,x)} \theta^*(X)(g, x) = X(g).$$

By the inverse function theorem, Π_G is a local diffeomorphism between L and G at any point of L. Furthermore, it follows from the right invariance of E that $\bar{R}_h$ sends leaves of $\mathcal{E}$ diffeomorphically onto other leaves, for each $h \in G$.

In what follows, we fix a coordinate chart $(u_1, \ldots, u_m, W)$ on a neighborhood of e in G such that $u_i(e) = 0$ for each i. We say that an open set W' containing e is a *cubical neighborhood* of e if $W' \subset W$ is a neighborhood of e and the image of W' under u_i is the interval $(-a, a)$, for some $a > 0$, so that W' is written as an open cube of the form $(-a, a)^m$. Notice that an arbitrary union of cubical neighborhoods of e is also a cubical neighborhood of e. Cubical neighborhoods can easily be constructed using the exponential map.

Following [21], we call a smooth map $\psi : V \times U \to G \times M$ an *auxiliary map* if V is a cubical neighborhood of e in G, U is an open set in M, ψ is a diffeomorphism onto its image of the form $\psi(g, x) = (g, \phi(g, x))$ for some smooth function $\phi : V \times U \to M$, and the following conditions hold:

1. For each $y \in U$, $g \mapsto \psi(g, y)$ maps V diffeomorphically onto the component of (e, y) in $L_y \cap \Pi_G^{-1}(V)$, where L_y is the leaf of $\mathcal{E}$ that contains (e, y).
2. For each $y \in U$, Π_G maps the component of (e, y) in $L_y \cap \Pi_G^{-1}(V^2)$ bijectively onto V^2.

It is a simple exercise to show that for each $x \in M$, one can always find a neighborhood U of x and a sufficiently small cubical neighborhood V of $e \in G$

such that there exists a unique auxiliary map of $V \times U$ into $G \times M$. Moreover, if we define $\psi^y := \psi(\cdot, y) : V \to G \times M$, then

$$\theta(X)(y) = (D\Pi_M)_{(e,y)}(D\psi^y)_e X(e)$$

for all $X \in \tilde{\mathfrak{g}}$ and $y \in U$. In fact, $(D\psi^y)_e X(e)$ is a vector in $E(e, y)$ that projects onto $X(e)$ under $(D\Pi_G)_{(e,y)}$, but $\theta^*(X)(e, y) = (X(e), \theta(X)(y))$ is the unique such vector.

Lemma 3.9.1 Let $\psi : V \times U \to G \times M$ and $\psi' : V' \times U' \to G \times M$ be two auxiliary maps, and write $\phi = \Pi_M \circ \psi$ and $\phi' = \Pi_M \circ \psi'$. Then the following hold.

1. ψ and ψ' agree on the intersection of their domains.
2. If $(h, x) \in V \times U$, then $\bar{R}_{h^{-1}}$ maps L_x diffeomorphically onto $L_{\phi(h,x)}$, and if $W \subset G$ then $\bar{R}_{h^{-1}}$ maps each component of $L_x \cap \Pi_G^{-1}(W)$ diffeomorphically onto a component of $L_{\phi(h,x)} \cap \Pi_G^{-1}(Wh^{-1})$.
3. If (h, x), $(g, \phi(h, x))$, and (gh, x) belong to $V \times U$, then

$$\phi(gh, x) = \phi(g, \phi(h, x)).$$

4. The conclusion of 3 also holds if (h, x) and (gh, x) belong to $V \times U$ and $(g, \phi(h, x))$ belongs to $V' \times U'$.
5. Let $D = (V \times U) \cup (V' \times U')$. Then there is a unique map $\bar{\phi}$ of D into M such that $\bar{\phi}$ agrees with ϕ on $V \times U$ and agrees with ϕ' on $V' \times U'$. Moreover, $\bar{\phi}(e, x) = x$ for all $x \in U \cup U'$, and if (h, x), $(g, \phi(h, x))$, and (gh, x) belong to D, then

$$\bar{\phi}(gh, x) = \bar{\phi}(g, \bar{\phi}(h, x)).$$

Proof. Properties 1 and 2 are easy. Notice that 3 already gives a local action of a weaker sort. It is, in fact, the classical form of Lie's second fundamental theorem. We verify 3 and leave the other statements as an exercise to the reader. The entire proof can be found in [21, p. 54–6].

By definition, $(g, \phi(g, \phi(h, x))) = \psi(g, \phi(h, x))$ belongs to the component of $(e, \phi(h, x))$ in $L_{\phi(h,x)} \cap \Pi_G^{-1}(V)$, which is a subset of the component of $(e, \phi(h, x))$ in $L_{\phi(h,x)} \cap \Pi_G^{-1}(V^2h^{-1})$ (notice that $V = Vhh^{-1} \subset V^2h^{-1}$). It follows from item 2 that $(gh, \phi(g, \phi(h, x)))$ belongs to the component of $(h, \phi(h, x))$ in $L_x \cap \Pi_G^{-1}(V^2)$. On the other hand, $(h, \phi(h, x)) = \psi(h, x)$ belongs to the component of (e, x) in $L_x \cap \Pi_G^{-1}(V)$, which is contained in the component of (e, x) in $L_x \cap \Pi_G^{-1}(V^2)$, whence $(gh, \phi(g, \phi(h, x)))$ belongs to the component of (e, x) in $L_x \cap \Pi_G^{-1}(V^2)$. Also $(gh, \phi(gh, x)) = \psi(gh, x)$ lies in the component of (e, x) in $L_x \cap \Pi_G^{-1}(V)$, which is contained in the component of (e, x) in $L_x \cap \Pi_G^{-1}(V^2)$. Since Π_G maps the component of (e, x) in $L_x \cap \Pi_G^{-1}(V^2)$ bijectively, and $\Pi_G(gh, \phi(gh, x)) = gh = \Pi_G(gh, \phi(g, \phi(h, x)))$, property 3 follows. $\quad\square$

Lemma 3.9.2 Let $\{\psi_\alpha : V_\alpha \times U_\alpha \to G \times M\}_{\alpha \in I}$ be the collection of all auxiliary maps associated to θ, and let $D = \bigcup_{\alpha \in I}(V_\alpha \times U_\alpha)$. Then there is a unique smooth map $\phi : D \to M$ such that ϕ is a local G-action on M with infinitesimal generator θ that agrees with $\Pi_M \circ \psi_\alpha$ for all α.

Proof. For each $x \in M$, let I_x be the subset of I consisting of all indices α such that $x \in U_\alpha$, and set $D_x = \bigcup_{\alpha \in I_x} V_\alpha$. I_x is not empty since auxiliary maps exist. Moreover, each V is a cubical neighborhood of e, so the same is true for D_x. In particular, each D_x is a connected neighborhood of e. As each V_α and U_α is open, so is D. The existence and uniqueness of ϕ and the fact that ϕ is a local G-action on M is now immediate from item 5 of the lemma.

Let $x \in M$ and let $\psi : V \times U \to G \times M$ be an auxiliary map with $x \in U$. Defining $\psi^x : V \to G \times M$ by $\psi^x(g) = \psi(g, x)$ yields $\Pi_M \circ \psi^x = \phi^x|_V$. Since V is a neighborhood of e, it follows that $(D\phi^x)_e = (D\Pi_M)_{\psi(g,x)} \circ (D\psi^x)_e$. Let $\check{\phi}$ be the infinitesimal generator of ϕ. Then for any given $X \in \tilde{\mathfrak{g}}$, it follows from item 1 of the previous lemma that

$$\check{\phi}(X)(x) = (D\phi^x)_e X(e) = ((D\Pi_M)_{\psi(g,x)} \circ (D\psi^x)_e)X(e) = \theta(X)(x).$$

Therefore, $\check{\phi} = \theta$. $\quad\square$

The next theorem is an immediate consequence of the previous lemma.

Theorem 3.9.3 If $\theta : \tilde{\mathfrak{g}} \to \mathfrak{X}(M)$ is an infinitesimal G-action on M, then θ is the infinitesimal generator of a local G-action on M.

We now ask, when is an infinitesimal G-action θ the infinitesimal generator of a global G-action? It will be seen that a sufficient condition is that θ be complete and G be a connected simply connected Lie group. The next lemma shows that a global G-action exists if there is a local G-action with a "uniform" domain.

Lemma 3.9.4 Let θ be an infinitesimal G-action on a smooth manifold M, where G is a connected Lie group, and suppose there exists a local G-action $\phi : D \to M$ with infinitesimal generator θ such that D contains $V \times M$ for some open neighborhood V of e. Then for each $x \in M$, $\Pi_G : L_x \to G$ is a covering map (in particular, it is surjective.) If G is, moreover, simply connected, then there is a global G-action on M with infinitesimal generator θ.

Proof. We first show that for every positive integer n, $V^n \subset \Pi_G(L_x)$. For $n = 1$ this is immediate since Π_G maps the connected component of (e, x) in $L_x \cap \Pi_G^{-1}(V)$ onto V. Suppose $V^{n-1} \subset \Pi_G(L_x)$ and let $g \in V^{n-1}$. Then there is $y \in M$ such that $(g, y) \in L_x$. But $\bar{R}_{g^{-1}}(L_x) = L_y$, since $\bar{R}_{g^{-1}}(g, y) = (e, y)$, and $\Pi_G(L_y)$ contains V. Therefore

$$V g \subset (R_g \circ \Pi_G)(L_y) = (\Pi_G \circ \bar{R}_g)(L_y) = \Pi_G(L_x).$$

Therefore, $V^n \subset V V^{n-1} \subset \Pi_G(L_x)$, completing the induction. Since G is connected, $G = \bigcup_{n=1}^{\infty} V^n$, and Π_G maps L_x onto G.

We now show that L_x is a covering space of G. Since Π_G is onto G, it remains to show that for each $g \in G$ there is a neighborhood W of g such that Π_G maps each component of $L_x \cap \Pi_G^{-1}(W)$ diffeomorphically onto W.

Let U be an open neighborhood of e such that $U U^{-1} \subset V$. Then, if $h \in U^{-1}$, $U h$ is an open connected neighborhood of e contained in V. It is now a simple computation to show that $W = U g$ is the neighborhood of g we seek [21, p. 81].

If G is simply connected, the map $\Pi_G : L_x \rightarrow G$ is a diffeomorphism and the global G-action can now be defined as

$$\phi(g, x) = \left[\Pi_M \circ \left(\Pi_G \big|_{L_x} \right)^{-1} \right](g)$$

[21, proof of theorem XII, p. 74]. □

Lemma 3.9.5 Let $\theta : \tilde{\mathfrak{g}} \rightarrow \mathfrak{X}(M)$ be a $\mathfrak{g}$-action, and let $X_1, \ldots, X_k$ be any elements of $\tilde{\mathfrak{g}}$ such that $\theta(X_i)$ is a complete vector field for each i, $1 \leq i \leq k$. Let ϕ^i be the global flow of X_i. Then for each $x \in M$,

$$\Phi^x : (t_1, \ldots, t_k)$$
$$\in \mathbb{R}^k \mapsto \left(\exp(t_k X_k) \cdots \exp(t_1 X_1), \left(\phi_{t_k}^k \circ \cdots \circ \phi_{t_1}^1 \right)(x) \right) \in G \times M$$

is a smooth map from $\mathbb{R}^k$ into the leaf L_x. Moreover, if $\phi : D \rightarrow M$ is a local flow on M with infinitesimal generator Z such that for some $X \in \tilde{\mathfrak{g}}$

$$\sigma(t) := (\exp(tX), \phi_t(x)) \in L_x$$

for each $(t, x) \in D$, then $Z = \theta(X)$.

Proof. It is clear that the map Φ^x is smooth. The issue is to show that it maps into L_x. The case $k = 1$ is immediate. We assume that the claim holds for $k - 1$ and show that it is true for k.

Write $g = \exp(t_{k-1}X_{k-1}) \cdots \exp(t_1 X_1)$ and $y = (\phi_{t_{k-1}}^{k-1} \circ \cdots \circ \phi_{t_1}^1)(x)$. The induction hypothesis is that $(g, y) \in L_x$, and the proof will be complete if we show that $(\exp(t_k X_k)g, \phi_{t_k}^k(y)) \in L_x$ for all $t_k \in \mathbb{R}$. The result for $k = 1$ implies that $(\exp(t_k X_k), \phi_{t_k}^k(y)) \in L_y$ for all $t \in \mathbb{R}$, so $(\exp(t_k X_k)g, \phi_{t_k}^k(y))$ belongs to $\bar{R}_g L_y$. But $(e, y) = \bar{R}_{g^{-1}}(g, y) \in \bar{R}_{g^{-1}} L_x$ so $L_y = \bar{R}_{g^{-1}} L_x$, concluding the induction.

The last assertion is a consequence of the fact that $(D\Pi_G)_{(g,x)}$ maps $E(g, x)$ injectively onto $T_g G$. □

Lemma 3.9.6 Let $X, Y \in \mathfrak{g}$ bc such that $\theta(X), \theta(Y)$ are complete vector fields. Then $\theta(\mathrm{Ad}(\exp X)Y)$ is also complete.

Proof. Let ϕ and ψ be the global flows of $\theta(X)$ and $\theta(Y)$, respectively. Then $\eta_t := \phi_1 \circ \psi_t \circ \phi_{-1}$ is a global flow. For each $(t, x) \in \mathbb{R} \times M$,

$$(\exp(t\,\mathrm{Ad}(\exp X)Y), \eta_t(x)) = (\exp X \exp(tY) \exp(-X), (\phi_1 \circ \psi_t \circ \phi_{-1})(x))$$

belongs to L_x by the previous lemma, so by the last assertion of the same lemma $\phi_1 \circ \psi_t \circ \phi_{-1}$ is the global flow of $\theta(\mathrm{Ad}(\exp X)Y)$. □

Lemma 3.9.7 Let $\theta : \tilde{\mathfrak{g}} \to \mathfrak{X}(M)$ be an action of $\mathfrak{g}$ on M and G a Lie group with Lie algebra $\mathfrak{g}$. Let $X_1, \ldots, X_m$ be a basis for $\tilde{\mathfrak{g}}$ such that $\theta(X_1), \ldots, \theta(X_m)$ are complete vector fields. Then, there is a local G-action $\phi : D \to M$ whose infinitesimal generator is θ such that D contains $V \times M$ for some open neighborhood V of $e \in G$.

Proof. Let $(z_1, \ldots, z_m, V)$ be a coordinate chart about $e \in G$ with inverse given by a diffeomorphism $\varphi : (-1, 1)^m \to V$ such that

$$\varphi(z_1, \ldots, z_m) = \exp(z_1 X_1) \cdots \exp(z_m X_m).$$

By lemma 3.9.5, for each $x \in M$,

$$\Phi^x : g = \varphi(z_1, \ldots, z_m) \in V \mapsto \left(g, \left(\phi_{z_1}^1 \circ \cdots \circ \phi_{z_m}^m\right)(x)\right)$$

maps into L_x. Denoting by V^x the image of V under Φ^x, it is immediate that $\Pi_G : V^x \to V$ is a bijection. V^x is, moreover, open in L_x. Namely, given $p \in V^x$, let U be a neighborhood of p in L_x such that Π_G maps U diffeomorphically into G, and let W be a neighborhood of $\Pi_G(p)$ in V such that $\Phi^x(W) \subset U$. Then $\Phi^x(W) = (\Pi_G|_U)^{-1}(W)$ is an open set of L_x containing p and included in V^x. Since V^x is connected, it is the connected component of $L_x \cap \pi_G^{-1}(V)$ that contains $(e, x) = \Phi^x(e)$. Therefore, Π_G maps the connected component of $L_x \cap \pi_G^{-1}(V)$ that contains (e, x) diffeomorphically onto V. This shows that the auxiliary maps ψ_α of lemma 3.9.2 can be assumed to have domain $V \times U_\alpha$. Therefore, that lemma implies the existence of a local G-action whose domain D contains $V \times M$.	□

The following theorem is now immediate.

Theorem 3.9.8 Let $\theta : \tilde{\mathfrak{g}} \to \mathfrak{X}(M)$ be an infinitesimal G-action, where G is a connected simply connected Lie group with Lie algebra $\mathfrak{g}$. Suppose that $\tilde{\mathfrak{g}}$ has a basis $X_1, \ldots, X_m$ such that $\theta(X_i)$ is a complete vector field on M for each i, $1 \leq i \leq m$. Then there is a global action of G on M with infinitesimal generator θ.

If G is not simply connected the conclusion of the theorem may not hold. Consider, for example, a smooth vector field on a compact manifold M, which we may think of as defining an infinitesimal $\mathbb{T}^1$-action. The vector field integrates to an $\mathbb{R}$-action on M, but not to a $\mathbb{T}^1$-action if there are noncompact orbits.

Corollary 3.9.9 Let H be a connected and simply connected Lie group. Then, for any Lie group G and any Lie algebra homomorphism $\varphi : \mathfrak{h} \to \mathfrak{g}$, φ is induced from a unique Lie group homomorphism $\Phi : H \to G$.

Proof. The Lie algebra homomorphism φ induces a Lie algebra homomorphism $\theta : \tilde{\mathfrak{h}} \to \tilde{\mathfrak{g}} \subset \mathfrak{X}(G)$ (use the inverse map $i : G \to G$ to transform left-invariant vector fields into right-invariant vector fields). The corollary is now a consequence of the theorem.	□

4
Algebraic Actions

Actions of the kind considered in this chapter will play a subsidiary but important role in the rigidity results to be discussed in chapter 10. These are actions "defined by polynomial equations," and, as dynamical systems, they are very special in that none of the complicated behavior of orbits, such as nontrivial recurrence, can arise.

After introducing some of the basic notions in algebraic geometry we discuss, without proof, Rosenlicht's theorem, which shows that the space of orbits of an algebraic action is not far from being an algebraic variety.

4.1 Affine Varieties

A real (or complex) affine algebraic set V is a subset of $\mathbb{R}^n$ (or $\mathbb{C}^n$) consisting of the common zeros of a set of polynomials. We also need to consider projective (and quasi-projective) algebraic sets, which are subsets of real (or complex) projective space, and will be defined later. The term *algebraic variety* will usually refer to complex algebraic sets, although we also use the terms *real variety*, *real algebraic group*, and *real algebraic action* to refer to the corresponding real notions.

More precisely, let $\mathbb{F}$ be a subfield of $\mathbb{C}$, which for the most part will be either $\mathbb{R}$ or $\mathbb{C}$. Let $x = (x_1, \ldots, x_n)$ denote the coordinate functions of $\mathbb{C}^n$. Set $x^m := x_1^{m_1} \cdots x_n^{m_n}$, $m = (m_1, \ldots, m_n) \in \mathbb{Z}^n$, so that a polynomial in n variables with coefficients in $\mathbb{F}$ can be written in the form

$$f(x) = \sum_{m \in I} a_m x^m,$$

where I is a finite subset of $\mathbb{Z}^n$ and $a_m \in \mathbb{F}$ for each $m \in I$. We identify f with the function it defines on $\mathbb{C}^n$. The set of all such polynomials is denoted $\mathbb{F}[x] := \mathbb{F}[x_1, x_2, \ldots, x_n]$.

Let $J \subset \mathbb{F}[x]$ be any subset. We define the $\mathbb{F}$-algebraic set associated to J by

$$\mathcal{V}_{\mathbb{F}}(J) := \{p \in \mathbb{F}^n \mid f(p) = 0 \text{ for all } f \in J\}.$$

A set $V \subset \mathbb{F}^n$ such that $V = \mathcal{V}_{\mathbb{F}}(J)$ for some $J \subset \mathbb{F}[x]$ will be called an *affine* $\mathbb{F}$-*algebraic set*. When $\mathbb{F} = \mathbb{C}$ (resp., $\mathbb{R}$), we call V an *affine variety* (resp., *real affine variety*).

Given an arbitrary set $S \subset \mathbb{F}[x]$, we may consider the ideal generated by S, that is, the subalgebra of $\mathbb{F}[x]$ consisting of polynomials of the form

$$h = \sum_{i=1}^{l} \varphi_i f_i,$$

where $f_i \in S$ and $\varphi_i \in \mathbb{F}[x]$. Notice that $\mathcal{V}_{\mathbb{F}}(S) = \mathcal{V}_{\mathbb{F}}(J)$, where J is the ideal generated by S.

The set of $\mathbb{F}$-*points* of an affine variety $V \subset \mathbb{C}^n$ is $V(\mathbb{F}) := V_{\mathbb{F}} := V \cap \mathbb{F}^n$. In particular, the real points $V(\mathbb{R})$ can be defined as the fixed-point set in V of the automorphism $c : \mathbb{C}^n \to \mathbb{C}^n$ defined by taking complex conjugation at each coordinate.

An affine variety $V \subset \mathbb{C}^n$ is said to be *defined over* $\mathbb{F}$ if there exists $J \subset \mathbb{F}[x]$ such that $V = \mathcal{V}_{\mathbb{C}}(J)$, that is, V consists of common zeros of a family of polynomials with coefficients in $\mathbb{F}$.

Given any subset $Q \subset \mathbb{F}^n$, define

$$\mathcal{I}_{\mathbb{F}}(Q) := \{f \in \mathbb{F}[x] \mid f(p) = 0 \text{ for all } p \in Q\}.$$

Notice that $\mathcal{I}_{\mathbb{F}}(Q)$ is an ideal of $\mathbb{F}[x]$.

The *Zariski topology* of $\mathbb{F}^n$ is defined by declaring affine algebraic sets $\mathcal{V}_{\mathbb{F}}(J)$ to be the closed sets. A Zariski-closed subset W of an affine $\mathbb{F}$-algebraic set V is itself an affine $\mathbb{F}$-algebraic set.

Exercise 4.1.1 Verify that we do indeed obtain a topology. That is, show that

1. $\mathcal{V}(I_1) \cup \mathcal{V}(I_2) = \mathcal{V}(I_1 I_2)$;
2. $\bigcap_{\alpha \in A} \mathcal{V}(I_\alpha) = \mathcal{V}(I)$, where A is an arbitrary index set and I is the ideal generated by the union of the I_α;
3. $\mathcal{V}(\mathbb{F}[x]) = \emptyset$ and $\mathcal{V}(\emptyset) = \mathbb{F}^n$.

Exercise 4.1.2 Prove the following claims:

1. $V = \{p\}$ is a closed subset for each $p \in \mathbb{F}^n$.
2. $\mathbb{Z}$ is Zariski-dense in $\mathbb{C}$, that is, any polynomial that vanishes on $\mathbb{Z}$ must also vanish on $\mathbb{C}$.

Let V be an affine $\mathbb{F}$-algebraic set and $f \in \mathbb{F}[x]$. The open set

$$V_f := \{p \in V \mid f(p) \neq 0\}$$

in V is called a *principal open set*. It is a simple exercise to check that $V_f \cap V_g = V_{fg}$ and that the principal open sets form a basis for the Zariski topology of V. If $V = \mathcal{V}_{\mathbb{F}}(J)$ and $h \in \mathbb{F}[x]$, then V_h can be given the structure of an affine algebraic set by means of the embedding into $\mathbb{F}^{n+1}$ defined by

$$\{(p, p_{n+1}) \in \mathbb{F}^n \times \mathbb{F} \mid f(p) = 0, f \in J, \text{ and } h(p)p_{n+1} - 1 = 0\}.$$

In this way, the "polynomial functions" on V_h are given by the restrictions to V_h of rational functions of the form g/h^k, $g, h \in \mathbb{F}[x]$.

Given affine $\mathbb{F}$-algebraic sets $V = \mathcal{V}_{\mathbb{F}}(J_1) \subset \mathbb{F}^n$, $W = \mathcal{V}_{\mathbb{F}}(J_2) \subset \mathbb{F}^m$, their product $V \times W \subset \mathbb{F}^{m+n}$ is also an $\mathbb{F}$-algebraic set in a natural way. Namely, if $J_1 \subset \mathbb{F}[x]$ and $J_2 \subset \mathbb{F}[y]$, we can view J_1 and J_2 as subsets of $\mathbb{F}[x, y] = \mathbb{F}[x] \otimes \mathbb{F}[y]$. If J denotes the ideal generated by the union of J_1 and J_2, then $V \times W$ is the $\mathbb{F}$-algebraic set associated to J. Recall that each $h \in \mathbb{F}[x] \otimes \mathbb{F}[y]$ is a finite sum $h = \sum_k a_k f_k \otimes g_k$, $a_k \in \mathbb{F}$, $f_k \in \mathbb{F}[x]$, $g_k \in \mathbb{F}[y]$, and

$$(f_k \otimes g_k)(x, y) := f_k(x)g_k(y).$$

As a set, $V \times W$ is the standard Cartesian product, but the Zariski topology on $V \times W$ does not coincide with the product of the Zariski topologies.

Exercise 4.1.3 What are the Zariski-closed subsets of $\mathbb{C}$? Show that the product topology on $\mathbb{C} \times \mathbb{C}$, where each factor has the Zariski topology, is not equal to the Zariski topology of $\mathbb{C}^2$.

4.2 Projective Varieties

The n-dimensional projective space is the quotient

$$P^n(\mathbb{F}) = (\mathbb{F}^{n+1} - (0, \ldots, 0))/\mathbb{F}^*,$$

where the multiplicative group $\mathbb{F}^* := \mathbb{F} - \{0\}$ acts on $\mathbb{F}^{n+1} - (0, \ldots, 0)$ by scalar multiplication of vectors. We denote a point in $P^n(\mathbb{F})$ by an $(n+1)$-tuple $[X_0, \ldots, X_n]$. There are coordinate charts $\varphi_i : \mathbb{F}^n \to P^n(\mathbb{F})$, $i = 0, \ldots, n$, defined by

$$\varphi_i(x_1, \ldots, x_n) = [x_1, \ldots, x_i, 1, x_{i+1}, \ldots, x_n].$$

When $\mathbb{F} = \mathbb{R}$ (resp., $\mathbb{C}$), these charts give $P^n(\mathbb{F})$ the structure of a smooth real (resp., complex) compact manifold.

A homogeneous polynomial $f \in \mathbb{F}[X_0, \ldots, X_n]$ defines a subset of $P^n(\mathbb{F})$ as follows:

$$\bar{\mathcal{V}}_{\mathbb{F}}(f) := \{[X_0, \ldots, X_n] \in P^n(\mathbb{F}) \mid f(X_0, \ldots, X_n) = 0\}.$$

The definition makes sense since $f(\lambda X_0, \ldots, \lambda X_n) = \lambda^d f(X_0, \ldots, X_n)$, where d is the degree of f. To a set J of homogeneous polynomials in $\mathbb{F}[X_0, \ldots, X_n]$ we can associate a subset of $P^n(\mathbb{F})$ by

$$\bar{\mathcal{V}}_{\mathbb{F}}(J) = \bigcap_{f \in J} \bar{\mathcal{V}}_{\mathbb{F}}(f).$$

$\bar{\mathcal{V}}_{\mathbb{F}}(J)$ will be called a *projective* $\mathbb{F}$-*algebraic set*. When $\mathbb{F} = \mathbb{C}$, we call it a *projective variety*. Just as for affine varieties, the Zariski topology is defined so that the closed sets are the projective algebraic varieties. A Zariski locally closed subset of $P^n(\mathbb{F})$ (i.e., the intersection of a closed and an open set) is called a *quasi-projective* $\mathbb{F}$-*algebraic set* (or quasi-projective variety, if $\mathbb{F} = \mathbb{C}$).

The chart φ_i pulls back projective algebraic sets to affine algebraic sets. Given $\bar{\mathcal{V}}_{\mathbb{F}}(f_1, \ldots, f_m) \subset P^n(\mathbb{F})$, define

$$g_i(x_1, \ldots, x_n) := f_i(x_1, \ldots, x_{i-1}, 1, x_i, \ldots, x_n).$$

Then $\varphi_i^{-1}(\bar{\mathcal{V}}_{\mathbb{F}}(f_1, \ldots, f_m)) = \mathcal{V}(g_1, \ldots, g_n)$ is an affine algebraic set in $\mathbb{F}^n$. Conversely, we can associate to any affine algebraic set $V \subset \mathbb{F}^n$ a projective algebraic set $W \subset P^n(\mathbb{F})$ as follows. If $f \in \mathbb{F}[x_1, \ldots, x_n]$ we can write $f = \sum_{i=1}^d f_i$, where each f_i is homogeneous of degree i. Let

$$F(X_0, X_1, \ldots, X_n) = \sum_{i=0}^d X_0^{d-i} f_i(X_1, \ldots, X_n).$$

Then F is homogeneous of degree d and called the *homogenization* of f. Whenever $X_0 \neq 0$, we have

$$F(X_0, \ldots, X_n) = X_0^d f(X_1/X_0, X_2/X_0, \ldots, X_n/X_0).$$

If $S \subset \mathbb{F}^n$ and J is the set of all homogenizations F of some $f \in \mathcal{I}_{\mathbb{F}}(S)$, then J defines a projective algebraic set $P(S)$ such that S is mapped bijectively onto $P(S) \cap \varphi_0(\mathbb{F}^n)$ under φ_0. A similar correspondence holds using φ_i.

There is also a one-to-one correspondence between projective algebraic sets in $P^n(\mathbb{F})$ and nonempty algebraic sets in $\mathbb{F}^{n+1}$ invariant under scalar multiplication by $\mathbb{F}^*$. In fact, if S is a projective algebraic set determined by a set of homogeneous polynomials J, then $\mathcal{V}_{\mathbb{F}}(J) \subset \mathbb{F}^{n+1}$ is an affine algebraic set invariant under scalar multiplication by $\mathbb{F} - \{0\}$. Conversely, let $S \subset \mathbb{F}^{n+1}$ be an algebraic set invariant under scalar multiplication. Let $J \subset \mathbb{F}[x_0, \ldots, x_n]$ be the set of polynomials vanishing on S, which is generated by a finite family $\{f_1, \ldots, f_m\}$, and write $f_i = \sum_{j=0}^{d_i} f_{ij}$, where each f_{ij} is homogeneous of degree j. Then for each $x \in S$,

$$0 = f_i(\lambda x) = \sum_{j=0}^{d_i} f_{ij}(\lambda x) = \sum_{j=0}^{d_i} \lambda^j f_{ij}(x).$$

Since this equation holds for each i and $\lambda \in \mathbb{F}$ we must have $f_{ij} = 0$ for each i, j. Therefore $\mathcal{I}_{\mathbb{F}}(S)$ is generated by the homogeneous polynomials f_{ij} and defines a projective algebraic set in $P^n(\mathbb{F})$.

The notion of product of projective varieties requires the so-called Segre embedding

$$s : P^n(\mathbb{F}) \times P^m(\mathbb{F}) \to P^{nm+n+m}(\mathbb{F})$$

defined by $s([X_0, \ldots, X_n], [Y_0, \ldots, Y_m]) = [X_0 Y_0, \ldots, X_i Y_j, \ldots, X_n Y_m]$. The image of s is a projective algebraic set and s is one-to-one. Whenever we take a Cartesian product of projective varieties, it is understood that its Zariski topology is defined by means of the Segre embedding.

We denote by $G_{\mathbb{F}}(m, n)$ the Grassmannian of m-planes in $\mathbb{F}^n$, that is, the set of all m-dimensional subspaces in $\mathbb{F}^n$. It can be shown that $G_{\mathbb{F}}(m, n)$ is in a natural way the $\mathbb{F}$-points of a projective variety. The Grassmannian can be regarded as a subset of a projective space by means of the *Plücker embedding*

$$G_{\mathbb{F}}(m, n) \to P(\Lambda^m(\mathbb{F}^n)) = P^{\binom{n}{m}-1}(\mathbb{F}),$$

sending the plane $L \subset \mathbb{F}^n$ spanned by vectors $v_1, \ldots, v_m$ to the alternating tensor $v_1 \wedge \cdots \wedge v_m$, where s is the dimension of $\Lambda^m(\mathbb{F}^n)$.

Exercise 4.2.1 Show that $G = GL(n, \mathbb{F})$, $\mathbb{F} = \mathbb{R}, \mathbb{C}$, acts transitively on $G_{\mathbb{F}}(m, n)$ and describe the isotropy subgroup H of the point of $G_{\mathbb{F}}(m, n)$ corresponding to $\mathbb{F}^m \subset \mathbb{F}^n$. Therefore the homogeneous space G/H is compact and can be given the structure of a projective variety.

4.3 The Descending Chain Condition

We recall that $\mathbb{F}[x]$ is a *Noetherian ring*, that is, any increasing chain of ideals $I_1 \subset I_2 \subset \cdots$ eventually stabilizes: $I_k = I_{k+1} = \cdots$, for some positive integer k. This implies that any ideal of $\mathbb{F}[x]$ is generated by a finite number of polynomials. Therefore, any $V_{\mathbb{F}}(J)$, $J \subset \mathbb{F}[x]$, is the intersection of finitely many $V_{\mathbb{F}}(f)$, $f \in \mathbb{F}[x]$.

Proposition 4.3.1 Let A be a family of affine $\mathbb{F}$-algebraic sets in $\mathbb{F}^n$. Then A has a minimal member, that is, there exists $V \in A$ such that $V = W$ for all $W \in A$ that are contained in V. The same holds for a family of projective $\mathbb{F}$-algebraic sets in $P^n(\mathbb{F})$.

Proof. It suffices to show that any descending sequence of affine algebraic sets

$$\mathcal{V}(J_1) \supset \mathcal{V}(J_2) \supset \cdots \supset \mathcal{V}(J_r) \supset \cdots$$

eventually stabilizes. (The projective case can be reduced to the affine case by the correspondence between projective algebraic sets and $\mathbb{F}^*$-invariant affine algebraic sets in $\mathbb{F}^{n+1}$.) Equivalently, the ascending sequence of ideals J_i must stabilize. But the sum of the J_i is a finitely generated ideal, with generating set $\{g_l\}$, and each g_l must belong to some J_i. Therefore, for some sufficiently large i, $g_l \in J_i$ for all l. $\square$

An algebraic set V, affine or projective, is called *irreducible* if there does not exist a decomposition $V = V_1 \cup V_2$ with $V_1, V_2 \subset V$ being proper algebraic subsets. It is an easy consequence of the proposition that each V decomposes in a unique way as a finite union of irreducible algebraic subsets

$$V = V_1 \cup \cdots \cup V_l$$

such that V_i is not contained in V_j for each i and j. The V_i are called the *irreducible components* of V.

4.4 Functions and Morphisms

Let $V \subset \mathbb{F}^n$ be an irreducible affine algebraic set and $I = \mathcal{I}_{\mathbb{F}}(V)$ its ideal. The set of *polynomial functions* on V is the quotient algebra $\mathbb{F}[V] := \mathbb{F}[x]/I$, and its elements can be identified in a natural way with the restriction to V of polynomials in $\mathbb{F}[x]$.

The *field of fractions*, or *function field*, $\mathbb{F}(V)$ of $\mathbb{F}[V]$ is the set of all *rational functions* $f = g/h$, $g, h \in \mathbb{F}[V]$, h not identically 0. It is indeed a field since V is irreducible. A rational function f is not really a function on V since h may vanish on a subset of V. However, the domain of f, that is, the subset of V where h does not vanish, is a nonempty principal open set, and it is a dense subset of V since V is assumed to be irreducible. A rational function f is said to be *regular at a point* p if p is contained in the domain of f.

If $W \subset \mathbb{F}^m$ is another algebraic set, we say that a map $f : V \to W$ is a *polynomial map* if there exist m polynomials $f_i \in \mathbb{F}[x_1, \ldots, x_n], i = 1, \ldots, m$, such that $f(x) = (f_1(x), \ldots, f_m(x))$ for all $x \in V$.

A *rational map* $F : V \to W$ is a partially defined map given by rational functions, $F = (F_1, \ldots, F_m)$. The domain of F is, by definition, the intersection of the domains of all the F_i. Any point in the domain of F is called a *regular point* of F.

We say that a rational map $F : V \to W$ is *dominant* if the image under F of the domain of F is dense in W in the Zariski topology. When $F : V \to W$ and $G : W \to U$ are rational maps and F is dominant, then $G \circ F$ is also a rational map.

Let U be a Zariski-open subset of the irreducible $\mathbb{F}$-algebraic set V. A *morphism* $F : U \to W$ is the restriction of a rational map from V to W such that U is contained in the domain of F. If V and W are affine varieties ($\mathbb{C}$-algebraic sets) and $F : V \to W$ is a morphism (hence a rational map with domain V), then it can be shown by applying Hilbert's Nullstellensatz [30] that F is a polynomial map.

4.5 Nonsingular Points

Let $V \subset \mathbb{F}^n$ be an affine algebraic set, $\mathbb{F} = \mathbb{R}$ or $\mathbb{C}$, and fix $p = (a_1, \ldots, a_n)$ in V. For each $f \in \mathbb{F}[x]$, write

$$df_p(x - p) = \sum_{i=1}^{n} \frac{\partial f}{\partial x_i}(p)(x_i - a_i).$$

The *(algebraic) tangent space* of V at p is the linear subspace

$$T_p V = \{v \in \mathbb{F}^n \mid df_p v = 0 \text{ for all } f \in \mathcal{I}_{\mathbb{F}}(V)\}.$$

Lemma 4.5.1 The function $V \to \mathbb{N}$ defined by $p \mapsto \dim T_p V$ is an upper semicontinuous function (in the Zariski topology of V). In other words, for each integer m, the subset $V(m) = \{p \in V \mid \dim T_p V \geq m\}$ is closed.

Proof. Let $f_1, \ldots, f_k$ be a set of generators of $\mathcal{I}_{\mathbb{F}}(V)$. Since each $g \in \mathcal{I}_{\mathbb{F}}(V)$ can be written as $g = \sum_{i=1}^{k} \phi_i f_i$, we have $dg_p = \sum_{i=1}^{k} \phi(p)(df_i)_p$. Therefore

$$T_p V = \bigcap_{i=1}^{k} \ker((df_i)_p) \subset \mathbb{F}^n.$$

A point p belongs to $V(m)$ if and only if the matrix $F(p) := (\frac{\partial f_i}{\partial x_j}(p))$, $1 \leq i \leq k$, $1 \leq j \leq n$, has rank at most $n - m$ and this holds if and only if every $(n - m + 1) \times (n - m + 1)$ minor of $F(p)$ vanishes. But each minor is a polynomial, so $V(m)$ is an algebraic set. $\qquad\square$

Proposition 4.5.2 There exists an integer m and a nonempty open subset $M_0 \subset V$ such that $\dim T_p V = m$ for all $p \in M_0$ and $\dim T_p V \geq m$ for all $p \in V$.

Proof. Let m be the minimum value of $\dim T_p V$ among the $p \in V$. Then $V(m-1) = \emptyset$, $V(m) = V$, and $V(m+1)$ is a proper closed subset of V. Therefore $M_0 = V(m) - V(m+1)$ is open and nonempty. $\qquad\square$

If V is irreducible, the proposition implies that on an open and dense (in the Zariski topology) subset $M_0 \subset V$, the dimension of $T_p V$ is constant and assumes the least value, m, of $\dim T_p V$ among the $p \in V$. That value is called the *dimension* of V, $\dim V$. We say that a point p is *nonsingular* if $\dim T_p V = \dim V$ and that it is *singular* if $\dim T_p V > \dim V$. V is said to be *nonsingular* or *smooth* if all its points are nonsingular.

It can be shown by a simple argument (see [35]) that for each $p \in M_0$, there is an open (in the ordinary Hausdorff topology of $\mathbb{F}^n$) neighborhood $U \subset \mathbb{F}^n$ of p, small enough that it does not intersect the complement of M_0, and polynomials $h_1, \ldots, h_{n-m}$, depending on p, such that $(dh_i)_q$ are linearly independent for all $q \in U$ and $V \cap U$ is the set of common zeros in U of the h_i. By the implicit function theorem it then follows that M_0 is a locally closed smooth submanifold of $\mathbb{F}^n$. Notice that at each point of M_0 the algebraic tangent space and the ordinary tangent space coincide.

Since the complement of the set of nonsingular points is an algebraic set, $V_1 = V - M_0$, we may apply the proposition to V_1, decomposing it into a nonempty open subset M_1 of nonsingular points (in V_1) and a proper algebraic subset of V_1, which we call V_2. Proceeding by induction, we obtain the following proposition.

Proposition 4.5.3 Let $V \subset \mathbb{F}^n$ be an irreducible affine algebraic set ($\mathbb{F} = \mathbb{R}$ or $\mathbb{C}$). Then V decomposes into a disjoint union of locally closed smooth embedded submanifolds $M_i \subset \mathbb{F}^n$, $i = 0, 1, \ldots, s$. Define $V_0 = V$. Then the union $V_i := M_i \cup \cdots \cup M_s$, for each $1 \leq i \leq s$, is a proper algebraic subset of V_{i-1}. Let m_i denote the constant dimension of $T_q V$ on M_i. For each $0 \leq i \leq s$ and $p \in M_i$ there exist $h_1, \ldots, h_{n-m_i}$ in the function field of V and a neighborhood U of p such that the differentials $(dh_i)_x$, $i = 1, \ldots, n - m_i$, are linearly independent and

$$U \cap V = \{q \in U \mid h_i(q) = 0 \text{ for each } i = 1, \ldots, n - m_i\}.$$

Although we have limited the foregoing discussion to affine algebraic sets, the notions of nonsingular points and dimension also apply to projective algebraic sets, by using affine open neighborhoods (such as the image of the affine charts

φ_i introduced earlier) of each point p of a projective variety V. It turns out that the facts discussed here do not depend on the affine neighborhood chosen.

4.6 Real Points of Complex Varieties

Let $V \subset \mathbb{R}^n$ be an affine real algebraic variety with ideal $I = \mathcal{I}_\mathbb{R}(V) \subset \mathbb{R}[x]$. Then the Zariski closure of V (in $\mathbb{C}$) is the affine variety

$$\bar{V} = \{ p \in \mathbb{C}^n \mid f(p) = 0 \text{ for all } f \in I \}.$$

This is the unique smallest complex variety containing V. Since for each affine $\mathbb{F}$-algebraic set W we have $W = \mathcal{V}_\mathbb{F}(\mathcal{I}_\mathbb{F}(W))$, it follows that V is the set of real points of its Zariski closure in $\mathbb{C}^n$, that is, $V = \bar{V}(\mathbb{R})$.

Recall that $c : \mathbb{C}^n \to \mathbb{C}^n$ is the involution of $\mathbb{C}^n$ induced by complex conjugation. Define f^c, for each $f \in \mathbb{C}[x]$, by $f^c = c \circ f \circ c$ and notice that $f^c = f$ for all $f \in \mathbb{R}[x]$. In particular, c restricts to an involution of $\bar{V}$ and V is the set of fixed points of c.

Define $\bar{I} = \mathcal{I}_\mathbb{C}(\bar{V})$. Then $I = \bar{I} \cap \mathbb{R}[x]$, $\bar{I}$ is the ideal of $\mathbb{C}[x]$ generated by I, and I is the set of all $(f + f^c)/2$, $f \in \bar{I}$.

Exercise 4.6.1 Let V be an affine real algebraic set. If $\bar{V}$ decomposes into irreducible components $V_1^*, \ldots, V_k^*$, and V_i, $i = 1, \ldots, k$, is the real part of V_i^*, show that the V_i are the irreducible components of V and that $\bar{V}_i = V_i^*$.

Proposition 4.6.2 At each point $p \in V$, $\dim_\mathbb{R} T_p V = \dim_\mathbb{C} T_p \bar{V}$. In particular, $T_p \bar{V}$ is the complexification of $T_p V$. Furthermore, if $\bar{V} = M_0 \cup V_1$ is the decomposition of $\bar{V}$ into the nonsingular and singular points, then $V = M_0(\mathbb{R}) \cup V_1(\mathbb{R})$ is the decomposition of V into nonsingular and singular points.

Proof. First notice that if W is an $\mathbb{F}$-algebraic set with ideal $J \subset \mathbb{F}[x]$, $\mathbb{F} = \mathbb{C}$ or $\mathbb{R}$, then $\dim_\mathbb{F} T_p W = n - r_\mathbb{F}(V, p)$, where $r_\mathbb{F}(V, p)$ is the $\mathbb{F}$-dimension of the linear subspace of the dual of $\mathbb{F}^n$ spanned by all differentials df_p, $f \in J$. The proposition follows from the fact that I generates $\bar{I}$ and that if $f_i \in I$ and $h_i \in \mathbb{C}[x]$, then $d(\sum_i h_i f_i)_p = \sum_i h_i(p)(df_i)_p$. The remaining claims follow easily from the characterization of nonsingular points in terms of the dimension of the algebraic tangent spaces. $\square$

4.7 The Ordinary Topology of a Variety

Varieties can also be given a Hausdorff topology, namely, the relative topology as subsets of $\mathbb{C}^n$ or $P^n(\mathbb{C})$ with their manifold topologies. We refer to that as the *ordinary topology* of the variety. It is clear that real or complex algebraic sets are locally compact and that projective varieties are compact in the ordinary topology. The next result is due to H. Whitney [35]. See also [26, theorem 3.5].

Proposition 4.7.1 Irreducible complex varieties are connected in the ordinary topology. If V is a real or complex variety, then it has a finite number of connected components, also in the ordinary topology.

Proof. For the first assertion, see [30, chapter 7]. A consequence of this first claim is that an irreducible affine variety of dimension 0 is a single point, whence any zero-dimensional variety is a finite set.

We only give here the argument for real affine varieties, for the second assertion.

In order to obtain a contradiction, suppose that the proposition is false. We may assume V is the smallest variety with an infinite number of connected components (i.e., minimal under inclusion among the varieties with an infinite number of components). If this is not already the case we can choose a proper subvariety V_1 with an infinite number of connected components, then a proper subvariety V_2 of V_1 with the same property, and so forth. But the sequence would have to terminate by the descending chain condition, and the last set would be minimal with the property of having infinitely many connected components.

Consider the splitting $V = M \cup V_1$ into singular and nonsingular points. Then V_1 has at most finitely many connected components, by the minimality of V, whence M has infinitely many components, which must be smooth closed submanifolds of M with same dimension m. Moreover, $m \geq 1$, since any zero-dimensional variety is a finite set, as noted in the first paragraph. Let $M_1, M_2, \ldots$ denote the connected components of M.

Choose a point $q = (a_1, \ldots, a_n) \in \mathbb{R}^n$ not equidistant from all points of M_1 and set $g(x) = \sum_{i=1}^{n}(x_i - a_i)^2$. For arbitrary polynomials $f_1, \ldots, f_r$, $r = n - m$, and each $\mu = (\mu_1, \ldots, \mu_{r+1}) \in \mathbb{N}^{r+1}$, consider the Jacobian

$$\Phi_\mu(f_1, \ldots, f_r) = \frac{\partial(f_1, \ldots, f_r, g)}{\partial(\mu_1, \ldots, \mu_r, \mu_{r+1})}.$$

Let I be the ideal of real polynomials vanishing on V. Let V' be the real variety defined by the set of all $\Phi_\mu(f_1, \ldots, f_r)$, $f_i \in I$, together with all the polynomials in I. Then V' is a real subvariety of V.

Let p_i be a point of M_i closest to q. Then $dg_{p_i} v = 0$ for all $v \in T_{p_i} M_i$; hence $(df_1)_{p_i}, \ldots, (df_r)_{p_i}, dg_{p_i}$ are linearly dependent elements of the dual of $T_{p_i} M_i$; if the f_k are in I. Therefore the $\Phi_\mu(f_1, \ldots, f_r)$ are 0 at p_i, whence $p_i \in V'$. This implies that V' also contains an infinite number of connected components.

On the other hand, g is not constant on M_1; thus there is a point $p \in M_1$ and a vector $v \in T_p M_1$ such that $dg_p v \neq 0$. There are also polynomials $f_1, \ldots, f_r$ in I with $(df_1)_p, \ldots, (df_r)_p$ linearly independent. But then, for some μ, $\Phi_\mu(f_1, \ldots, f_r)$ does not vanish at p, and p is not in V'. Therefore V' is a proper subvariety of V, which is a contradiction due to the minimality of V. $\quad\square$

Proposition 4.7.2 Let V be a smooth irreducible variety defined over $\mathbb{R}$ such that $V(\mathbb{R})$ is nonempty. Then, if $U \subset V$ is any nonempty Zariski-open set defined over $\mathbb{R}$, the set of $\mathbb{R}$-points $U(\mathbb{R})$ is dense in $V(\mathbb{R})$ in the ordinary topology. Furthermore, any nonempty open (in the ordinary topology) subset $F \subset V(\mathbb{R})$ is Zariski-dense in V. In particular, $V(\mathbb{R})$ is Zariski-dense in V.

Proof. We give below only a sketch of the proof. The reader will find more details in [26, lemma 3.2].

All assertions in the proposition reduce to proving that $U \cap F$ is nonempty, where $U \subset V$ is Zariski-open and $F \subset V(\mathbb{R})$ is open in the ordinary topology. Let $X = V - U$. For any $p \in F$ it is possible to find an affine neighborhood W of p and a nonzero $f \in \mathbb{R}[W]$ such that f vanishes on $X \cap W$. If F does not intersect U, then f vanishes identically on F.

The idea is to regard f as an analytic function near p using an analytic coordinate system of a neighborhood of p in V, and show that if f vanishes on F it must vanish identically, a contradiction. $\quad\square$

4.8 Homogeneous Spaces

Denote by $M(n, \mathbb{F})$ the linear space of n-by-n matrices with coefficients in $\mathbb{F}$. A *linear algebraic group* G is a subgroup of $GL(n, \mathbb{C})$ that is also an affine variety in $M(n, \mathbb{C})$. Therefore its elements are the common roots of a subset of $\mathbb{C}[x] := \mathbb{C}[x_{11}, \ldots, x_{ij}, \ldots, x_{nn}]$ of polynomials in the indeterminates $x := (x_{ij})$, the n^2 coordinate entries of $M(n, \mathbb{C})$.

We say that a linear algebraic group G is defined over $\mathbb{F}$ if G is a variety defined over $\mathbb{F}$ and the group operations (inversion and multiplication) are morphisms defined over $\mathbb{F}$. The set of real points of a linear algebraic group defined over $\mathbb{R}$ will be referred to as a *real algebraic group*.

The group $GL(n, \mathbb{C})$ is itself a linear algebraic group defined over $\mathbb{Q}$ since it can be identified with the subgroup of $GL(n + 1, \mathbb{C})$ of matrices of the form $\left(\begin{smallmatrix} A & 0 \\ 0 & \delta \end{smallmatrix}\right)$ such that $1 - \delta \det A = 0$. The group

$$SL(n, \mathbb{C}) = \{A \in GL(n, \mathbb{C}) \mid \det A = 1\}$$

is also defined over $\mathbb{Q}$. The same is true for all the classical groups we will study in chapter 5.

Let G be a linear algebraic group. Since any point of G can be mapped to any other point by an automorphism of G, say by left translation (and since a nonempty variety always has at least one nonsingular point), it follows that G is a smooth variety, hence a Lie group. It can also be shown that the irreducible components of G agree with the connected components. (This is a consequence of G being a smooth variety.) When G and the group operations are defined over $\mathbb{R}$, the set of real points $G_{\mathbb{R}}$ is also a Lie group, and by the previous section it has a finite number of connected components.

Let H be a Zariski-closed subgroup of the linear algebraic group G. Then the quotient G/H can be given the structure of a quasi-projective variety, by a natural embedding into a projective space. The precise statement is as follows (see [5, 5.1]).

Theorem 4.8.1 (Chevalley) Let G be a linear algebraic group and H a Zariski-closed subgroup, both defined over $\mathbb{F}$. Then there exists a faithful representation $\rho : G \to GL(n, \mathbb{C})$ defined over $\mathbb{F}$ and a one-dimensional subspace $L \subset \mathbb{C}^n$ also defined over $\mathbb{F}$ such that $H = \{g \in G \mid gL = L\}$.

The linear representation ρ naturally induces an algebraic action of G on the projective space $P^{n-1}(\mathbb{C})$. Let $p \in P^{n-1}(\mathbb{C})$ be the point corresponding to L. Then the preceding theorem yields a geometric realization of G/H as the orbit Gp. This orbit can be shown to be a quasi-projective variety (i.e., an open subset of a projective variety) defined over $\mathbb{F}$.

Rather than give the proof in detail, we illustrate the main argument with a simple example. Let G be the one-dimensional multiplicative group $GL(1, \mathbb{C})$ $= \mathbb{C}^*$ and $H = \{-1, +1\}$. Then H is the variety corresponding to the ideal $I \subset \mathbb{C}[x]$ generated by the single polynomial $f_0(x) = x^2 - 1$. G acts on its algebra of polynomials $\mathbb{C}[G]$ by means of the (infinite-dimensional) linear representation $R : G \to GL(\mathbb{C}[G])$ defined by

$$R_g(f)(g') = f(g^{-1}g'),$$

where $f \in \mathbb{C}[G]$ and $g, g' \in G$.

It can be shown that the linear span of a generating set of I lies in a finite-dimensional subspace W of $\mathbb{C}[G]$ invariant under all R_g. In our example, it is easily seen that $W = \{ax^2 + b \mid a, b \in \mathbb{C}\}$ is such a subspace. Set

$$L := W \cap I = \{a(x^2 - 1) \mid a \in \mathbb{C}\}.$$

Then L is H-invariant. Regard $P^1(\mathbb{C})$ as the space of one-dimensional subspaces of W and denote a point of $P^1(\mathbb{C})$ by $[a, b]$. Thus the induced action of

$u \in \mathbb{C}^*$ on $P^1(\mathbb{C})$ becomes $u \cdot [a, b] = [u^2 a, b]$, L is identified with the point $p_0 := [1, -1]$, and we have $H = \{u \in \mathbb{C}^* \mid u \cdot p_0 = p_0\}$.
As a result, we obtain the identification

$$\mathbb{C}^*/\{-1, +1\} \cong \{[u^2, -1] \mid u \in \mathbb{C}^*\} \subset P^1(\mathbb{C}).$$

It is apparent, from what has preceded, that the set of real points of the quotient is

$$(\mathbb{C}^*/\{-1, +1\})_{\mathbb{R}} \cong \{[x, -1] \mid x \in \mathbb{R}^*\},$$

which is not connected. Therefore, even though the action of G on G/H is transitive, the action of $G(\mathbb{R}) = \mathbb{R}^*$ on $(G/H)(\mathbb{R})$ is not. (We have in this case two orbits: the positive reals and the negative reals.) The next proposition shows that $G(\mathbb{R})$ has at most finitely many orbits on $(G/H)(\mathbb{R})$.

Proposition 4.8.2 Let G be a linear algebraic group defined over $\mathbb{R}$ and H a Zariski-closed subgroup also defined over $\mathbb{R}$. Then $(G/H)(\mathbb{R})$ is the union of a finite number of $G(\mathbb{R})$-orbits and each orbit is a union of connected components of $(G/H)(\mathbb{R})$. In particular, if $(G/H)(\mathbb{R})$ is connected, there is exactly one orbit.

Proof. Let p be any point of $(G/H)(\mathbb{R})$ and let $\tau : G \to G/H$ be the orbit map, defined by $\tau(g) = g \cdot p$. Then τ is a smooth surjective morphism defined over $\mathbb{R}$ between smooth varieties defined over $\mathbb{R}$. Moreover, the differential $D_g \tau$ is surjective for all $g \in G$. It follows from the comments in section 4.6 that, for each $g \in G(\mathbb{R})$, the differential

$$D_g \tau : T_g G(\mathbb{R}) \to T_{g \cdot p}(G/H)(\mathbb{R})$$

is also surjective. Hence the restriction of τ to $G(\mathbb{R})$ is a submersion into $(G/H)(\mathbb{R})$, hence an open map. But the image of this map is the $G(\mathbb{R})$-orbit of p. Therefore, $G(\mathbb{R})$-orbits are open. They are also closed, since the complement of an orbit is a union of orbits. Consequently, each orbit is a union of connected components of $(G/H)(\mathbb{R})$. Their number is finite by Whitney's theorem, discussed in section 4.7. □

4.9 Rosenlicht's Theorem

Let V be an affine, projective, or, more generally, quasi-projective variety defined over $\mathbb{R}$, and let H be a linear algebraic group also defined over $\mathbb{R}$. An *algebraic action* (defined over $\mathbb{R}$) of H on V is a morphism (defined over $\mathbb{R}$)

$$\Phi : H \times V \to V$$

such that Φ is a group action on V.

The following theorem contains the key property of algebraic actions that will be needed later in the book. We refer the reader to [29] for a more general statement and the proof. See also [15] for a different, although very sketchy, proof.

Theorem 4.9.1 (Rosenlicht) Let H, V, and Φ be as before. Then there exists a variety W and a rational map $\tau : V \to W$, both defined over $\mathbb{R}$, such that the following holds. The image under τ of its domain of definition in V is Zariski-dense in W, and there exist dense open subsets $W' \subset W$, $V' \subset V$, defined over $\mathbb{R}$, such that V' is H-invariant and $W' = V'/H$; that is $\tau|_{V'}$ is a surjective morphism from V' to W' inducing a bijection between H-orbits in V' and points of W'. Moreover, V', W', and $\tau|_{V'}$ are smooth.

By applying Rosenlicht's theorem to the complement of $M_0 := V'$ in V, we obtain a smooth open H-invariant subset M_1 of that complement such that the orbit space M_1/H is also a smooth variety. Applying the theorem inductively we arrive at the following corollary.

Corollary 4.9.2 There exists a finite decomposition of V into H-invariant sets

$$V = M_0 \cup M_1 \cup \cdots \cup M_l$$

such that, for each $0 \le i \le s$, the union $M_i \cup \cdots \cup M_l$ is Zariski-closed in V and contains M_i as a Zariski-open subset. Moreover, for each i, M_i/H has the structure of a smooth variety defined over $\mathbb{R}$. In particular, H-orbits in M_l are closed sets.

Another immediate consequence of the theorem is that the H-orbits are locally closed and embedded in V. By applying proposition 4.8.2 we obtain that the orbits of the associated real action are also locally closed and embedded. This, together with theorem 1.2.2, implies the next corollary.

Corollary 4.9.3 Let $\Phi : H \times V \to V$ be a real algebraic action, defined by restriction to the real points of an algebraic action defined over $\mathbb{R}$. Then each orbit of Φ is locally closed and is an embedded submanifold of V. In particular, real algebraic actions are tame.

By a *real algebraic one-parameter group* we mean a real algebraic group isomorphic to either $GL(1, \mathbb{R}) \cong \mathbb{R}^*$ or the additive group $\mathbb{R}$.

Corollary 4.9.4 Let $\Phi : H \times V \to V$ be a real algebraic action, where H is a one-parameter real algebraic subgroup. Then any recurrent point is a fixed point.

Proof. Since orbits are embedded, any recurrent point $x \in V$ must actually be a periodic point. In that case, the isotropy group of x is the set of real points of a Zariski-closed infinite algebraic subgroup of a one-dimensional group; therefore $H_x = H$ and x is a fixed point. □

Given an algebraic group H, let H' be the smallest algebraic subgroup of H containing all the real one-parameter subgroups – a normal subgroup. We say that H is *generated by real algebraic one-parameter subgroups* if $H' = H$.

Corollary 4.9.5 Let $\Phi : H \times V \to V$ be a real algebraic action and suppose that H is generated by real algebraic one-parameter subgroups. Let μ be an H-invariant probability measure on V. Then μ is supported on the set of H-fixed points.

Proof. This is an immediate consequence of the previous corollary and Poincaré recurrence. □

Exercise 4.9.6 For each indicated subgroup G of $GL(2, \mathbb{R})$, describe all the orbits of the linear action on $\mathbb{R}^2$. Verify in each case that the orbits are locally closed embedded submanifolds of $\mathbb{R}^2$ and describe the quotient space $\mathbb{R}^2 / G$. Do the same for the natural action that each G induces on $P^1(\mathbb{R})$.

(a) $\left\{ \begin{pmatrix} \lambda & 0 \\ 0 & \lambda \end{pmatrix} \middle| \lambda \in \mathbb{R} - \{0\} \right\}$;

(b) $\left\{ \begin{pmatrix} \lambda & 0 \\ 0 & \lambda^{-1} \end{pmatrix} \middle| \lambda \in \mathbb{R} - \{0\} \right\}$;

(c) $\left\{ \begin{pmatrix} 1 & u \\ 0 & 1 \end{pmatrix} \middle| u \in \mathbb{R} \right\}$;

(d) $\left\{ \begin{pmatrix} 1 & u \\ 0 & \lambda \end{pmatrix} \middle| u \in \mathbb{R}, \lambda \in \mathbb{R} - \{0\} \right\}$;

(e) $\left\{ \begin{pmatrix} \lambda & u \\ 0 & \lambda^{-1} \end{pmatrix} \middle| u \in \mathbb{R}, \lambda \in \mathbb{R} - \{0\} \right\}$.

Recall that a discrete subgroup Γ of a Lie group G is called a *lattice* if G/Γ admits a probability measure invariant under the action of G on the quotient by left translations.

Theorem 4.9.7 (Borel density theorem) Let Γ be a lattice in a real algebraic group G. Suppose that G is generated by real algebraic one-parameter subgroups. Then Γ is Zariski-dense in G.

Proof. Let L denote the Zariski closure of Γ in G. The natural projection $p : G/\Gamma \to G/L$ commutes with left multiplication by elements in G, so the push-forward measure $\nu := p_*\mu$ is a G-invariant probability measure on G/L. By the last corollary, ν must be supported on the set of fixed points for the G-action on G/L. Consequently, fixed points exist and since that action is transitive we conclude that $G = L$. $\qquad\square$

4.10 The Rough Structure of Algebraic Groups

We described in this section a number of general facts concerning the structure of algebraic groups. The reader will find more details on the topics discussed here in [5] or [26].

An diagonalizable matrix of $GL(n, \mathbb{C})$ is also called a *semisimple element* of $GL(n, \mathbb{C})$. A *unipotent element* u of $GL(n, \mathbb{C})$ is a matrix such that $(u-I)^k = 0$ for some positive integer k, where I is the identity matrix. Any $g \in GL(n, \mathbb{C})$ has a *Jordan decomposition*, that is, a splitting into semisimple and unipotent parts, $g = g_s g_u$, such that $g_s g_u = g_u g_s$. It can be shown that if G is an algebraic subgroup of $GL(n, \mathbb{C})$, then the semisimple and unipotent parts of any $g \in G$ are also in G. Moreover, if $\rho : G \to H$ is a homomorphism of algebraic groups that is also a morphism of varieties, semisimple and unipotent elements of G are sent to semisimple and unipotent elements of H, respectively.

A linear algebraic group U is said to be *unipotent* if all of its elements are unipotent. For example, the one-dimensional group G_a consisting of matrices of the form $\left(\begin{smallmatrix} 1 & x \\ 0 & 1 \end{smallmatrix}\right)$ is unipotent. (The subscript stands for "additive.")

Notice that for any unipotent element $g \in GL(n, \mathbb{C})$,

$$(g - I)^n = 0$$

so the Taylor series of the function log, applied to unipotent matrices, is a polynomial of degree $n - 1$:

$$\log u = (u - I) - \frac{(u - I)^2}{2} + \frac{(u - I)^3}{3} - \cdots + (-1)^n \frac{(u - I)^{n-1}}{n - 1}.$$

If $U \subset GL(n, \mathbb{C})$ is unipotent, $\log : U \to u$ gives an isomorphism of varieties from U onto its Lie algebra, which consists of nilpotent elements (if $X \in u$, then X is an $n \times n$ complex matrix such that $X^n = 0$). The inverse map is the truncated exponential map:

$$\exp X = I + X + \frac{X^2}{2!} + \cdots + \frac{X^{n-1}}{(n-1)!}.$$

In particular, U is connected and simply connected.

It can be shown that if U is a unipotent subgroup of $GL(n, \mathbb{F})$, where $\mathbb{F}$ is either $\mathbb{R}$ or $\mathbb{C}$, there exists a matrix $g \in GL(n, \mathbb{F})$ such that gUg^{-1} is contained in the group of upper triangular matrices with 1s on the diagonal. It follows that there is a series of subgroups

$$U = U_0 \supset U_1 \supset \cdots \supset U_n = \{I\}$$

such that U_i / U_{i+1} is isomorphic as an algebraic group to G_{a}, for $i = 0, \ldots,$ $n - 1$.

Let G be any group. Recall that for $g, h \in G$, we defined the commutator of g and h by $(g, h) = h^{-1}g^{-1}hg$. The subgroup of G generated by all the commutators (g, h), $g, h \in G$, will be denoted by (G, G). It is clear that (G, G) is a normal subgroup of G. Also define inductively the groups $G^{(1)} := G_{(1)} := (G, G)$, $G^{(i+1)} := (G^{(i)}, G^{(i)})$, $G_{(i+1)} := (G, G_{(i)})$. G is called *solvable* if the series

$$G \supset G^{(1)} \supset G^{(2)} \supset \cdots \supset G^{(m+1)}$$

eventually terminates in $\{I\}$. G is called *nilpotent* if the series

$$G \supset G_{(1)} \supset G_{(2)} \supset \cdots \supset G_{(m+1)}$$

eventually terminates in $\{I\}$. Each $G^{(i)}$ (resp., $G_{(i)}$) turns out to be a closed (normal) subgroup of $G^{(i-1)}$ (resp., $G_{(i-1)}$).

Let G be now a connected linear algebraic group (defined over $\mathbb{R}$). It can be shown that there exists a unique maximal connected solvable normal subgroup of G, which is also an algebraic group (defined over $\mathbb{R}$). It is called the *radical* of G, and will be denoted by $R(G)$. It is also possible to define the (unique) maximal connected unipotent normal subgroup of G. It is called the *unipotent radical* of G and will be denoted by $R_{\mathrm{u}}(G)$. This is also an algebraic group (defined over $\mathbb{R}$). Moreover, $R_{\mathrm{u}}(G)$ is the subgroup of $R(G)$ consisting of the unipotent elements.

G is said to be *reductive* if $R_{\mathrm{u}}(G) = \{I\}$. It is called a *semisimple group* if $R(G) = \{I\}$. (We will give in chapter 7 a different definition of reductive and semisimple Lie groups, which will easily imply the definitions given here. The reader should be aware that the standard definitions of these concepts are the ones given here. The equivalence of the definitions depends on results of Mostow that will not be discussed in this book. See [26].)

Theorem 4.10.1 (Levi decomposition) Let G be a connected linear real alge-braic group defined over $\mathbb{R}$. Then there exists a reductive connected real al-gebraic subgroup H of G, also defined over $\mathbb{R}$, such that $G = HR_u(G)$, that is, G is the semidirect product of H and the unipotent radical. Moreover, any reductive subgroup of G defined over $\mathbb{R}$ is conjugate to a subgroup of H by an element of the real group $R_u(G)_{\mathbb{R}}$.

The radical $R(H)$ of H is the connected component of the center of H, denoted here by Z; (H, H) is a connected semisimple algebraic group defined over $\mathbb{R}$, and H decomposes into an almost direct product: $H = (H, H)Z$ (i.e., H and Z commute and $H \cap Z$ is finite).

5

The Classical Groups

We study here the main examples of semisimple Lie groups: the special linear groups, unitary groups, orthogonal groups, and symplectic groups, and their various "real forms." These are called the *classical groups*. The general theory of semisimple Lie groups will be addressed in chapter 7.

5.1 *GL(n, ℝ)*, *GL(n, ℂ)*, and *GL(n, ℍ)*

We have already encountered the *general linear group* $GL(n, \mathbb{R})$. It consists of all invertible $n \times n$ matrices with real entries and group operations given by matrix multiplication and matrix inversion. More generally, if V is any finite-dimensional real vector space, we denote by $GL(V)$ the set of invertible linear maps from V onto itself. $GL(V)$ is an abstract group with multiplication given by composition of linear maps. It is also a smooth manifold in a natural way since it is an open subset of the vector space $\text{End}(V)$ of all linear maps of V into itself. Notice that $GL(\mathbb{R}^n)$ is naturally isomorphic to $GL(n, \mathbb{R})$; namely, if $(e_1, \ldots, e_n)$ denotes the standard basis of $\mathbb{R}^n$, then to any $g \in GL(\mathbb{R}^n)$ we associate the matrix $A = (a_{ij})$ such that $ge_i = \sum_j e_j a_{ji}$.

The Lie algebra $\mathfrak{g}$ of a Lie group G was defined in chapter 3 as the linear space of all left-invariant vector fields on G. If $G \subset GL(n, \mathbb{R})$, $\mathfrak{g}$ has an especially simple description, as follows. First identify $\mathfrak{g}$, as a vector space, with the subspace $T_e G \subset \mathfrak{gl}_n(\mathbb{R}) := T_e(GL(n, \mathbb{R}))$, where the latter is the space of all real n-by-n matrices. We claim that the Lie bracket corresponds under this identification to the commutator of matrices $AB - BA$. The precise statement is given in the next exercise.

Since $GL(n, \mathbb{R})$ is an open subset of the vector space $\mathfrak{gl}_n(\mathbb{R})$, its tangent bundle can be written as $T(GL(n, \mathbb{R})) = GL(n, \mathbb{R}) \times \mathfrak{gl}_n(\mathbb{R})$. Any vector field X on $GL(n, \mathbb{R})$ is then of the form $X(g) = (g, A(g))$, for some smooth function $A : GL(n, \mathbb{R}) \to \mathfrak{gl}_n(\mathbb{R})$. It is immediate that

$$D(L_h)_g X(g) = (hg, hA(g)),$$

so X is a left-invariant vector field exactly when $A(g) = gA_0$ for some $A_0 \in \mathfrak{gl}_n(\mathbb{R})$ uniquely determined by X. The flow lines $\Phi_t(g) = gc(t)$ of such a left-invariant X must satisfy the differential equation $\frac{d}{dt}\Phi_t(g) = X(\Phi_t(g)) = (\Phi_t(g), \Phi_t(g)A_0)$, so $c(t)$ is the unique solution to $c'(t) = c(t)A_0$ with initial condition $c(0) = I$. The solution to this initial-value problem is well known:

$$c(t) = e^{tA_0} = \sum_{k=0}^{\infty} \frac{t^k}{k!} A_0^k,$$

and the series converges uniformly on any compact subset of $\mathbb{R}$ to a smooth matrix-valued function of t. Therefore, $\Phi_t(g) = ge^{tA_0}$. This conclusion is needed for the next exercise.

Exercise 5.1.1 Given $A, B \in \mathfrak{gl}(n, \mathbb{R})$, let X, Y be the left-invariant vector fields on $GL(n, \mathbb{R})$ such that $X(g) = (g, gA)$ and $Y(g) = (g, gB)$. Then the bracket $[X, Y]$ is the left-invariant vector field Z such that $Z(g) = (g, g[A, B])$, where $[A, B] = AB - BA$. Conclude the same for the Lie algebra $\mathfrak{g}$ of a Lie subgroup G of $GL(n, \mathbb{R})$, where $\mathfrak{g}$ is now identified with $T_e G \subset \mathfrak{gl}_n(\mathbb{R})$. (Hint: It suffices to show that $([X, Y]f)(e) = \frac{d}{dt}|_{t=0} f(e^{t[A,B]})$ for functions f that are the restriction to $GL(n, \mathbb{R})$ of linear functions on the full space of n-by-n matrices.)

Although the groups considered in this book are mostly real linear groups, that is, subgroups of $GL(V)$ for some real vector space V, we will often define them in terms of matrices with complex coefficients. The next exercise shows that (closed) subgroups of the group $GL(n, \mathbb{C})$ of invertible complex matrices of order n can naturally be viewed as (closed) subgroups of $GL(2n, \mathbb{R})$.

Exercise 5.1.2 Associate to each column vector $z = (z_1, \ldots, z_n)^t$ in $\mathbb{C}^n$, with $z_k = x_k + iy_k$, the vector $\eta(z) = (x_1, \ldots, x_n, y_1, \ldots, y_n)^t \in \mathbb{R}^{2n}$. The correspondence $z \mapsto \eta(z)$ defines an $\mathbb{R}$-linear isomorphism between $\mathbb{C}^n$ and $\mathbb{R}^{2n}$. Denote also by η the map that associates to each $A_1 + iA_2 \in GL(n, \mathbb{C})$ the matrix $\left(\begin{smallmatrix} A_1 & -A_2 \\ A_2 & A_1 \end{smallmatrix}\right)$ in $GL(2n, \mathbb{R})$. Show that (the latter) η is an injective homomorphism and that $\eta(Az) = \eta(A)\eta(z)$ for all $A \in GL(n, \mathbb{C})$ and $z \in \mathbb{C}^n$. If A^* denotes the conjugate transpose of A, show that $\eta(A^*) = \eta(A)^t$. Also show that $GL(n, \mathbb{C})$ corresponds under η to the subgroup of all $A \in GL(2n, \mathbb{R})$ such that $AJ = JA$, where J is the block matrix $\left(\begin{smallmatrix} 0 & I \\ -I & 0 \end{smallmatrix}\right)$ and I is the n-by-n identity matrix.

It turns out, as will be seen later, that connected semisimple linear groups only contain matrices with determinant 1. They are, therefore, subgroups of the *special linear group*

$$SL(n, \mathbb{C}) = \{A \in GL(n, \mathbb{C}) \mid \det A = 1\}.$$

Besides $\mathbb{C}$, it will also be convenient to use the skew field of quaternions $\mathbb{H}$ in order to define some of the groups given subsequently.

We recall some basic facts concerning the quaternions. $\mathbb{H}$ is the $\mathbb{R}$-vector space $\mathbb{R}^4$ with multiplication defined on the basis $1 = (1, 0, 0, 0)$, $i = (0, 1, 0, 0)$, $j = (0, 0, 1, 0)$, $k = (0, 0, 0, 1)$ by the relations

$$i^2 = j^2 = k^2 = ijk = -1, \qquad ij = -ji = k,$$
$$jk = -kj = i, \qquad ki = -ik = j$$

and 1 is the multiplicative identity. As usual, we omit 1 and write an element $q \in \mathbb{H}$ as $q = x_0 + x_1 i + x_2 j + x_3 k$. $\mathbb{H}$ has a natural involution given by $\bar{q} = x_0 - x_1 i - x_2 j - x_3 k$, and $q\bar{q} = \bar{q}q$ is the square of the Euclidean norm of the vector $q \in \mathbb{R}^4$. Moreover, for $p, q \in \mathbb{H}$, one has $\overline{pq} = \bar{q}\,\bar{p}$. The multiplicative inverse of $q \neq 0$ is given by $\bar{q}/(q\bar{q})$.

One should also notice that $\mathbb{H}$ can be written as $\mathbb{H} = \mathbb{C} \oplus j\mathbb{C}$. We obtain an identification of $\mathbb{H}$ with $\mathbb{C}^2$ via the $\mathbb{R}$-linear mapping

$$x_0 + x_1 i + x_2 j + x_3 k = x_0 + ix_1 + j(x_2 - ix_3) \mapsto (x_0 + ix_1, x_2 - ix_3).$$

Accordingly, the $\mathbb{R}$-vector space $\mathbb{H}^n$ of column vectors with entries in $\mathbb{H}$ may be identified with $\mathbb{C}^n \oplus \mathbb{C}^n$ by mapping an element $u = z_1 + jz_2 \in \mathbb{H}^n$ to $\eta(u) := \binom{z_1}{z_2} \in \mathbb{C}^n \oplus \mathbb{C}^n$. Notice that η is an isomorphism of real vector spaces.

We denote by $\mathbb{F}$ any one of $\mathbb{R}$, $\mathbb{C}$, or $\mathbb{H}$ and set $V = \mathbb{F}^n$, the space of column vectors of size n with entries in $\mathbb{F}$. V can be regarded as a vector space over $\mathbb{F}$, where scalar multiplication is defined from the right and denoted $v\lambda$, for $v \in V$ and $\lambda \in \mathbb{F}$. We denote by $V_{\mathbb{R}}$ the subspace $\mathbb{R}^n$, and by $(e_1, \ldots, e_n)$ the standard basis of $V_{\mathbb{R}}$, which is, of course, also a basis of V over $\mathbb{F}$.

$GL(V)$ is, by definition, the group of all invertible $\mathbb{F}$-linear maps from V onto itself. Given $\varphi \in GL(V)$, we write

$$\varphi(e_i) = \sum_j e_j a_{ji}.$$

If to $v = \sum_i e_i z_i$, $z_i \in \mathbb{F}$, one associates the column vector $z = (z_1, \ldots, z_n)^t$, then $\varphi(v)$ corresponds to the matrix multiplication Az, with $A = (a_{ij})$. Notice that $z \mapsto Az$ is indeed $\mathbb{F}$-linear since scalar multiplication acts on the right. The correspondence $\varphi \mapsto A$ defines a group isomorphism between $GL(V)$ and $GL(n, \mathbb{F})$.

One may regard the group $GL(n, \mathbb{H})$ as a subgroup of $GL(2n, \mathbb{C})$. The explicit embedding is given in the next exercise.

Exercise 5.1.3 Define the map $\eta : GL(n, \mathbb{H}) \to GL(2n, \mathbb{C})$ that associates to each $A = A_1 + jA_2 \in GL(n, \mathbb{H})$ the matrix $\left(\begin{smallmatrix} A_1 & -\bar{A}_2 \\ A_2 & \bar{A}_1 \end{smallmatrix} \right)$. Show that η is a group isomorphism onto its image. Verify that $\eta(A^*) = \eta(A)^*$, where A^* denotes matrix transpose followed by the involution $a_{ij} \mapsto \overline{a_{ij}}$ of the entries of A. Check that $\eta(Au) = \eta(A)\eta(u)$, for $A \in GL(n, \mathbb{H})$ and $u \in \mathbb{H}^n$, where $\eta(u)$ has already been defined. Show that scalar (right) multiplication by j corresponds under η to the $\mathbb{R}$-linear map

$$\kappa : \begin{pmatrix} z_1 \\ z_2 \end{pmatrix} \mapsto \begin{pmatrix} 0 & -I \\ I & 0 \end{pmatrix} \begin{pmatrix} \bar{z}_1 \\ \bar{z}_2 \end{pmatrix}.$$

Conclude that $GL(n, \mathbb{H})$ can be identified under η with the subgroup of $GL(2n, \mathbb{C})$ consisting of all A in $GL(2n, \mathbb{C})$ that commute with κ, that is, the subgroup $\{A \in GL(2n, \mathbb{C}) \mid AJ = J\bar{A}\}$, where

$$J = J_{2n} = \begin{pmatrix} 0 & -I \\ I & 0 \end{pmatrix} = \eta(jI).$$

The group $GL(n, \mathbb{H})$, viewed as a subgroup of $GL(2n, \mathbb{C})$, is often denoted

$$U^*(2n) := \{A \in GL(2n, \mathbb{C}) \mid AJ = J\bar{A}\}.$$

We denote by $SL(n, \mathbb{H})$ the subgroup of $GL(n, \mathbb{H})$ that is mapped by η onto

$$SU^*(2n) := U^*(2n) \cap SL(2n, \mathbb{C}).$$

5.2 Automorphism Groups of Bilinear Forms

We now define subgroups of $GL(n, \mathbb{F})$ that preserve bilinear forms on V. Let $Q : V \times V \to \mathbb{F}$ be an $\mathbb{R}$-bilinear map satisfying the property

$$Q(u\lambda, v\mu) = \bar{\lambda} Q(u, v)\mu$$

for all $u, v \in V$ and $\mu, \lambda \in \mathbb{F}$. We also make the assumption that Q is defined over the real numbers, that is, $Q(u, v)$ is real whenever u and v are vectors in $V_{\mathbb{R}}$. We now define the group of $\mathbb{F}$-linear isomorphisms of V that preserve Q:

$$G_Q := \{\varphi \in GL(V) \mid \varphi^* Q = Q\},$$

where $(\varphi^* Q)(u, v) := Q(\varphi(u), \varphi(v))$. Clearly, G_Q is a closed subgroup of $GL(V)$, hence a Lie group.

Exercise 5.2.1 Show that under the natural identification of $GL(V)$ with $GL(n, \mathbb{F})$ obtained by means of the standard basis of $V = \mathbb{F}^n$, G_Q corresponds to the subgroup

$$\{A \in GL(n, \mathbb{F}) \mid A^*MA = M\},$$

where $M = (Q(e_i, e_j))$.

For example, define for $p + q = n$ the form $Q_{p,q}(u, v) := \bar{u}^t I_{p,q} v$, where $I_{p,q} := \begin{pmatrix} -I_p & 0 \\ 0 & I_q \end{pmatrix}$. The subgroup $G_{p,q} := G_{Q_{p,q}}$ of $GL(n, \mathbb{F})$ can then be written as

$$G_{p,q} = \{A \in GL(n, \mathbb{F}) \mid A^* I_{p,q} A = I_{p,q}\}.$$

According to whether $\mathbb{F}$ is $\mathbb{R}$, $\mathbb{C}$, or $\mathbb{H}$, the group $G_{p,q}$ is denoted, respectively,

$$O(p, q) = \{A \in GL(n, \mathbb{R}) \mid A^t I_{p,q} A = I_{p,q}\},$$
$$U(p, q) = \{A \in GL(n, \mathbb{C}) \mid A^* I_{p,q} A = I_{p,q}\},$$
$$Sp(p, q) = \{A \in GL(n, \mathbb{H}) \mid A^* I_{p,q} A = I_{p,q}\}.$$

The last group is also denoted $U(p, q)_{\mathbb{H}}$. We will refer to these three classes of groups as the *unitary groups* over $\mathbb{R}$, $\mathbb{C}$, and $\mathbb{H}$, respectively, with signature $p - q$. The first will also be called the *orthogonal* group (with signature $p - q$). We also denote

$$SO(p, q) = O(p, q) \cap SL(n, \mathbb{R}),$$
$$SU(p, q) = U(p, q) \cap SL(n, \mathbb{C}),$$
$$SO(n) = SO(n, 0) = SO(0, n),$$
$$SU(n) = SU(n, 0) = SU(0, n),$$
$$Sp(n) = Sp(n, 0) = Sp(0, n).$$

It will follow from the next exercise that $Sp(p, q)$, when viewed as a subgroup of $GL(2n, \mathbb{C})$, is already contained in the special linear group. In order to see how $Sp(p, q)$ can be expressed as a subgroup of $GL(2n, \mathbb{C})$, one should first notice that the Hermitian quaternionic form $Q = Q_{p,q}$ decomposes as $Q = Q_1 + jQ_2$, where Q_1 is a Hermitian complex form and Q_2 is a skew-symmetric complex form. In fact, a simple computation shows that

$$Q(u, v) = \overline{\eta(u)}^t K_{p,q} \eta(v) - j\eta(u)^t C J C^{-1} \eta(v),$$

where $K_{p,q} = \begin{pmatrix} I_{p,q} & 0 \\ 0 & I_{p,q} \end{pmatrix} = \eta(I_{p,q})$ and $C = \begin{pmatrix} I & 0 \\ 0 & I_{p,q} \end{pmatrix}$.

Notice that C and $K_{p,q}$ commute. Therefore, after making a change of basis of $\mathbb{C}^{2n}$ by means of the orthogonal matrix C, we can write $Sp(p, q)$ in the form given subsequently. From now on, we adopt the following as a definition:

$$Sp(p, q) := \left\{ A \in GL(2n, \mathbb{C}) \mid A^* K_{p,q} A = K_{p,q} \text{ and } A^t J A = J \right\}.$$

The expression $A^t J A = J$ also characterizes the complex symplectic group:

$$Sp(2n, \mathbb{C}) := \{A \in GL(2n, \mathbb{C}) \mid A^t J A = J\},$$

so $Sp(p, q)$ is a subgroup of $Sp(2n, \mathbb{C})$. We have, in particular,

$$Sp(n) = U(2n) \cap Sp(2n, \mathbb{C}).$$

We also define the real symplectic group as

$$Sp(2n, \mathbb{R}) := Sp(2n, \mathbb{C}) \cap GL(2n, \mathbb{R}).$$

Exercise 5.2.2 Show that the elements of $Sp(2n, \mathbb{C})$ are matrices of determinant 1. You can do this as follows. First observe that $\omega(u, v) := u^t J v$ defines a nondegenerate alternating 2-form on $\mathbb{C}^{2n}$, so $\omega^n := \omega \wedge \cdots \wedge \omega$ is a nonzero alternating $2n$-form. By definition, an element $A \in Sp(2n, \mathbb{C})$, regarded as a linear mapping on $\mathbb{C}^{2n}$, leaves ω invariant under pull-back of forms. The same must then be true for ω^n. But invariance of ω^n under pull-back by A precisely means that the determinant of A is 1.

The complex pseudo-orthogonal groups are defined by taking $\mathbb{F} = \mathbb{C}$ and letting Q be symmetric rather than Hermitian. Therefore,

$$O_{p,q}(\mathbb{C}) := \{A \in GL(n, \mathbb{C}) \mid A^t I_{p,q} A = I_{p,q}\},$$
$$SO_{p,q}(\mathbb{C}) := O_{p,q}(\mathbb{C}) \cap SL_{p+q}(\mathbb{C}).$$

Therefore, $SO(p, q) = SO_{p,q}(\mathbb{R}) = SO_{p,q}(\mathbb{C}) \cap GL(2n, \mathbb{R})$ (similarly for $O(p, q)$).

Exercise 5.2.3 Define the block-diagonal unitary matrix $C = \text{diag}[i I_p, I_q]$, which satisfies $C^2 = I_{p,q}$. Show that

$$C SO_{p,q}(\mathbb{C}) C^{-1} = SO_{p+q}(\mathbb{C}).$$

Therefore the complex pseudo-orthogonal groups are not distinguished, up to isomorphism, by the signature of the quadratic form. Note, however, that the isomorphism does not preserve the real subspace $\mathbb{R}^n$ of $\mathbb{C}^n$.

Finally, we consider the case of a skew-Hermitian quaternionic form Q. Define $Q(u, v) = \bar{u}^t J v$. The corresponding group is denoted

$$O^*(2n) := \{A \in GL(n, \mathbb{H}) \mid A^* J A = J\}.$$

Observe that n should be an even integer. Embedding $O^*(2n)$ into $GL(2n, \mathbb{C})$ via the homomorphism η, and identifying the group with its isomorphic image, we obtain (recall that $\eta(B^*) = \eta(B)^*$)

$$O^*(2n) = U^*(2n) \cap \{A \in GL(2n, \mathbb{C}) \mid A^*\eta(J)A = \eta(J)\}$$
$$= \{A \in GL(2n, \mathbb{C}) \mid A^*\eta(J)A = \eta(J) \text{ and } A^t\eta(J)JA = \eta(J)J\}.$$

The symbol $\eta(J)J$ actually stands for $\eta(J_n)J_{2n}$, where J_n has blocks of size $n/2$ and J_{2n} has blocks of size n. We have also used the fact that $A \in U^*(2n)$ is characterized by the identity $AJ = J\bar{A}$, so that

$$\eta(J)J = A^*\eta(J)AJ = A^*\eta(J)J\bar{A},$$

and the fact that $\eta(J)J$ is a real matrix.

Exercise 5.2.4   Define a unitary block matrix with blocks of size $n/2$ by

$$C = \frac{1}{\sqrt{2}} \begin{pmatrix} -iI & 0 & 0 & -iI \\ 0 & I & I & 0 \\ 0 & -iI & -iI & 0 \\ I & 0 & 0 & -I \end{pmatrix}.$$

Show that $CO^*(2n)C^{-1} = \{A \in GL(2n, \mathbb{C}) \mid A^* J A = J\} \cap O_{2n}(\mathbb{C})$.

Having in mind the previous exercise, we change our definition of $O^*(2n)$ to the following simpler expression:

$$O^*(2n) := \{A \in GL(2n, \mathbb{C}) \mid A^* J A = J\} \cap O_{2n}(\mathbb{C}).$$

We also write

$$SO^*(2n) := O^*(2n) \cap SL(2n, \mathbb{C}).$$

The definitions of the classical groups are summarized here. Recall that I_n is the identity matrix of size n and

$$J_{2n} := \begin{pmatrix} 0 & -I_n \\ I_n & 0 \end{pmatrix}, \quad K_{p,q} := \begin{pmatrix} I_{p,q} & 0 \\ 0 & I_{p,q} \end{pmatrix}, \quad I_{p,q} := \begin{pmatrix} -I_p & 0 \\ 0 & I_q \end{pmatrix}.$$

$$SL(n, \mathbb{C}) = \{A \in GL(n, \mathbb{C}) \mid \det A = 1\},$$

$$SL(n, \mathbb{R}) = \{A \in GL(n, \mathbb{R}) \mid \det A = 1\},$$

$$Sp(2n, \mathbb{C}) = \{A \in GL(2n, \mathbb{C}) \mid A^t J_{2n} A = J_{2n}\},$$

$$Sp(2n, \mathbb{R}) = \{A \in GL(2n, \mathbb{R}) \mid A^t J_{2n} A = J_{2n}\},$$

$$SO^*(2n) = \{A \in SL(2n, \mathbb{C}) \mid A^t A = I_{2n}, \ A^* J_{2n} A = J_{2n}\},$$

$$SU^*(2n) = \{A \in SL(2n, \mathbb{C}) \mid A J_{2n} = J_{2n} \bar{A}\},$$

$$SO_n(\mathbb{C}) = \{A \in SL(n, \mathbb{C}) \mid A^t A = I_n\},$$

$$SO(p, q) = \{A \in SL(n, \mathbb{R}) \mid A^t I_{p,q} A = I_{p,q}\},$$

$$SU(p, q) = \{A \in SL(n, \mathbb{C}) \mid A^* I_{p,q} A = I_{p,q}\},$$

$$Sp(p, q) = \big\{A \in GL(2(p+q), \mathbb{C}) \mid A^* K_{p,q} A = K_{p,q},$$
$$A^t J_{2(p+q)} A = J_{2(p+q)}\big\}.$$

The exercises to follow indicate how the groups already defined account for other groups defined by more general bilinear mappings on a vector space V over $\mathbb{R}$, $\mathbb{C}$, or $\mathbb{H}$. But first, we need to introduce one more class of subgroups of the general linear group. Consider for a given $\mathbb{F}$-linear subspace W of V the group

$$G_W = \{\varphi \in GL(V) \mid \varphi(W) = W\}.$$

By choosing a basis $(e_1, \dots, e_n)$ of V such that $(e_1, \dots, e_k)$ is a basis of W, we can describe G_W as the subgroup of $GL(n, \mathbb{F})$ consisting of matrices that have block form

$$\begin{pmatrix} A & B \\ 0 & C \end{pmatrix},$$

where A and C are nonsingular matrices of order k and $n - k$, respectively, and B is arbitrary. More generally, given a nested sequence of subspaces

$$\{0\} \subset W_1 \subset \cdots \subset W_m \subset V,$$

such that W_s has dimension n_s, the subgroup of $GL(V)$ that stabilizes all the subspaces W_s corresponds (by choosing a basis $(e_1, \dots, e_n)$ of V such that $(e_1, \dots, e_{n_s})$ is a basis of W_s) to the subgroup of $GL(n, \mathbb{F})$ consisting of all matrices with block form

$$\begin{pmatrix} A_{11} & \cdots & * \\ \vdots & \ddots & \vdots \\ 0 & \cdots & A_{mm} \end{pmatrix},$$

where the diagonal blocks are nonsingular matrices of order $n_1, \ldots, n_m$.

Exercise 5.2.5 Given an $\mathbb{R}$-bilinear form $Q : V \times V \to \mathbb{F}$ on V, define the Hermitian and skew-Hermitian parts of Q as follows:

$$Q_{\pm}(u, v) := \frac{Q(u, v) \pm \overline{Q(v, u)}}{2},$$

so that $Q = Q_+ + Q_-$ and $Q_{\pm}(u, v) = \pm \overline{Q_{\pm}(v, u)}$. Show that

$$G_Q = G_{Q_+} \cap G_{Q_-}$$

(G_Q is defined at the beginning of section 5.2). Show the same for the symmetric and antisymmetric forms, which are defined as before but without the conjugate bar.

Exercise 5.2.6 Define the subspace $N_Q := \{v \in V \mid Q(v, \cdot) = 0\}$. Q is said to be nondegenerate if N_Q is the zero space. Show that N_Q is G_Q stable, in the sense that any $\varphi \in G_Q$ maps N_Q onto itself. Also show that (under the assumption that Q is defined over $\mathbb{R}$) there exists a subspace W, $W \subset \mathbb{R}^n \subset V$, that linearly spans N_Q over $\mathbb{F}$. Express G_Q in terms of a basis $e_1', \ldots, e_n'$ (for $\mathbb{R}^n$) such that $e_1', \ldots, e_k'$ is a basis for N_Q, and explain how this group can be described in terms of the nondegenerate bilinear form that Q induces on V/N_Q.

The preceding exercises imply that when given a group defined by a bilinear form, we can, in a sense, reduce it to the case of a nondegenerate form that is one of the following: Hermitian symmetric, Hermitian skew-symmetric, symmetric, or skew-symmetric.

Exercise 5.2.7 Let $Q : V \times V \to \mathbb{F}$ be a nondegenerate Hermitian symmetric $\mathbb{R}$-bilinear form satisfying $Q(u\lambda, v\mu) = \bar{\lambda} Q(u, v)\mu$ for all $u, v \in V$ and $\lambda, \mu \in \mathbb{F}$. Also assume that Q is real (i.e., $Q(u, v)$ is real for all $u, v \in \mathbb{R}^n$). Show that there is a basis

$$\{u_1, \ldots, u_k, w_1, \ldots, w_{n-2k}, v_k, \ldots, v_1\} \subset \mathbb{R}^n$$

such that $Q(u_i, u_j) = Q(v_i, v_j) = 0$ and $Q(u_i, v_j) = \epsilon \delta_{ij} = Q(w_i, w_j)$, where $\epsilon \in \{-1, 1\}$ and $\delta_{ij} = 1$ if $i = j$ and 0 otherwise.

The matrix of ϵQ expressed in terms of the basis given by the previous exercise is

$$L_1 := \begin{pmatrix} 0 & 0 & F \\ 0 & I & 0 \\ F & 0 & 0 \end{pmatrix}, \qquad F := \begin{pmatrix} 0 & \cdots & 1 \\ \vdots & & \vdots \\ 1 & \cdots & 0 \end{pmatrix},$$

where the middle block is the identity matrix of size $n - 2k$ and F is a square block of size k with 1s on the SW-NE diagonal and 0s elsewhere.

Exercise 5.2.8 Keeping the notation of the previous exercise, define a new basis for V given by

$$\{\breve{u}_1, \ldots, \breve{u}_k, w_1, \ldots, w_{n-2k}, \breve{v}_k, \ldots, \breve{v}_1\},$$

where $\breve{u}_i = (u_i + v_i)/\sqrt{2}$ and $\breve{v}_i = (u_i - v_i)/\sqrt{2}$. With respect to this new basis, show that Q has associated matrix given by $L = \text{diag}[1, \ldots, 1, -1, \ldots, -1]$, where the eigenvalue -1 has multiplicity k.

Exercise 5.2.9 Suppose now $\dim_{\mathbb{F}} V = n = 2m$ and let Q be a nondegenerate skew-Hermitian $\mathbb{R}$-bilinear form satisfying $Q(u\lambda, v\mu) = \bar{\lambda} Q(u, v)\mu$ for all $u, v \in V$ and $\lambda, \mu \in \mathbb{F}$. Also assume that Q is real. Show that there is a basis

$$\{u_1, \ldots, u_m, v_1, \ldots, v_m\} \subset \mathbb{R}^n$$

such that $Q(u_i, u_j) = Q(v_i, v_j) = 0$ and $Q(u_i, v_j) = -\delta_{ij}$. Therefore the matrix of Q in terms of this basis is J. It is sometimes convenient to consider the basis

$$\{v_1, \ldots, v_m, u_m, \ldots, u_1\},$$

in which case the matrix of Q becomes

$$J' := \begin{pmatrix} 0 & -F \\ F & 0 \end{pmatrix},$$

where F is as defined earlier.

5.3 The Lie Algebras

The Lie algebras of the foregoing groups can be easily calculated. In fact, we have:

Exercise 5.3.1 Given a group $G = \{A \in GL(n, \mathbb{C}) \mid A^\tau L A = L\}$, where A^τ stands for either A^t or A^*, show that its Lie algebra is given by

$$\mathfrak{g} = \{X \in \mathfrak{gl}_n(\mathbb{C}) \mid X^\tau L + LX = 0\}.$$

The space of all $n \times n$ complex matrices is the Lie algebra of $GL(n, \mathbb{C})$ and is denoted $\mathfrak{gl}_n(\mathbb{C})$.

Using the exercise, one finds the Lie algebras enumerated subsequently. We denote the Lie algebra of a group by the corresponding lowercase gothic letters.

$$\mathfrak{sl}_n(\mathbb{C}) = \{X \in \mathfrak{gl}_n(\mathbb{C}) \mid \operatorname{Tr} X = 0\},$$

$$\mathfrak{sl}_n(\mathbb{R}) = \{X \in \mathfrak{gl}_n(\mathbb{R}) \mid \operatorname{Tr} X = 0\},$$

$$\mathfrak{sp}_{2n}(\mathbb{C}) = \left\{ \begin{pmatrix} X_{11} & X_{12} \\ X_{21} & -\bar{X}_{11} \end{pmatrix} \in \mathfrak{gl}_{2n}(\mathbb{C}) \;\middle|\; \begin{array}{ll} X_{11} & n \text{ by } n \\ X_{12} = X_{12}^t & n \text{ by } n \\ X_{21} = X_{21}^t & n \text{ by } n \end{array} \right\},$$

$$\mathfrak{sp}_{2n}(\mathbb{R}) = \left\{ \begin{pmatrix} X_{11} & X_{12} \\ X_{21} & -X_{11} \end{pmatrix} \in \mathfrak{gl}_{2n}(\mathbb{R}) \;\middle|\; \begin{array}{ll} X_{11} & n \text{ by } n \\ X_{12} = X_{12}^t & n \text{ by } n \\ X_{21} = X_{21}^t & n \text{ by } n \end{array} \right\},$$

$$\mathfrak{so}_n(\mathbb{C}) = \{X \in \mathfrak{gl}_n(\mathbb{C}) \mid X^t = -X\},$$

$$\mathfrak{su}^*(2n) = \left\{ \begin{pmatrix} X_{11} & X_{12} \\ -\bar{X}_{12} & \bar{X}_{11} \end{pmatrix} \in \mathfrak{sl}_{2n}(\mathbb{C}) \;\middle|\; X_{11}, X_{12} \quad n \text{ by } n \right\},$$

$$\mathfrak{so}^*(2n) = \left\{ \begin{pmatrix} X_{11} & X_{12} \\ -\bar{X}_{12} & \bar{X}_{11} \end{pmatrix} \in \mathfrak{sl}_{2n}(\mathbb{C}) \;\middle|\; \begin{array}{ll} X_{11}, X_{12} & n \text{ by } n \\ X_{11}^* = -X_{11} & n \text{ by } n \\ X_{12}^* = X_{11} & n \text{ by } n \end{array} \right\},$$

$$\mathfrak{so}(p,q) = \left\{ \begin{pmatrix} X_{11} & X_{12} \\ X_{12}^t & X_{22} \end{pmatrix} \in \mathfrak{sl}_n(\mathbb{R}) \;\middle|\; \begin{array}{ll} X_{11}^t = -X_{11} & p \text{ by } p \\ X_{22}^t = -X_{22} & q \text{ by } q \end{array} \right\},$$

$$\mathfrak{su}(p,q) = \left\{ \begin{pmatrix} X_{11} & X_{12} \\ X_{12}^* & X_{22} \end{pmatrix} \in \mathfrak{sl}_n(\mathbb{C}) \;\middle|\; \begin{array}{ll} X_{11}^* = -X_{11} & p \text{ by } p \\ X_{22}^* = -X_{22} & q \text{ by } q \end{array} \right\},$$

$$\mathfrak{sp}(p,q) = \left\{ \begin{pmatrix} X_{11} & X_{12} & X_{13} & X_{14} \\ X_{12}^* & X_{22} & X_{14}^t & X_{24} \\ -\bar{X}_{13} & \bar{X}_{14} & \bar{X}_{11} & -\bar{X}_{12} \\ X_{14}^* & -\bar{X}_{24} & -X_{12}^t & \bar{X}_{22} \end{pmatrix} \right.$$

$$\left. \in \mathfrak{sl}_{2n}(\mathbb{C}) \;\middle|\; \begin{array}{ll} X_{11}^* = -X_{11} & p \text{ by } p \\ X_{22}^* = -X_{22} & q \text{ by } q \\ X_{13}^t = -X_{13} & p \text{ by } p \\ X_{24}^t = -X_{24} & q \text{ by } q \end{array} \right\}.$$

6

Geometric Structures

We introduce the language of principal bundles, reductions, and geometric structures on manifolds, and explore the notion of invariant geometric structures on G-spaces. A useful concept, called the *algebraic hull* of a G-space, is also introduced. Loosely speaking, the algebraic hull of a G-action is the "maximal" geometric structure of "algebraic type" preserved by the action.

The main results of the chapter are the "reduction lemmas" given in section 6.4, the definition and basic properties of the algebraic hull given in section 6.5, and the results of section 6.7. They will not be needed until chapter 10. (Some of the definitions given in sections 6.1 and 6.2 are also used in section 9.3.) The reader who prefers to skip to chapters 7 and 8, and return to this one later, can do so without loss of continuity.

6.1 Principal Bundles

A smooth *fiber bundle* consists of manifolds E, M, S and a smooth mapping $p : E \to M$ for which the following holds. Each $x \in M$ has an open neighborhood U such that $E|_U := p^{-1}(U)$ is diffeomorphic to $U \times S$ via a diffeomorphism that respects fibers; that is, there is a diffeomorphism

$$\varphi : E|_U \to U \times S$$

such that, on $E|_U$, $p_1 \circ \varphi = p$, where p_1 is the natural projection onto the first factor. E is called the *total space*, M is the *base*, S is the *standard fiber*, and p is a surjective submersion called the *projection*. The map φ is a *local trivialization* of E over U. We will often refer to E itself, rather than (E, p, M, S), as being the bundle, unless there is some possibility for confusion. The set $p^{-1}(x)$ is the fiber of E above $x \in M$ and will be denoted E_x.

Given a collection of trivializations $(\varphi_\alpha, U_\alpha)$ such that $\{U_\alpha\}$ is an open cover of M, then

$$\left(\varphi_\alpha \circ \varphi_\beta^{-1}\right)(x, s) = (x, \varphi_{\alpha\beta}(x, s)),$$

where $\varphi_{\alpha\beta} : (U_\alpha \cap U_\beta) \times S \to S$ is smooth and $s \mapsto \varphi_{\alpha\beta}(x, s)$ is a diffeomorphism of S for each $x \in U_{\alpha\beta} := U_\alpha \cap U_\beta$. The mappings $\varphi_{\alpha\beta}(x) := \varphi_{\alpha\beta}(x, \cdot)$ of S are the *transition functions* of the bundle. They satisfy the *cocycle condition*: $\varphi_{\alpha\gamma}(x) \circ \varphi_{\gamma\beta}(x) = \varphi_{\alpha\beta}(x)$ for each $x \in U_\alpha \cap U_\beta \cap U_\gamma$, and $\varphi_{\alpha\alpha}(x)$ is the identity map for each $x \in U_\alpha$. We refer to $\{(\varphi_\alpha, U_\alpha)\}$ as a *trivializing atlas* for E.

Let $f : N \to M$ be a smooth mapping between smooth manifolds and suppose that $p : E \to M$ is the projection map of a smooth fiber bundle with standard fiber S. The *pull-back* of E by f, denoted f^*E, is the fiber bundle with base N, standard fiber S, total space

$$f^*E := \{(x, u) \in N \times E \mid f(x) = p(u)\},$$

and projection $f^*p : (x, u) \mapsto x$. It can be shown that f^*E is indeed a smooth fiber bundle.

A *section* of E is a map $\sigma : M \to E$ such that $p \circ \sigma = \mathrm{id}_M$. The section is smooth (resp., C^r, continuous, measurable) if σ is a smooth (resp., C^r, continuous, measurable) map.

We will also need to consider fiber bundles that are not smooth. A fiber bundle is said to be C^r, $r \geq 1$, if it admits a trivializing atlas for which the trivializing functions (denoted by φ as before) are C^r differentiable. Similarly, E is a C^0 or *topological bundle* if the trivializing functions are continuous.

It will also be convenient later to consider *measurable bundles*. The fiber S will always be a smooth manifold (often a real variety), but we allow M and E to be measurable spaces. (M will typically be a smooth manifold since our measurable bundles will arise as possibly discontinuous subbundles of smooth bundles.) We say that E is a measurable bundle if it admits a global measurable trivialization; in other words, there exists a measurable isomorphism $\varphi : E \to M \times S$ that preserves fibers. Notice that smooth bundles admit global measurable trivializations, so they are measurable bundles with respect to the underlying Borel measurable structure. Because of this fact, applying the language of fiber bundles in the measurable situation will only be a matter of convenience.

Exercise 6.1.1 Show that a topological bundle E over a topological manifold M can be measurably trivialized. Moreover, if μ is any σ-finite measure on M, show that a global trivialization $\varphi : E \to M \times S$ can be chosen so that there is an open set $U \subset M$ of full μ-measure such that φ is continuous on E_U. (Hint: The idea is to patch together local trivializations, making sure that the boundaries of the trivializing neighborhoods have measure 0. The following remark will be useful. If $B(r, x)$ is the ball of radius r at x with respect to some metric on M, there must be a value r' arbitrarily close to r such that the boundary of $B(r', x)$ has μ-measure 0.)

Let H be a Lie group, and let E be a fiber bundle over a manifold M with projection p and standard fiber S. E is called an H-*bundle* if there is a smooth action $\Phi : H \times S \to S$ and a family $\{\varphi_\alpha\}$ of smooth trivializations of E over an open cover $\{U_\alpha\}$ of M such that the transition maps take values in H. More precisely, for each pair α, β with $U_\alpha \cap U_\beta \neq \emptyset$ and each $x \in U_\alpha \cap U_\beta$, $\varphi_{\alpha\beta}(x) = \Phi_h \in \mathrm{Diff}(S)$, for some $h \in H$.

The tangent bundle of M is an example of a $GL(n, \mathbb{R})$-bundle, where n is the dimension of M. In this case, the standard fiber is $\mathbb{R}^n$ with the linear action of $GL(n, \mathbb{R})$, and the transition functions are the Jacobian matrices of coordinate changes. More generally, a *vector bundle* over M is a $GL(V)$-bundle, for a vector space V that is also the standard fiber.

Vector bundles can be combined by the familiar operations on vector spaces, such as the direct sum and tensor product. Thus, given vector bundles E_1 and E_2 over a manifold M, with standard fibers V_1 and V_2, it is possible to define their direct sum $E_1 \oplus E_2$ and tensor product $E_1 \otimes E_2$. We will return to this in the section on associate bundles.

A *principal bundle* is an H-bundle with standard fiber H such that the action of H on itself is by left translation. Each principal H-bundle P admits a (right) H-action $P \times H \to P$ defined as follows. If $\sigma \in P|_U$ and $\varphi : P|_U \to U \times H$ is a trivialization of P over a neighborhood of $x = p(\sigma)$, and $h \in H$, then $\varphi(\sigma h) := \varphi(\sigma)h$, where the action on the right-hand side is given by $(x, h_1)h := (x, h_1 h)$. This is well defined since the transition functions are left translation by elements in H, and these commute with right translation.

Exercise 6.1.2 Use theorem 3.6.4 to show that if H is a closed subgroup of a Lie group G, then the natural projection $p : G \to G/H$ makes G a principal H-bundle over G/H.

It can be shown that if $p : P \to M$ is a surjective submersion and H is a Lie group that acts freely on P from the right such that the orbits are exactly the fibers of p, then P is a principal H-bundle.

Exercise 6.1.3 Let G be a Lie group, Γ a discrete subgroup of G, and $\rho : G \to L$ a smooth homomorphism into a Lie group L. Define $P = (G \times L)/\Gamma$, where Γ acts on the product by $(g, l) \cdot \gamma = (g\gamma, \rho(\gamma)^{-1}l)$. Show that there is a well-defined smooth projection $p : P \to M = G/\Gamma$ and a right action of L on P making P a principal L-bundle over M.

Let P and P' be, respectively, principal H- and H'-bundles. We say that a map $l : P \to P'$ is a *homomorphism of principal bundles* if there is a homomorphism $\psi : H \to H'$ such that for each $\sigma \in P$ and each $h \in H$,

$$l(\sigma h) = l(\sigma)\psi(h).$$

Notice that a homomorphism of principal bundles respects fibers, whence it defines a map $\bar{l} : M \to M'$. When $\bar{l}$ is the identity and ψ is the inclusion of a subgroup H of H', we say that P is an H-*reduction* of P'. Examples of reductions are given in the next section. The map l may be smooth, C^r, continuous, or simply measurable. Accordingly, we may talk about smooth, C^r, continuous, or measurable reductions.

We introduce now a fundamental example of a principal bundle, called the frame bundle. A *frame* at $x \in M$ is a linear isomorphism $\sigma : \mathbb{R}^n \to T_x M$. Notice that $GL(n, \mathbb{R})$ acts freely and transitively on the right on the set of all frames at x by

$$(\sigma, A) \mapsto \sigma \circ A,$$

where $A \in GL(n, \mathbb{R})$ and for each $v \in \mathbb{R}^n$, $(\sigma \circ A)v = \sigma(Av)$. Here, Av denotes matrix multiplication of the column vector v by A. Let P be the set of all frames at all $x \in M$. It is a simple exercise to verify that P defines a smooth principal $GL(n, \mathbb{R})$-bundle, called the *frame bundle of order* 1, or simply the *frame bundle* of M.

More generally, let $p : E \to M$ be the projection map of a vector bundle over M, whose standard fiber is an m-dimensional vector space V. We denote by $\mathcal{F}(E)$ the frame bundle of E, which is defined just as for $E = TM$. Notice that a frame at $x \in M$ now corresponds to a linear isomorphism $\sigma : V \to E_x$.

6.2 Geometric A-Structures

Let P be a principal H-bundle over a manifold M, and let $\varphi : H \times S \to S$ be a smooth action of H on a manifold S. Then H also acts on the right on the product $P \times S$ as follows:

$$(\sigma, s)h := (\sigma h, h^{-1}s).$$

The space of orbits $P \times_H S := (P \times S)/H$ can be shown to carry a unique smooth manifold structure for which the quotient map from $P \times S$ to $P \times_H S$

is a submersion and, with respect to the natural projection $\bar{p} : P \times_H S \to M$ that maps the coset $(\sigma, s)H$ to $p(\sigma)$, $P \times_H S$ is a fiber bundle over M with standard fiber S, called the *associated bundle* for the H-action on S.

When S is a vector space and the action of H on S comes from a linear representation $\rho : H \to GL(S)$, and P is a principal H-bundle, then the associated bundle for this action is a vector bundle.

If E is a vector bundle over M whose standard fiber is a vector space V of dimension m, then E may be recovered from its frame bundle $\mathcal{F}(E)$ by taking the associated $GL(V)$-bundle obtained by the obvious linear action ρ of $GL(V)$ on V. The dual bundle E^* comes from the dual representation of $GL(V)$ on V^* defined by

$$\rho^* : (A, \alpha) \in GL(V) \times V^* \mapsto \alpha \circ A^{-1} \in V^*.$$

By using the natural representation of $GL(V)$ on $(\bigotimes^r V^*) \otimes (\bigotimes^s V)$, which extends by linearity the correspondence

$$\alpha_1 \otimes \cdots \otimes \alpha_r \otimes v_1 \otimes \cdots \otimes v_s \mapsto \rho^*(\alpha_1) \otimes \cdots \otimes \rho^*(\alpha_r) \otimes \rho(v_1) \otimes \cdots \otimes \rho(v_s),$$

we obtain the tensor bundle

$$E^{(r,s)} := \left(\bigotimes^r E^* \right) \otimes \left(\bigotimes^s E \right).$$

Let F be another vector bundle over M, with standard fiber W. Then $E^* \otimes F$ is a vector bundle over M whose standard fiber is the vector space $\mathrm{End}(V, W)$ of all linear maps from V into W.

Now let P be a principal H-bundle over a manifold M with projection map p, let $H \times S \to S$ be a smooth action, and form the associated bundle $P \times_H S$. We call a section of $P \times_H S$ a *geometric structure* on M (of type S). The geometric structure is smooth (resp., C^r, continuous, measurable) if the section that defines it is a smooth (resp., C^r, continuous, measurable) map from M into $P \times_H S$. We say that a geometric structure is of *algebraic type*, or that it is an *A-structure*, if the H-action on S is real algebraic (see chapter 4).

For example, the tangent bundle TM is an associated $GL(n, \mathbb{R})$-bundle of the frame bundle over M, and a smooth section of TM is a smooth vector field on M. A pseudo-Riemannian metric on M of signature s can be described as a section of the associated bundle $\mathcal{F}(TM) \times_{GL(V)} B$, where $V = \mathbb{R}^n$ and $B \subset V^* \otimes V^*$ is the space of symmetric nondegenerate bilinear forms on V of signature s. The action of $GL(V)$ on B is given by

$$(A \cdot \beta)(u, v) = \beta(A^{-1}u, A^{-1}v),$$

for each $A \in GL(V)$, $\beta \in B$ and all $u, v \in V$. Vector fields and pseudo-Riemannian metrics are examples of A-structures. Other examples will be given later.

Let $C^r(P \times_H S)$ denote the space of C^r sections of $P \times_H S$. Here r stands for any degree of regularity: $0 \le r \le \infty$, or $r = $ meas. For uniformity of notation we say that a measurable function is C^{meas}. Similarly, let $C^r(P, S)^H$ denote the space of C^r mappings $\varphi : P \to S$ that are H-equivariant, that is,

$$\varphi(\sigma h) = h^{-1}\varphi(\sigma)$$

for all $h \in H$ and $\sigma \in P$.

Proposition 6.2.1 Let (P, p, M, H) be a smooth principal H-bundle over M, and let $H \times S \to S$ be a smooth action. Then there is a canonical bijection between $C^r(P \times_H S)$ and $C^r(P, S)^H$.

Proof. Let $\varphi \in C^r(P, S)^H$. We obtain a section σ_φ of $P \times_H S$ by the following diagram, where q is the quotient map defining $P \times_H S$:

$$
\begin{array}{ccc}
P & \xrightarrow{(\mathrm{id}_P, \varphi)} & P \times S \\
{\scriptstyle p}\downarrow & & \downarrow{\scriptstyle q} \\
M & \xrightarrow{\ \sigma_\varphi\ } & P \times_H S
\end{array}
$$

Notice that (id_P, φ) is H-equivariant, so it indeed descends through the quotient by H.

Conversely, let $\sigma \in C^r(P \times_H S)$ and define φ_σ as follows. For each $\xi \in P$, with $p(\xi) = x$, define $q_\xi : S \to (P \times_H S)_x$ by $q_\xi(s) := (\xi, s)H$. Notice that q_ξ is a smooth diffeomorphism. Also define a C^r map

$$\tau : P \times_M (P \times_H S) := \{(\xi, \eta) \in P \times (P \times_H S) \mid p(\xi) = \bar{p}(\eta)\} \to S$$

by $\tau(\xi, \eta) := q_\xi^{-1}(\eta)$. We now define for each $\xi \in P_x$

$$\varphi_\sigma(\xi) := \tau(\xi, (\xi, \sigma(x))H).$$

It is a simple exercise to check that $\sigma \mapsto \varphi_\sigma$ and $\varphi \mapsto \sigma_\varphi$ are inverse maps and that all maps are C^r. □

The next proposition shows that reductions of P can be viewed as geometric structures.

Proposition 6.2.2 Let P be, as before, a principal H-bundle over M, and let L be a closed subgroup of H. Then the C^r L-reductions of P correspond bijectively to C^r sections of the associated bundle $P \times_H H/L$, where H acts on H/L by left translation.

Proof. By the previous proposition, each section $\sigma \in C^r(P \times_H H/L)$ corresponds to a $\varphi_\sigma \in C^r(P, H/L)^H$, which is a surjective submersion since H acts transitively on H/L. Thus $Q := \varphi_\sigma^{-1}(eL)$, the preimage of the identity coset, is a submanifold of P that is stable under the action of L on P. The L-orbits of Q are the fibers of $p|_Q : Q \to M$ and one easily checks that Q is a principal L-bundle.

Conversely, let Q be a principal L-bundle over M that is a subbundle of P, that is, the inclusion map is a principal bundle homomorphism. In particular, notice that the action of L on P stabilizes Q. Consider the mapping

$$\tau : P \times_M Q := \{(\xi, \eta) \in P \times Q \mid p(\xi) = p(\eta)\} \to H$$

defined by $(\xi, \eta) \mapsto h$, where h is the unique element of H such that $\eta = \xi h$. Then τ is smooth and satisfies $\tau(\xi, \eta h) = \tau(\xi, \eta)h$ and $\tau(\xi h, \eta) = h^{-1}\tau(\xi, \eta)$ for each $(\tau, \eta) \in P \times_M Q$ and all $h \in H$. By this equivariance property the following diagram commutes:

$$
\begin{array}{ccc}
P \times_M Q & \xrightarrow{\;\tau\;} & H \\
\left\downarrow{\scriptstyle p_1}\right. & & \left\downarrow{\scriptstyle p}\right. \\
P & \xrightarrow{\;\sigma_\varphi\;} & H/L
\end{array}
$$

where p_1 is projection onto the first factor and the quotient map φ is equivariant, that is, $\varphi(h\xi) = h^{-1}\varphi(\xi)$ for each $\xi \in P$ and $h \in H$. By construction, the preimage of the identity coset, $\varphi^{-1}(eL)$, is precisely Q. The two maps just obtained, taking reductions of P to sections of the associated bundle and vice versa, are the inverse maps of each other. $\square$

The previous proposition shows that a geometric structure represented by an H-equivariant map $P \to S$ defines a reduction of P whenever the action of H on S is transitive. In fact, by identifying S with H/L, where L is the isotropy group of some point of S, H acts by left translation on S and the proposition applies. For example, we have seen earlier that a pseudo-Riemannian metric on M of signature s can be described by a $GL(n, \mathbb{R})$-equivariant map from the frame bundle $\mathcal{F}(TM)$ into the space B of all nondegenerate symmetric bilinear forms on $\mathbb{R}^n$ with signature s. The action of $GL(n, \mathbb{R})$ on B is transitive. In fact, any such bilinear form is the translate under some element of $GL(n, \mathbb{R})$ of

$$\beta_0(u, u) := u_1^2 + \cdots + u_k^2 - u_{k+1}^2 - \cdots - u_n^2,$$

where $s = 2k - n$. (This can be derived from the conclusion of exercise 5.2.7.) The isotropy group of β_0 is the orthogonal group of β_0.

Exercise 6.2.3 Describe a smooth vector subbundle with fiber dimension m of the tangent bundle of an n-dimensional manifold M both as a reduction of $\mathcal{F}(TM)$ and as a $GL(n, \mathbb{R})$-equivariant map from $\mathcal{F}(TM)$ into a Grassmannian variety.

6.3 Invariant Geometric Structures

By an *automorphism* of a principal H-bundle P over a manifold M we mean a smooth diffeomorphism l of P that commutes with the H-action, that is, $l(\xi h) = l(\xi)h$ for all $\xi \in P$ and $h \in H$.

The set of all automorphisms of P forms a subgroup $\mathrm{Aut}(P)$ of $\mathrm{Diff}(P)$. Each $\varphi \in \mathrm{Aut}(P)$ covers a diffeomorphism $\bar{\varphi}$ of M, that is, the following diagram commutes:

$$
\begin{array}{ccc}
P & \xrightarrow{\ \varphi\ } & P \\
{\scriptstyle p}\downarrow & & \downarrow{\scriptstyle p} \\
M & \xrightarrow[\ \bar{\varphi}\ ]{} & M
\end{array}
$$

(We note that many authors define an automorphism φ of P so that $\bar{\varphi} = \mathrm{id}_M$.) The action of $\mathrm{Aut}(P)$ on P induces an action on each associated bundle $P \times_H S$; namely, for each $\varphi \in \mathrm{Aut}(P)$ let $\tilde{\varphi}$ be the unique mapping for which the following diagram is commutative:

$$
\begin{array}{ccc}
P \times S & \xrightarrow{\ (\varphi,\mathrm{id}_S)\ } & P \times S \\
{\scriptstyle q}\downarrow & & \downarrow{\scriptstyle q} \\
P \times_H S & \xrightarrow[\ \tilde{\varphi}\ ]{} & P \times_H S
\end{array}
$$

Notice that $\tilde{\varphi}$ is indeed well defined since $\varphi(\xi h) = \varphi(\xi)h$ for all $\xi \in P$ and $h \in H$. For simplicity, we will denote the map on $P \times_H S$ by φ rather than $\tilde{\varphi}$.

$\mathrm{Aut}(P)$ also acts on geometric structures in a natural way. Namely, let σ be a geometric structure on P viewed as a section of the associated bundle $P \times_H S$. For each $\varphi \in \mathrm{Aut}(P)$ we define

$$(\varphi_* \sigma)(x) := \varphi(\sigma(\bar{\varphi}^{-1}(x)))$$

for all $x \in M$.

For example, let $\bar{\varphi}$ be a diffeomorphism of M and φ the diffeomorphism induced by $\bar{\varphi}$ on the frame bundle $\mathcal{F}(TM)$, that is, given $\xi \in \mathcal{F}(TM)_x$, $\varphi(\xi)$ is the frame at $\bar{\varphi}(x)$ such that $\varphi(\xi)(v) := D\bar{\varphi}_x \xi v$ for each $v \in \mathbb{R}^n$. Therefore, if σ is a vector field on M, hence a section of $TM = \mathcal{F}(TM) \times_{GL(n,\mathbb{R})} \mathbb{R}^n$, the action of φ on σ just defined corresponds to the usual push-forward operation on vector fields.

We define an action of $\mathrm{Aut}(P)$ on $C^r(P, S)^H$ by

$$\varphi \cdot \mathcal{G} := \mathcal{G} \circ \varphi^{-1}.$$

This is well defined since $\varphi \cdot \mathcal{G}$ is easily seen to be H-equivariant. When $\mathcal{G}$ is the H-equivariant map determined by σ, the actions on $C^r(P \times_H S)$ and on $C^r(P, S)^H$ agree under the correspondence obtained in proposition 6.2.1. More precisely,

$$\varphi_* \sigma_{\mathcal{G}} = \sigma_{\varphi \cdot \mathcal{G}},$$

where $\sigma_{\mathcal{G}}$ is the section of $P \times_H S$ associated to the H-equivariant map $\mathcal{G} : P \to S$. This is a straightforward verification and is left to the reader.

We say that $\mathcal{G}$ is *invariant* under $\varphi \in \mathrm{Aut}(P)$ if $\varphi \cdot \mathcal{G} = \mathcal{G}$. It is often the case that the automorphism φ of P is induced in a canonical way by a diffeomorphism $\bar{\varphi}$ of M. For example, any diffeomorphism of M induces an automorphism of $\mathcal{F}(TM)$ as pointed out earlier. In such cases we say that $\bar{\varphi}$ is an *isometry* of $\mathcal{G}$ whenever the induced automorphism φ fixes $\mathcal{G}$.

6.4 The Reduction Lemmas

We have seen that an L-reduction of the frame bundle over a smooth manifold M can be interpreted as a geometric structure on M, which we will sometimes call an L-*structure*. Each one of the classical groups studied earlier gives rise to some standard structure in differential geometry. We enumerate below a few of the most common examples. Notice that they are all examples of A-structures. In each case, M is a manifold of dimension n and $F(M) = \mathcal{F}(TM)$ denotes the frame bundle.

Example 1 Complete parallelism or $\{e\}$-structure. When L is the trivial subgroup $\{e\}$ of $GL(n, \mathbb{R})$, a C^r L-reduction of $F(M)$ is a C^r section of $F(M)$, that is, a C^r field of linear isomorphisms $\sigma(x) : \mathbb{R}^n \to T_x M$. A section σ of $F(M)$ determines a $GL(n, \mathbb{R})$-equivariant map $\mathcal{G}_\sigma : F(M) \to GL(n, \mathbb{R})$ that sends $\xi = \sigma(x)A \in F(M)_x$ to $\mathcal{G}_\sigma(\xi) = A^{-1}$.

Example 2 Nonvanishing n-form or $SL(n, \mathbb{R})$- structure. Let V denote the space of nonzero alternating n-forms on $\mathbb{R}^n$ and denote by μ_0 the form

$$\mu_0(u_1, \ldots, u_n) = \det(u_{ij}),$$

where u_{ij} are the entries of the matrix whose columns are the vectors $u_1, \ldots, u_n \in \mathbb{R}^n$. The general linear group acts transitively on V and the isotropy subgroup of μ_0 is $SL(n, \mathbb{R})$. A smooth $SL(n, \mathbb{R})$-reduction of $F(M)$ determines a smooth assignment of a nonzero n-form at each $T_x M$, a smooth nonvanishing n-form $x \mapsto \nu(x)$ on M. The equivariant map $\mathcal{G} : F(M) \to V$ determined by ν can be described as follows. To each $\xi \in F(M)_x$, $\mathcal{G}(\xi)$ is the n-form on $\mathbb{R}^n$ given by $\nu(x)(\xi\cdot, \ldots, \xi\cdot)$.

Example 3 Pseudo-Riemannian metric or $O(p, q)$-structure. The space V of nondegenerate symmetric bilinear forms on $\mathbb{R}^n$ of signature s is a transitive $GL(n, \mathbb{R})$-space, and it can be written as $GL(n, \mathbb{R})/O(p, n - p)$, where $s = 2p - n$ is the signature and $O(p, n - p)$ is the isotropy subgroup of

$$\beta_0(u, u) := u_1^2 + \cdots + u_p^2 - u_{p+1}^2 - \cdots - u_n^2.$$

A C^r $O(p, q)$-reduction of $F(M)$ corresponds to a C^r field $x \mapsto \langle \cdot, \cdot \rangle_x$, where $\langle \cdot, \cdot \rangle_x$ is a nondegenerate symmetric bilinear form of signature s on $T_x M$. The $GL(n, \mathbb{R})$-equivariant map associated to a pseudo-Riemannian metric is the map $\mathcal{G} : F(M) \to V$ such that $\mathcal{G}(\xi) = \langle \xi\cdot, \xi\cdot \rangle_{p(\xi)}$.

Example 4 Almost symplectic or $Sp(n, \mathbb{R})$-structure. Let V be space of all nondegenerate alternating bilinear forms on $\mathbb{R}^n$, $n = 2m$. Then V is a transitive $GL(n, \mathbb{R})$-space and $Sp(n, \mathbb{R})$ is the isotropy subgroup of

$$\omega_0(u, v) = u_1 v_{m+1} + \cdots + u_m v_{2m} - u_{m+1} v_1 - \cdots - u_{2m} v_m,$$

so V, which is the orbit of ω_0, is naturally identified with the homogeneous space $GL(2m, \mathbb{R})/Sp(2m, \mathbb{R})$. A smooth $Sp(2m, \mathbb{R})$-reduction of $F(M)$ is equivalent to smoothly assigning a nondegenerate alternating 2-form Ω_x to $T_x M$ for each $x \in M$. The equivariant map $\mathcal{G} : F(M) \to V$ associated to Ω is given by $\mathcal{G}(\xi) = \Omega(\xi\cdot, \xi\cdot)_{p(\xi)}$.

Example 5 Subbundles of TM or $GL(n, m, \mathbb{R})$-structure. Viewing $\mathbb{R}^m$ as the subspace of $\mathbb{R}^n$ of vectors with the last $n - m$ components equal to 0, we let $GL(n, m, \mathbb{R})$ denote the subgroup of $GL(n, \mathbb{R})$ consisting of invertible matrices that map $\mathbb{R}^m$ into itself. The group $GL(n, m, \mathbb{R})$ is the isotropy subgroup of $\mathbb{R}^n$ for the transitive action of $GL(n, \mathbb{R})$ on the Grassmannian variety V of m-dimensional subspaces of $\mathbb{R}^n$. A smooth $GL(n, m, \mathbb{R})$-reduction of $F(M)$ consists of a smooth field $x \mapsto D(x)$ of m-dimensional subspaces $D(x) \subset T_x M$. The equivariant map $\mathcal{G}$ associated to the field D sends each $\xi \in F(M)$ to the subspace $\xi^{-1} D(x) \subset \mathbb{R}^n$, with $x = p(\xi)$. Notice that if g is a diffeomorphism of M, then

$$(g_* D)(x) := Dg_{g^{-1}x} D(g^{-1}x) = \xi \mathcal{G}(g^{-1}\xi),$$

for any frame ξ at x.

Example 6 Almost complex or $GL(m, \mathbb{C})$-*structure*. Here $n = 2m$ and we view $GL(m, \mathbb{C})$ as a subgroup of $GL(2m, \mathbb{R})$ (see chapter 5). A smooth (continuous, measurable, etc.) $GL(m, \mathbb{C})$-structure on M is equivalent to a smooth (continuous, measurable, etc.) assignment, at each $x \in M$, of a linear map $J_x : T_x M \to T_x M$ such that $J_x^2 = -I$, where I is the identity map. In a similar way, one defines the notion of an *almost quaternionic* structure on M, which is a $GL(p, \mathbb{H})$-reduction of $F(M)$, $4p = n$.

The next example includes all of the previous examples.

Example 7 L-structure defined by a tensor. Let ρ be a linear representation of $GL(n, \mathbb{R})$ on W, $\rho : GL(n, \mathbb{R}) \to GL(W)$. A tensor field of type W is a section of the associated bundle $F(M) \times_{GL(n,\mathbb{R})} W$. For example, when $W = (\bigotimes^r \mathbb{R}^{n*}) \otimes (\bigotimes^s \mathbb{R}^n)$ and ρ is the representation induced by the standard representation of $GL(n, \mathbb{R})$ on $\mathbb{R}^n$, we obtain the bundle $T^{(r,s)} M$ of tensors of type (r, s). A section of $T^{(r,s)} M$ is called a tensor field of type (r, s). Notice that a vector field is a section of $T^{(0,1)} M$. In general, having a tensor field of type W on M is equivalent to having a $GL(n, \mathbb{R})$-equivariant map $\mathcal{G} : F(M) \to W$, where equivariance in this case is defined by

$$\mathcal{G}(\xi A) = \rho(A)^{-1} \mathcal{G}(\xi)$$

for all $A \in GL(n, \mathbb{R})$ and $\xi \in F(M)$.

Not every tensor field of type W can be regarded as an L-structure for some $L \subset GL(n, \mathbb{R})$. In general, a geometric A-structure of type V, defined by an H-equivariant map $\mathcal{G} : P \to V$, does not define an L-reduction of the principal H-bundle P since the range of values of $\mathcal{G}$ need not be contained in a single H-orbit in V. In the examples given earlier, $GL(n, \mathbb{R})$ acted transitively on V, so we were able to define a reduction in each case.

In example 7, if $\mathcal{G}$ takes values in a single $GL(n, \mathbb{R})$-orbit, that is, if

$$\mathcal{G} : F(M) \to \rho(GL(n, \mathbb{R}))w_0$$

for some $w_0 \in W$, then $\mathcal{G}^{-1}(w_0)$ is an L-reduction of $F(M)$ for the closed subgroup

$$L = \{g \in GL(n, \mathbb{R}) \mid \rho(g)w_0 = w_0\}.$$

The next proposition shows that if a geometric A-structure ω admits a group of isometries that acts on M topologically transitively, then ω is an L-structure, at least over an open and dense subset of M.

Proposition 6.4.1 (First reduction lemma) Let V be a real algebraic H-space and P a principal H-bundle over a manifold M. Let $\mathcal{G} : P \to V$ be a $C^r, r \geq 0$, H-equivariant map, and suppose that a group G of automorphisms of P acts topologically transitively on M. Suppose moreover that $\mathcal{G}$ is G-invariant. Then there exists an open and dense G-invariant subset U of M such that $\mathcal{G}$ maps $P|_U$ onto a single H-orbit, $H \cdot v_0 \subset V$, for some $v_0 \in V$. The set $\mathcal{G}^{-1}(v_0) \subset P$ is a C^r G-invariant L-reduction of P, where $L \subset H$ is the isotropy subgroup of v_0. If $H \cdot v_0$ is a closed subset of V, then $U = M$.

Proof. Suppose that $x_0 \in M$ has a dense G-orbit in M and let $\xi_0 \in P_x$ be any point in the fiber of P above x_0. Set $v_0 = \mathcal{G}(\xi_0)$ and denote by W the closure of the H-orbit of v_0 in V. Since the G-orbit of x_0 is dense in M, the $G \times H$-orbit of ξ_0 is also dense in P, and maps into $H \cdot v_0$. Therefore $\mathcal{G}$ maps P into W. By corollary 4.9.3, $H \cdot v_0$ is open in W, so $\mathcal{G}^{-1}(H \cdot v_0)$ is an open and dense subset of P. This set is saturated by H-orbits since $\mathcal{G}$ is H-equivariant, whence it is of the form $P|_U$ for some open and dense subset $U \subset M$. Moreover, U is G-invariant since $\mathcal{G}$ is itself G-invariant. If $H \cdot v_0$ is closed in V, then $W = H \cdot v_0$, so $U = M$. Once we know that $\mathcal{G}$ maps into a single H-orbit, it follows from the earlier discussion in this chapter that we obtain a reduction as claimed. $\square$

The previous result has the following measurable counterpart, which should be compared with R. Zimmer's *cocycle reduction lemma* [36].

Proposition 6.4.2 (Second reduction lemma) Let V be a real algebraic H-space and P a measurable principal H-bundle over a second countable metrizable space M. Let $\mathcal{G} : P \to V$ be a measurable H-equivariant map, and suppose that a group G of automorphisms of P acts ergodically on M with respect to a quasi-invariant measure μ and leaves $\mathcal{G}$ invariant. Then, there exists a G-invariant measurable conull subset U of M such that $\mathcal{G}$ maps $P|_U$ into a single H-orbit $H \cdot v_0$ in V. The preimage of v_0 under $\mathcal{G}$ defines a measurable, G-invariant L-reduction of $P|_U$.

Proof. The H-equivariance of $\mathcal{G}$ implies that $\mathcal{G}$ induces a G-invariant measurable map $\bar{\mathcal{G}} : M \to V/H$. We have seen that the H-action on V is tame, since it is an algebraic action. It follows from proposition 2.3.5 that $\bar{\mathcal{G}}$ is constant a.e.; therefore $\mathcal{G}$ sends a G-invariant set $P|_U$, with $\mu(M - U) = 0$, into a single orbit in V. □

6.5 The Algebraic Hull

We discuss now a very useful invariant of G-spaces, introduced by R. Zimmer, called the *algebraic hull*. The concept comes in C^r and measurable versions. We first consider the measurable case, using a more geometric language than in [36].

Suppose that a group G acts by automorphisms of a principal H-bundle P and that the G-action on the base M preserves a measure class represented by a probability measure μ. We say that $Q \subset P$ is a G-invariant *measurable L-reduction* of P if Q is a measurable L-reduction of $P|_U$ for some G-invariant μ-conull measurable subset $U \subset M$ and the G-action on P restricts to a G-action on Q.

Proposition 6.5.1 (Zimmer) Let M be a second countable metrizable G-space with a quasi-invariant probability measure μ. Suppose that the action is ergodic with respect to μ. Let H be a real algebraic group and let P be a measurable principal H-bundle on which G acts by bundle automorphisms over the G-action on M. Then the following hold:

1. There exists a real algebraic subgroup $L \subset H$ and a G-invariant measurable L-reduction $Q \subset P$ such that Q is minimal; that is, Q does not admit a measurable G-invariant L'-reduction for a proper real algebraic subgroup L' of L.
2. If Q_1 and Q_2 are G-invariant reductions with groups L_1 and L_2, respectively, satisfying the preceding minimality property, then there is $h \in H$ such that $L' = hLh^{-1}$ and $Q_2 = Q_1h^{-1}$.
3. Any G-invariant measurable L'-reduction of P, for real algebraic L', contains a G-invariant measurable L''-reduction, where L'' is a conjugate in H of the minimal L obtained in item 1.

Proof. Suppose that we have a nested sequence of invariant reductions $Q_1 \supset Q_2 \supset \cdots$ with groups $L_1 \supset L_2 \supset \cdots$. The groups L_i form a descending chain of real algebraic groups. By the descending chain condition (section 4.3), the sequence must stabilize at a finite level, so a minimal reduction must exist.

The uniqueness claimed in item 2 can be seen as follows. A G-invariant L_i-reduction Q_i yields a G-invariant H-equivariant map

$$\mathcal{G}_i : P \to H/L_i.$$

Taking the product $\mathcal{G}_1 \times \mathcal{G}_2$, we obtain a G-invariant H-equivariant map

$$\mathcal{G} : P \to H/L_1 \times H/L_2.$$

The right-hand side is an H-space for the natural product action. Applying the reduction lemma to $\mathcal{G}$, we conclude that $\mathcal{G}$ maps $P|_U$ onto a single H-orbit in $H/L_1 \times H/L_2$, where U is a conull subset of M. We denote that orbit by $H \cdot (h_1 L_1, h_2 L_2)$. The isotropy group of $(h_1 L_1, h_2 L_2)$ is

$$L = \{ h \in H \mid h h_1 L_1 = h_1 L_1, h h_2 L_2 = h_2 L_2 \}$$

and we have a G-invariant measurable L-reduction Q of P. Notice that $L \subset h_1 L_1 h_1^{-1} \cap h_2 L_2 h_2^{-1}$. L cannot be a proper subgroup of $h_i L_i h_i^{-1}$ since, otherwise, $Q h_i$ would define a proper reduction of Q_i, contradicting the minimality of Q_i. Therefore, $Q h_i = Q_i$, $i = 1, 2$, proving 2. Notice that the same argument also shows 3.   □

We give next the C^r counterpart of the previous result.

Proposition 6.5.2 Let P be a principal H-bundle over a manifold M. Suppose that a group G acts by bundle automorphisms of P such that the action on M is topologically transitive. Then, for each $r \geq 0$, there exists a real algebraic subgroup $L \subset H$ and a G-invariant C^r L-reduction $Q \subset P|_U$, over a G-invariant dense open subset $U \subset M$, such that Q is minimal in the same sense already defined in the previous proposition. Moreover, the foregoing properties 2 and 3 also hold here after replacing "measurable" by "C^r" and taking into account that all reductions are only defined over a G-invariant open and dense subset of M.

Proof. This is along the lines of the previous proof, using the C^r form of the reduction lemma.   □

The conjugacy class of the group L obtained here is called the C^r (resp., measurable) *algebraic hull* of the G-action on P. By abuse of language, we sometimes call L itself the algebraic hull.

Exercise 6.5.3 Let Γ be a discrete subgroup of a Lie group G and $\rho : G \to L$ a smooth homomorphism into a real algebraic group L. Form the principal L-bundle $p : P = (G \times L)/\Gamma \to M = G/\Gamma$ as in exercise 6.1.3. If $\rho(\Gamma)$ is Zariski-dense in L, show that the C^r algebraic hull of the G-action on P is L. (If $g \in G$ and (g_0, l_0) represents an element $\xi \in P$, then $g\xi$ is the element represented by (gg_0, l_0).)

We derive a few basic facts about the algebraic hull.

Lemma 6.5.4 We assume here the same conditions and notations of proposition 6.5.1 (or 6.5.2, if $r \geq 0$). Let the algebraic hull H be a real algebraic subgroup of $GL(V)$ for a real vector space V and suppose that V_0 is an H-invariant linear subspace of V. Set $V_1 = V/V_0$ and consider the natural homomorphisms $\pi_i : H \to GL(V_i)$, where, for each $h \in H$, $\pi_0(h)$ is the restriction of h to V_0 and $\pi_1(h)$ is the induced linear map on V_1. We form the associated $GL(V_i)$-principal bundle $P_i = (P \times GL(V_i))/H, i = 0, 1$, which has a natural G-action by automorphisms induced from the G-action on P. Then the C^r algebraic hull for the G-action on P_i is $\pi_i(H)$.

Proof. Let L_i denote a representative of the C^r algebraic hull for the G-action on P_i. Notice that we may assume that $L_i \subset \pi_i(H)$. Let Q_i be a G-invariant C^r reduction of P_i with group L_i and define a map $l_i : P \to P_i$ by $l_i(\xi) = [\xi, e]$, where e is the identity element in $GL(V_i)$. It is immediate that l_i is a surjective homomorphism of principal bundles (in the sense defined in section 6.1, where π_i takes the role of the homomorphism ψ in the definition of homomorphism). Also notice that l_i commutes with the G-actions on P and P_i. It follows that $l_i^{-1}(Q_i)$ is a G-invariant C^r reduction of P with group $\pi_i^{-1}(L_i)$, a real algebraic subgroup of H. Therefore, $\pi_i^{-1}(L_i) = H$ by the definition of the algebraic hull, so $L_i = \pi_i(H)$, as claimed. $\square$

Proposition 6.5.5 We continue to assume the conditions of proposition 6.5.1 (or 6.5.2, if $r \geq 0$) and that H is a real algebraic subgroup of $GL(V)$ for a real vector space V. Suppose, moreover, that H is the set of real points of a connected real algebraic subgroup of $GL(V \otimes \mathbb{C})$. Then we can find a sequence of H-invariant subspaces

$$V_0 = \{0\} \subset V_1 \subset \cdots \subset V_k = V$$

such that the following holds. Let $\pi_i : H \to GL(V_i / V_{i-1})$ be the natural homomorphism, $i = 1, \ldots, k$, and denote by P_i the principal $\pi_i(H)$-bundle induced by π_i, as in the beginning of section 6.2. Then $\pi_i(H)$ is a reductive group and is the algebraic hull of the natural G-action on P_i.

Proof. According to the Levi decomposition, theorem 4.10.1, H is the semidirect product of a reductive group L and its unipotent radical U, which is a normal subgroup of H. We also recall from section 4.10 that there exists a nonzero element $x \in V$ such that $ux = x$ for all $u \in U$. Therefore, the linear subspace

$$V_1 = \{x \in V \mid ux = x \text{ for all } u \in U\}$$

is not zero. Since U is a normal subgroup of H, V_1 is H-invariant. In fact, given any $h \in H$ and $x \in V_1$, we have for each $u \in U$

$$uhx = hh^{-1}uhx = hx.$$

Denote by π_1 the homomorphism of H into $GL(V_1)$ obtained by restricting each $h \in H$ to V_1. By the previous lemma, $\pi_1(H)$ is the algebraic hull for the G-action on the associated $GL(V_1)$-bundle defined via π_1. Moreover, $\pi_1(H) = \pi_1(L)$ is reductive since U acts trivially on V_1 and the image of L under π_1 is reductive. (This last claim is an easy consequence of general results from chapter 7. We can avoid those results from chapter 7 by the following argument. By passing to a proper nonzero subspace of V_1 if necessary, we may assume that V_1 does not contain a nonzero subspace invariant under H. If $\pi_1(H)$ were not reductive, where π_1 is now the restriction homomorphism for this smaller V_1, we could find just as before a nonzero H-invariant subspace of V_1 on which the unipotent radical of $\pi_1(H)$ acts trivially. But, by the choice of V_1, this subspace should coincide with V_1, so the unipotent radical of $\pi_1(H)$ is actually trivial.)

We can now repeat the previous argument for the induced homomorphism $\bar{\pi} : H \to GL(V/V_1)$, so as to obtain an H-invariant subspace $V_2 \subset V$ properly containing V_1 such that the group $\pi_2(H)$, defined by the restriction of $\bar{\pi}$ to V_2/V_1, is reductive and is the algebraic hull of the G-action on the principal $\pi_2(H)$-bundle induced by π_2. The proposition follows by applying this argument a finite number of times.   □

In the previous proposition, it was assumed that H is the set of real points of a connected algebraic group. It turns out that this property of H always holds after passing to a *finite extension* of the G-action on M. Before making sense of this claim, we must introduce some notation. The discussion will be given in the C^r, $r \geq 0$, case. The measurable case is treated in a similar way.

Let $p : P \to M$ be a C^r principal H-bundle over a manifold M and G a group that acts on P by automorphisms. Suppose that $H = \bar{H}(\mathbb{R})$, where $\bar{H}$ is a linear algebraic group. Denote by $\bar{H}^0$ the connected component of $\bar{H}$ containing the identity element, and define $\mathcal{L} := H/\bar{H}^0(\mathbb{R})$, a finite set. H naturally acts on $\mathcal{L}$ by left multiplication. The associated bundle $\tilde{M} := (P \times \mathcal{L})/H$ is defined, as usual, as the quotient space for the action of H on $P \times \mathcal{L}$ given by $(\xi, l) \cdot h := (\xi h, hl)$. ($\mathcal{L}$ is a finite set by the results of sections 4.7 and 4.8. Notice that $\mathcal{L}$ naturally injects into $(\bar{H}/\bar{H}^0)_{\mathbb{R}}$, which is finite.) The induced G-action on $\tilde{M}$ is $g[\xi, l] = [g\xi, l]$, where $[\xi, l]$ is the orbit $(\xi, l)H$. Denote by $\pi : \tilde{M} \to M$ the projection defined by $\pi[\xi, l] = p(\xi)$ and consider the pull-back $\tilde{P} := \pi^* P$, which, we recall, is the H-bundle defined by

$$\tilde{P} := \{(\tilde{x}, \xi) \in \tilde{M} \times P \mid \pi(\tilde{x}) = p(\xi)\}$$

(see exercise 6.1.4). The right action of H on $\tilde{P}$ is given by $[\tilde{x}, \xi]h = [\tilde{x}, \xi h]$, and the left action of G on $\tilde{P}$ is $g(\tilde{x}, \xi) = (g\tilde{x}, g\xi)$.

Write $H' := \bar{H}^0(\mathbb{R})$. The space $\tilde{M}$ is a finite covering of M and can be thought of as the space of H'-orbits in P. In fact, notice that to each $[\xi_1, hH'] \in \tilde{M}$ we can associate the H'-orbit $\xi_1 hH' \subset P$ and the correspondence is bijective. Therefore, we have in a canonical way an H'-reduction $Q \subset \tilde{P}$. Namely,

$$Q := \{(\tilde{x}, \xi) \in \tilde{P} \mid \tilde{x} = [\xi_1, hH'] \text{ and } \xi_1 hH' = \xi H'\},$$

and it is easy to check that Q is a C^r fiber bundle. This is also a G-invariant reduction since $\xi_1 hH' = \xi H'$ if and only if $g\xi_1 hH' = g\xi H'$. We call the G-action on $\tilde{P}$ a *finite extension* of the G-action on P. The preceding remarks can be rephrased by saying that by passing to a finite extension, we obtain a G-invariant reduction of the bundle with a Zariski-connected group, that is, with a group H' that is the set of real points of a connected algebraic group.

Proposition 6.5.6 We assume the same conditions of proposition 6.5.1 or proposition 6.5.2, according to whether $r = $ meas. or $r \geq 0$, respectively, and use the notation introduced in the preceding paragraphs. Then the G-action on $\tilde{M}$ is ergodic in the measurable case and topologically transitive in the $r \geq 0$ case. Moreover, the C^r algebraic hull for the G-action $\tilde{P}$ is H'. In other words, by passing to a finite ergodic (or topologically transitive) extension, the algebraic hull becomes Zariski-connected.

Proof. We discuss only the $r \geq 0$ case, the measurable case being similar. (see [36, 9.2.6]). We begin by proving that the G-action on $\tilde{M}$ is topologically transitive. Let $\tilde{U} \subset \tilde{M}$ be an open nonempty G-invariant subset. It will be shown that $\tilde{U}$ is dense. Let $U \subset M$ be the image of $\tilde{U}$ under the projection $\pi : \tilde{M} \to M$. Notice that U is open and G-invariant, whence it is also dense since the G-action on M is topologically transitive.

Let $\Phi : \tilde{P} \to H/H'$ be the G-invariant H-equivariant map that defines the H'-reduction $Q \subset \tilde{P}$, according to propositions 6.2.1 and 6.2.2. Denote by $\pi_2 : \tilde{P} \subset \tilde{M} \times P \to P$ the projection map onto the second factor.

For each $\xi \in P$, let $m(\xi)$ be the number of distinct values of $\Phi(\xi')$, where ξ' ranges over the set of elements in $\tilde{P}|_{\tilde{U}}$ such that $\pi_2(\xi') = \xi$. Since $m(\xi h) = m(\xi)$ for all $h \in H$, m is actually a function on M. It is also G-invariant, since Φ is G-invariant. Let m_0 be the maximum value of the function m. Since Φ is continuous, there is an open nonempty G-invariant (hence dense) subset U' of U where $m = m_0$. By considering $\tilde{U} \cap \pi^{-1}(U')$ instead of $\tilde{U}$ and U' instead of U, we may assume that the function m is identically equal to m_0 over an open dense G-invariant subset of M, which we still denote by U.

The set of all subsets of H/H' of cardinality m_0 can be identified with a subset of $V := (H/H' \times \cdots \times H/H')/S(m_0)$, where $S(m_0)$ is the symmetric group on m_0 letters, acting by permutations on the m_0-fold product. Define a map $\bar{\Phi} : P|_U \to V$ by assigning to each $\xi \in P|_U$ the set of values of $\Phi(\xi')$ such that $\xi' \in \tilde{P}|_{\tilde{U}}$ and $\pi_2(\xi') = \xi$. It is immediate that $\bar{\Phi}$ is continuous, G-invariant, and H-equivariant.

The quotient H/H' can be regarded in a natural way as a subset of $(\bar{H}/\bar{H}^0)(\mathbb{R})$, where $\bar{H}$ is the Zariski closure of H and $\bar{H}^0$ is the connected component of e in $\bar{H}$. By corollary 4.9.3, the action of $H \times S(m_0)$ on the m_0-fold product of H/H_0 is tame, so we can apply the reduction lemma (proposition 6.4.1) to conclude that there exists an open dense G-invariant subset of M, which we still denote by U, such that $\bar{\Phi}(\xi)$ takes values in a single H-orbit in V, for all $\xi \in P|_U$. This yields a C^r L-reduction of P (over an open dense subset of M), where L is the real algebraic subgroup of H that stabilizes a subset of H/H' of cardinality m_0. Since H is already the C^r algebraic hull of the action, we conclude that $L = H$, so m_0 is the cardinality of H/H', which is the cardinality of the fibers of the covering map π. It follows that $\tilde{U} = \pi^{-1}(U)$; hence this is a dense set, as claimed. This proves that the G-action on $\tilde{M}$ is topologically transitive.

It remains to show that the C^r algebraic hull for the G-action on $\tilde{P}$ is H'. Let L be a real algebraic subgroup of H' and assume that there is a G-invariant C^r reduction of $\tilde{P}$ over a G-invariant open dense subset $\tilde{W} \subset \tilde{M}$. This reduction is associated to a G-invariant H-equivariant map $\Psi : \tilde{P}|_{\tilde{W}} \to H/L$. Denoting by $\mathcal{F}$ the space of all functions from the finite set H/H' into H/L, we define a map $\bar{\Psi}: P|_W \to \mathcal{F}$, where W is the image of $\tilde{W}$ under the open map π, by

$$\bar{\Psi}(\xi)(hH') = \Psi([\xi, hH'], \xi).$$

This map is easily seen to be G-invariant and H-equivariant. The set $\mathcal{F}$ can be viewed as a subset of the m_0-fold product of H/L factored out by the action of $S(m_0)$, where m_0 is the cardinality of H/H'. Arguing as before, we obtain a C^r reduction of P (over some open dense G-invariant subset of M) whose group is the isotropy subgroup in H of some function $f: H/H' \to H/L$. Since H is already the algebraic hull, we conclude that H stabilizes a finite subset of H/L, so L is a finite index subgroup of H. This implies $H' \subset L$, and therefore $L = H'$ and H' is indeed the algebraic hull for the action on $\tilde{P}$.   □

The following terminology will be used in the next proposition. We say that an action is C^r-*transitive* if it is topologically transitive for $r \geq 0$ or ergodic (with respect to a measure class) for $r = $ meas. Given an action of a group G by automorphisms of a vector bundle $p: E \to M$, we can naturally define an action on the frame bundle $\mathcal{F}(E)$ of E, and if the action on M is C^r-transitive, we can define its algebraic hull H. We also will refer to H as the algebraic hull for the G-action on E. A subset of M will be said to be C^r-*full* if it is open and dense for $r \geq 0$ or is conull for $r = $ meas.

The expression "by passing to a finite C^r-transitive extension, the G-action will have such-and-such property" will mean the following. There exists a finite covering $\pi: \tilde{M} \to M$ and a C^r-transitive action of G on $\tilde{M}$ commuting with π such that, after replacing the original action with the natural G-action on the pull-back bundle π^*E, such-and-such property will hold.

Proposition 6.5.7 Let $p: E \to M$ be a smooth vector bundle over a manifold M, and let G be a Lie group that acts on E by bundle automorphisms such that the action on M is C^r-transitive. Then after passing to a finite C^r extension the following holds. There exists a C^r-full G-invariant subset $U \subset M$ and G-invariant C^r subbundles E_i of $E|_U$ such that

$$0 = E_0 \subset E_1 \subset \cdots \subset E_m = E$$

and the C^r algebraic hull for the natural G-action on E_{i+1}/E_i is a Zariski-connected reductive group, for $i = 0, \ldots, m-1$.

Proof. This is an immediate consequence of proposition 6.5.5. We leave to the reader the exercise of translating between the two propositions.   □

6.6   Applications to Simple Lie Groups

It may seem at first that the concept of measurable algebraic hull is rather trivial since we only ask for measurable reductions. This is not the case, however, as the next lemma (and the example given after it) shows.

Lemma 6.6.1 Let M be a G-space with an ergodic G-invariant probability measure μ. Let $\rho : G \to GL(V)$ be a representation of G on the finite-dimensional (real) vector space V. Denote by H the Zariski closure of $\rho(G)$ in $GL(V)$ and suppose that $\rho(G)$ is a subgroup of finite index in H. We assume moreover that H is generated by algebraic one-parameter subgroups. Then H is the algebraic hull of the G-action by bundle automorphisms of the (trivial) principal H-bundle $P = M \times H$ given by $g(x, h) := (gx, \rho(g)h)$.

Proof. Let $L \subset H$ denote the algebraic hull, and let Q be a G-invariant measurable L-reduction of P. The reduction is, in effect, a G-invariant measurable assignment of an L-orbit (for the right-translation L-action on H) at each $x \in M$, that is, a G-invariant measurable section of the fiber bundle P/L, whose standard fiber is H/L. In the present situation, where P is already a product, having an L-reduction is equivalent to having a measurable map $\phi : M \to H/L$ such that for each $x \in M$, $g(x, \phi(x)) = (gx, \phi(gx))$. Therefore, ϕ has the property

$$\phi(gx) = \rho(g)\phi(x)$$

for all $g \in G$ and all $x \in M$. The probability measure $\phi_*\mu$ on H/L is $\rho(G)$-invariant since $\phi_*\mu = \phi_*g_*\mu = \rho(g)_*\phi_*\mu$ for each $g \in G$. By averaging $\phi_*\mu$ over the finite group $H/\rho(G)$ we obtain an H-invariant probability measure on H/L. We can now apply theorem 4.9.7 (the Borel density theorem) to conclude that $H = L$.   □

A Lie group G will be called *simple* if its Lie algebra $\mathfrak{g}$ is simple, that is, is nonabelian and does not contain any nontrivial ideals. Part of the theory of (semi)simple Lie groups will be developed in the next chapter. It turns out that most of the classical groups of chapter 5 – whose Lie algebras appear in the list at the end of that chapter – are simple with finite center. The only exceptions are $SO(4)$, $SO(2, 2)$, $SO^*(4)$, and their complexification $SO(4, \mathbb{C})$. (These are still semisimple groups, and their Lie algebras are products of simple Lie algebras.)

Each noncompact classical group is a real algebraic group and contains a real algebraic one-parameter subgroup; for example, the diagonal subgroup of

$SL(n, \mathbb{R})$ containing elements of the form

$$\mathrm{diag}[a, b, 1, \ldots, 1]$$

such that $ab = 1$ is an algebraic one-parameter group. It will be seen in the next chapter that any noncompact simple real algebraic group contains an algebraically embedded copy of a group locally isomorphic to $SL(2, \mathbb{R})$, whence also contains a real algebraic one-parameter subgroup.

If G is a real algebraic group and $L \subset G$ is a real algebraic one-parameter subgroup, then any conjugate of L in G is also a real algebraic one-parameter subgroup. Therefore the real algebraic subgroup of G generated by the one-parameter subgroups is a normal subgroup. If G contains a one-parameter subgroup and is a simple group, then G is generated by the one-parameter subgroups.

It will also be shown later that the adjoint representation

$$\mathrm{Ad} : G \to GL(\mathfrak{g})$$

for any (semi)simple Lie group G has the following property: $\mathrm{Ad}(G)$ is a finite-index subgroup of its Zariski closure in $GL(\mathfrak{g})$. Therefore the preceding lemma has the following corollary.

Corollary 6.6.2 [38] Let G be a noncompact connected simple Lie group and let M be a G-space with an ergodic G-invariant probability measure μ. Denote by H the Zariski closure of $\mathrm{Ad}(G)$ in $GL(\mathfrak{g})$. Then H is the algebraic hull of the G-action by bundle automorphisms on the (trivial) principal H-bundle $P = M \times H$ by $g(x, h) := (gx, \mathrm{Ad}(g)h)$.

The following is a very useful result, also from [38].

Theorem 6.6.3 (Zimmer) Let G be a noncompact connected simple Lie group, and let M be an ergodic G-space with a G-invariant probability measure. Denote by G_x the isotropy subgroup of $x \in M$. Then either $G_x = G$ at almost every $x \in M$ or G_x is discrete at almost every x.

Proof. Let $\mathfrak{g}_x$ be the Lie algebra of G_x and $\mathrm{Gr}(\mathfrak{g})$ the union of the Grassmannian varieties of subspaces of $\mathfrak{g}$. Define $\phi : M \to \mathrm{Gr}(\mathfrak{g})$ by $\phi(x) = \mathfrak{g}_x$. Then ϕ is easily seen to be measurable and for each $g \in G$ and $x \in M$

$$\phi(gx) = \mathrm{Ad}(g)\phi(x).$$

If μ is the G-invariant probability measure on M, $\phi_*\mu$ is an $\mathrm{Ad}(G)$-invariant probability measure on $\mathrm{Gr}(\mathfrak{g})$. Since $\mathrm{Ad}(G)$ is a finite-index subgroup of its

Zariski closure H in $GL(\mathfrak{g})$, we obtain, as in the proof of the previous lemma, an H-invariant probability measure on $\mathrm{Gr}(\mathfrak{g})$. That invariant measure must be supported on the set of H-fixed points theorem (4.9.7), so over a set of full measure in M, the map ϕ takes values in the set of H-fixed points. In particular, there is a conull G-invariant subset $S \subset M$ such that $\phi|_S$ is a G-invariant function. By ergodicity, ϕ is constant almost everywhere. Call $\mathfrak{l}$ the constant value of ϕ. Then $\mathfrak{l} = \mathrm{Ad}(g)\mathfrak{l}$ for all $g \in G$, whence $\mathfrak{l}$ is an ideal of $\mathfrak{g}$. But the only ideals of $\mathfrak{g}$ are $\mathfrak{g}$ and 0, so G_x is either G or a discrete subgroup at almost every x. ☐

Theorem 6.6.4 (Zimmer) Let G be a connected noncompact simple Lie group acting nontrivially on a compact n-dimensional manifold M. Suppose that the action preserves an H-structure on M, where H is a real algebraic subgroup of $GL(n, \mathbb{R})$ consisting of matrices of determinant ± 1. Then there is a Lie algebra embedding $\pi : \mathfrak{g} \to \mathfrak{h}$ such that the representation π of $\mathfrak{g}$ on $\mathbb{R}^n$ contains $\mathrm{ad}(\mathfrak{g})$ as a subrepresentation.

Proof. We have pointed out before that if an action preserves an H-structure on M such that H consists of matrices of determinant 1, then it also preserves a nonvanishing alternating n-form on M. Similarly, if we allow the elements of H to have determinant either 1 or -1, then the action preserves a nonvanishing n-form that is only well defined up to sign, that is, a *volume density*. This nevertheless allows us to define a smooth G-invariant measure on M. The total measure of M is finite since M is compact, so after normalization we may assume that M admits a G-invariant probability measure μ whose support is the entire M.

For each ergodic component of μ we can apply the previous theorem and conclude that G_x is either discrete or equal to G for μ-a.e. $x \in M$. If $G_x = G$ for μ-a.e. x, we have by continuity that the action is trivial, contrary to the hypothesis. Therefore, there must be a G-invariant measurable subset $S \subset M$ of positive μ-measure such that G_x is discrete for all $x \subset S$.

At each $x \in S$, the differential of the orbit map $\tau_x : G \to M$, $\tau_x(g) := gx$, yields an identification of the tangent space at x of the G-orbit of x with the Lie algebra of G, as indicated by the following arrow:

$$F_x := (D\tau_x)_e : \mathfrak{g} \xrightarrow{\cong} V_x := T_x(G \cdot x).$$

Moreover, with respect to this identification, the derivative action of each $g \in G$ on the G-invariant subbundle V of TM with fibers V_x is given by $\mathrm{Ad}(g)$, that is, $Dg_x : V_x \to V_{gx}$ and

$$F_{gx} \circ \mathrm{Ad}(g) = Dg_x \circ F_x.$$

Let m be the dimension of $\mathfrak{g}$ and view $\mathfrak{g}$ as the subspace $\mathbb{R}^m \subset \mathbb{R}^n$, corresponding to setting equal to 0 the last $n - m$ coordinates of $\mathbb{R}^n$. Let H_1 denote the image of G in $GL(m, \mathbb{R})$ under the adjoint representation. Then by the preceding discussion, we obtain over S a measurable $\bar{H}$-reduction of the frame bundle $F(M)|_S$, where $\bar{H}_1$ is a subgroup of $GL(n, m, \mathbb{R})$ that restricts to H on the invariant subspace $\mathbb{R}^n$. We can now apply corollary 6.6.2 to conclude that the algebraic hull of the action contains $\mathrm{Ad}(G)$. But some conjugate of the algebraic hull is contained in H. Since the Lie algebra of $\mathrm{Ad}(G)$ is isomorphic to $\mathfrak{g}$, we obtain that some conjugate of $\mathfrak{g}$ is a Lie subalgebra of $\mathfrak{h}$. $\square$

It can also be shown (see [38]) that if $\mathfrak{g}$ is the Lie algebra of a noncompact simple Lie group and $\pi : \mathfrak{g} \to \mathfrak{gl}(V)$ is a Lie algebra homomorphism such that on V there is a nondegenerate symmetric bilinear form of signature $2 - \dim V$, invariant under $\pi(\mathfrak{g})$, and $\mathrm{ad}(\mathfrak{g})$ is a subrepresentation of π, then $\mathfrak{g} = \mathfrak{sl}(2, \mathbb{R})$. Therefore, by the previous theorem, if a connected noncompact simple Lie group G acts nontrivially on a compact manifold preserving a Lorentz metric (i.e., a pseudo-Riemannian metric of signature $2 - \dim V$), then G is locally isomorphic to $SL(2, \mathbb{R})$.

Theorem 6.6.3 has the following interesting consequence. Suppose that G, as in the theorem, acts smoothly on a manifold M. Let μ be a G-invariant Borel probability measure on M that assigns positive measure to nonempty open sets, and that with respect to μ the action is ergodic (and nontrivial). Then the set U consisting of $x \in M$ such that G_x is discrete is an open, G-invariant, conull subset. It follows that the G-orbits of points in U are the leaves of a smooth foliation of U, and that the leaves have the same dimension as G. Moreover, the tangent bundle F of this foliation has a smooth trivialization, obtained by identifying the tangent space $T_x(G \cdot x)$ with the Lie algebra of G, for each x. With respect to this trivialization, the action of G on F is given by the adjoint representation of G, as indicated in the proof of theorem 6.6.4. It is a deep problem whether the G-action on the quotient bundle TM/F can similarly be described by a linear representation of G. We will return to this problem in chapter 10.

We finish the section with a third version of the reduction lemma, in a convenient form for later use.

Proposition 6.6.5 (Third reduction lemma) Let P be a principal H-bundle over M, where H is a real algebraic group, and let V be a real algebraic

H-space. Let S be a real algebraic one-parameter group also acting algebraically on V such that the H and S actions commute. Let T be a one-parameter
group of bundle automorphisms of P (hence the T-action commutes with the
H-action on P), and let $\mathcal{G} : P|_U \to V$ be an $H \times T$-equivariant map defined
over some subset $U \subset M$, where equivariance means in this case that there is a
smooth homomorphism $\rho : T \to S$ such that $\mathcal{G}(l\xi h) = h^{-1}\rho(l)\mathcal{G}(\xi)$ for each
$l \in T, h \in H$, and $\xi \in P$. Suppose moreover that one of the following holds:

1. T acts topologically transitively on M, U is a nonempty T-invariant open
   set in M, and $\mathcal{G}$ is continuous.
2. T acts ergodically on M with respect to a T-invariant ergodic probability
   measure μ, U is a measurable T-invariant conull set, and $\mathcal{G}$ is measurable.

Then, in each of these two cases we have, respectively:

1. There exists an open dense T-invariant subset $U' \subset U$ such that $\mathcal{G}|_{P_{U'}}$ takes
   values in a single H-orbit in V.
2. There is a measurable T-invariant subset $U' \subset U$ of full measure such that
   $\mathcal{G}|_{P_{U'}}$ takes values in a single H-orbit in V.

Proof. The main difference between this more complicated statement and the
two earlier versions of the reduction lemma is that here $\mathcal{G}$ is not necessarily
T-invariant but only equivariant. We give the proof in the topological case and
leave the ergodic case for the reader.

We let $V_i, i = 1, \ldots, m$, be the smooth varieties given by Rosenlicht stratifi-
cation for the action of $H \times S$ on V. Notice that $\mathcal{G}$, restricted to some open and
dense subset of the form $P|_{U'}$, $U' \subset U$, must take values in a single stratum
V_i. In fact, for each $x \in U$ with dense T-orbit in M and each $\xi \in P$ in the fiber
above x, the image under $\mathcal{G}$ of the $T \times H$-orbit of ξ is entirely contained in
some V_i, since these strata are $T \times H$-invariant. But V_i is open in its closure so,
by continuity, $\mathcal{G}^{-1}(V_i)$ is open and $T \times H$-invariant and contains a dense subset
of $P|_U$. Therefore, $\mathcal{G}^{-1}(V_i)$ is of the form $P|_{U'}$ as claimed. (In the measurable
case, all that we need is that the action of $H \times S$ on V is tame.)

Consequently, the restriction of $\mathcal{G}$ to $P|_{U'}$ descends through the bundle pro-
jection to define a continuous T-equivariant map $\bar{\mathcal{G}}$ from U' into a smooth real
algebraic variety $W_i = V_i/H$. If S acts trivially on W_i (or if ρ is the trivial
homomorphism), then $\bar{\mathcal{G}}$ is a T-invariant map, hence constant on an open dense
T-invariant set by topological transitivity, concluding the proof in this case.

Suppose that S acts nontrivially on W_i and that ρ is nontrivial, whence $\rho(T)$
is unbounded in S. If x is a T-recurrent point in M, by equivariance, $\bar{\mathcal{G}}(x)$ is an
S-recurrent point in W_i, hence an S-fixed point by corollary 4.9.4. We recall

that if the T-action on M is topologically transitive, there must be a dense set of recurrent points, which are all sent to the (closed) set of S-fixed points in W_i. Therefore, all points in the image of $\bar{\mathcal{G}}|_{U'}$ are S-fixed points and the result follows from what was said in the previous paragraph. (In the measurable case, we use the fact that $\bar{\mathcal{G}}_*\mu$ is an S-invariant probability measure on V, whence must be supported on the set of S-fixed points, by the Borel density theorem. By ergodicity, $\bar{\mathcal{G}}$ is then constant μ-almost everywhere.) $\square$

Corollary 6.6.6 Let V be a real algebraic S-space, where S is a real algebraic one-parameter group. Let T be a one-parameter group of homeomorphisms of a topological space M acting topologically transitively. Suppose that $\phi : U \to V$ is a continuous map defined on an open dense T-invariant subset $U \subset M$ and is T-equivariant, that is, there is a smooth homomorphism $\rho : T \to S$ such that $\phi(lx) = \rho(l)\phi(x)$ for each $l \in T$ and each $x \in U$. Then, ϕ is constant and its value is an S-fixed point. The same result, with the obvious modifications (as in the proposition), holds for ergodic actions.

Proof. Set $H = S$ and $P = M \times S$. T acts on P by $l(x, s') = (lx, s')$ and S acts on the right: $(x, s')s = (x, s's)$. Define $\mathcal{G}(x, s) := s^{-1}\phi(x)$. Since S is abelian, the conditions of the proposition are satisfied here, and the claim follows. $\square$

6.7 Measures as Geometric Structures

Let P be a principal H-bundle over M, V an H-space, and T a group of auto-morphisms of P such that $T \times P \to P$ is a continuous action. A T-invariant geometric structure of type V has been defined before as a T-invariant section of the associated bundle $P \times_H V$, and we have seen how invariant structures some-times yield invariant reductions of P. Thus, for example, a measurable field of m-dimensional planes on the n-dimensional manifold M can be described as a measurable section η of the associated bundle $F(M) \times_{GL(n,\mathbb{R})} V$, where V is the Grassmannian variety of m-planes in $\mathbb{R}^n$, and $\eta(x)$ is interpreted as an m-dimensional subspace of $T_x M$ at each $x \in M$. If such a measurable plane field is invariant under T and if the action of T on M is ergodic with respect to some invariant probability measure, then we obtain a measurable T-invariant L-reduction of P, where L is the subgroup of $GL(n, \mathbb{R})$ that stabilizes $\mathbb{R}^m$.

There is a fruitful probabilistic generalization of this notion of geometric structure, which will be used later when discussing rigidity theorems, and is implicit in Zimmer's proof of the cocycle superrigidity theorem [36]. Suppose that instead of, say, a field of m-dimensional planes that are exactly specified at each $x \in M$ we have a "field of probability distributions" so that at each

x, one is given a probability measure μ_x on the Grassmannian variety V_x of m-dimensional subspaces of $T_x M$; that is, the m-plane at x is only specified "up to probability μ_x." The field of probabilities is said to be invariant if the following holds: Let τ be an element of T and let the induced map on $P \times_H V$ also be denoted by τ, so that for each $x \in M$, $\tau : V_x \to V_{\tau x}$. Then the invariance condition is that $\tau_* \mu_x = \mu_{\tau x}$ for each x and τ. When μ_x is the Dirac measure supported at a single point of V_x we recover the "deterministic" notion of a plane field.

We give now some of the details. Let V be a compact metric space and $\mathcal{M}(V)_1$ the space of Borel probability measures on V. We have indicated in chapter 2 that $\mathcal{M}(V)_1$ is a compact convex metrizable space and that the group of homeomorphisms of V acts on $\mathcal{M}(V)_1$ via a continuous action by homeomorphisms. Form the associated bundle $P \times_H \mathcal{M}(V)_1$. Then a Borel measurable section of this bundle is our field of probabilities, $x \mapsto \mu_x$, where each μ_x is a probability measure on the fiber of $P \times_H V$ above x. Such a section can also be represented by a measurable H-equivariant map $\mathcal{G} : P \to \mathcal{M}(V)_1$.

Lemma 6.7.1 Let V be a compact metric space with a continuous action of a Lie group H, and let $p : P \times_H V \to M$ be the associated bundle, P being a (topological) principal H-bundle over a manifold M. We also denote by p the projection from P to M. Suppose that τ is a bundle automorphism of P such that the group T generated by τ acts on P such that the $\mathbb{Z}$-action is continuous. Suppose furthermore that T leaves invariant a Borel probability measure ν on $P \times_H V$ and that the T-invariant measure $\mu = p_* \nu$ on M is ergodic. Then there exist a T-invariant conull subset $\bar{A} \subset P$ and a T-invariant measurable section η of $P|_{\bar{A}} \times_H \mathcal{M}(V)_1$.

Proof. Let $x \mapsto \nu_x$ be the disintegration of ν with respect to the bundle projection p (recall theorem 2.1.9). Thus for each Borel subset A of $P \times_H V$

$$\nu(A) = \int_M \nu_x(A) \, d\mu(x).$$

For each $\tau \in T$, $\tau_* \nu$ disintegrates according to the formula

$$
\begin{aligned}
\tau_* \nu(A) &= \nu(\tau^{-1}(A)) \\
&= \int_M \nu_x(\tau^{-1}(A)) \, d\mu(x) \\
&= \int_M \nu_{\tau^{-1}x}(\tau^{-1}(A)) \, d\mu(x) \\
&= \int_M \tau_* \nu_{\tau^{-1}x}(A) \, d\mu(x),
\end{aligned}
$$

where we have used the T-invariance of μ. By uniqueness of the disintegration we have that $\tau_* \nu_x = \nu_{\tau x}$ for μ-a.e. $x \in M$.

We claim that the function $f(x) := \nu_x(p^{-1}(x))$ is equal to 1 for μ-a.e. x. This is because for μ-a.e. $x \in M$, $f(\tau x) = f(x)$. Therefore f is constant μ-a.e. by ergodicity, due to proposition 2.3.5. The constant is 1 since μ and ν are probability measures.

Recall that each $\xi \in P$ can be regarded as a homeomorphism from V onto $(P \times_H V)_x$, where x is the base point of ξ. Define a measurable map

$$\mathcal{G} : P \to \mathcal{M}(V)_1$$

by $\mathcal{G}(\xi) := (\xi^{-1})_* \nu_x$, for $x = p(\xi)$. Then $\mathcal{G}$ is a $T \times H$-map in the following sense: There is a conull set $\mathcal{A} \subset M$ such that for each $\xi \in P|_{\mathcal{A}}$ and each $h \in H$ we have $\mathcal{G}(\tau \xi h) = h_*^{-1} \mathcal{G}(\xi)$. Define

$$\bar{\mathcal{A}} = \bigcap_{m \in \mathbb{Z}} \tau^m(\mathcal{A}).$$

Then $\bar{\mathcal{A}}$ is a conull T-invariant set and for each $\xi \in P|_{\bar{\mathcal{A}}}$, each $h \in H$, and each $m \in \mathbb{Z}$

$$\mathcal{G}(\tau^m \xi h) = h_*^{-1} \mathcal{G}(\xi).$$

We have already seen that the existence of such a map $\mathcal{G}$ is equivalent to the existence of the T-invariant section we seek. $\square$

A key property that was needed in the reduction lemmas is that the H-action on V is tame. When V is a homogeneous space of real algebraic groups, the induced action of H on $\mathcal{M}(V)_1$ is also tame by the next lemma. We refer the reader to Zimmer's book [36, 3.2.17, 3.2.18] for a detailed proof of a somewhat more general result. A sketch of the proof is given at the end of this section.

Lemma 6.7.2 If B is a real algebraic subgroup of a real algebraic group H such that H/B is compact, then the H-action on $\mathcal{M}(H/B)_1$ is tame. Furthermore, for any $\nu_0 \in \mathcal{M}(H/B)_1$, the isotropy group H_{ν_0} of ν_0 is a real algebraic group.

The next proposition will be needed in chapter 10.

Proposition 6.7.3 Let P be a topological principal H-bundle over a manifold M with projection map p. Suppose that τ is a bundle automorphism of P such that the group T generated by τ acts on P such that the $\mathbb{Z}$-action is continuous and the induced action on M is ergodic with respect to a T-invariant probability measure μ. Suppose that H is a real algebraic group, and let B be a real algebraic subgroup of H such that the quotient $V := H/B$ is compact. Form the associated bundle $P_V := P \times_H V$, where H acts on V by left translation. Then there exists a T-invariant measurable L-reduction of $P|_A$, where A is a T-invariant measurable subset of full μ-measure and $L = H_{v_0}$ is the isotropy subgroup of a measure $v_0 \in \mathcal{M}(V)_1$.

Proof. Since the fibers of P_V are compact, we can apply exercise 2.1.8 to conclude that T preserves a probability measure v on P_V that projects to μ under the bundle map p. Using lemma 6.7.1 we obtain a T-invariant H-equivariant map $\mathcal{G} : P \to \mathcal{M}(V)_1$. Since, by the previous lemma, the action of H on $\mathcal{M}(V)_1$ is tame and T acts ergodically on M, the induced map $\bar{\mathcal{G}} : M \to \mathcal{M}(V)_1/H$ is constant μ-almost everywhere. Therefore, we can find a μ-conull set $A \subset M$ such that the restriction of $\mathcal{G}$ to $P|_A$ takes values in a single H-orbit in $\mathcal{M}(V)_1$. By theorem 1.2.2, each H-orbit in $\mathcal{M}(V)_1$ is homeomorphic to H/H_{v_0}, for some $v_0 \in \mathcal{M}(V)_1$. Therefore the reduction we seek is given by $\mathcal{G}^{-1}(eH_{v_0})$, where eH_{v_0} is the coset of the identity element in H/H_{v_0}.   □

We discuss now some aspects of the proof of lemma 6.7.2. By Chevalley's theorem, theorem 4.8.1, H/B can be regarded as a Zariski-closed H-orbit in a projective space $P^n(\mathbb{R})$. This allows us to consider $\mathcal{M}(H/B)_1$ as a closed H-invariant subspace of $\mathcal{M}(P^n(\mathbb{R}))_1$, consisting of H-invariant probability measures whose supports are contained in that H-orbit.

The standard linear action of $GL(n + 1, \mathbb{R})$ on $\mathbb{R}^{n+1}$ induces a real algebraic action of the (real algebraic) group $PGL(n + 1, \mathbb{R}) := GL(n + 1, \mathbb{R})/\mathbb{R}^*$ on $P^n(\mathbb{R})$. Let $v_0 \in \mathcal{M}(P^n(\mathbb{R}))_1$. If we show the lemma for the action of $PGL(n + 1, \mathbb{R})$ on $\mathcal{M}(P^n(\mathbb{R}))_1$, then the proof for H would reduce to the observation that the H-action on the orbit $PGL(n + 1, \mathbb{R}) \cdot v_0$ is tame. But by theorem 1.2.2, $PGL(n + 1, \mathbb{R}) \cdot v_0$ can be identified with the real variety $PGL(n+1, \mathbb{R})/PGL(n+1, \mathbb{R})_{v_0}$, and H acts on the quotient algebraically via a linear representation in $PGL(n + 1, \mathbb{R})$. But this action is tame by corollary 4.9.3. The conclusion that H_{v_0} is real algebraic is also a consequence of the facts that $PGL(n + 1, \mathbb{R})_{v_0}$ is real algebraic and that the representation of H into $PGL(n + 1, \mathbb{R})$ from Chevalley's theorem is real algebraic.

We are left with showing that the lemma holds for the natural action of $PGL(n + 1, \mathbb{R})$ on $\mathcal{M}(P^n(\mathbb{R}))_1$. The key remark is the following lemma due to Furstenberg. By a projective subspace $[V]$ we mean the projection in $P^n(\mathbb{R})$ of a linear subspace $V \subset \mathbb{R}^n$.

Lemma 6.7.4 (Furstenberg) Let $[g_m]$ be a sequence in $PGL(n + 1, \mathbb{R})$, and $\mu, \nu \in \mathcal{M}(P^n(\mathbb{R}))$, such that $\lim_{m \to \infty} [g_m]_* \mu = \nu$. Then one of the following holds:

1. Either $\{[g_m] \mid m = 1, 2, \ldots\}$ is bounded, that is, is contained in a compact set in $PGL(n + 1, \mathbb{R})$; or
2. there exist proper independent linear subspaces $V, W \subset \mathbb{R}^{n+1}$ such that the support of ν is contained in $[V] \cup [W]$. Furthermore,

$$\dim V + \dim W = n + 1.$$

Proof. Let $\| \cdot \|$ be a matrix norm on $GL(n + 1, \mathbb{R})$. By multiplying g_m by a scalar, which does not change the class $[g_m]$, we may assume that $\|g_m\| = 1$ for each m. Assume that $\{[g_m] \mid m = 1, 2, \ldots\}$ is not bounded. By passing to a subsequence we may suppose that $[g_m]$ goes to infinity, that is, that it eventually leaves every compact set in the group. On the other hand, since the g_m are contained in a bounded set in the space of all square matrices of size $n + 1$, we can pass to another subsequence so that $g_m \to g$ for a singular but nonzero matrix g. The kernel N of g and the image $R = g\mathbb{R}^{n+1}$ are proper subspaces of $\mathbb{R}^{n+1}$. By the compactness of the Grassmannian varieties, we can pass to yet another subsequence so that $g_m[N]$ converges to $[V]$ for some linear subspace V of the same dimension as N. Notice that if $x \in P^n(\mathbb{R}) - [N]$, then $g_m x \to gx \in R$. We can write $\mu = \mu_1 + \mu_2$, where $\mu_1(P^n(\mathbb{R}) - [N]) = 0$ and $\mu_2([N]) = 0$. By the compactness of the space of probability measures, we can pass to another subsequence so that $[g_m]_* \mu_i \to \nu_i$, $i = 1, 2$. In particular, $[g_m]_* \mu$ converges to $\nu = \nu_1 + \nu_2$. Since μ_1 is supported on $[N]$ and $[g_m N]$ converges to $[V]$, we have that ν_1 is supported on $[V]$. It only remains to show that ν_2 is supported on $[R]$. The proof will be finished if we show that the ν_2-integral of an arbitrary continuous function f on $P^n(\mathbb{R})$ that vanishes on $[R]$ is 0. This is what the following computation shows

$$\int_{P^n(\mathbb{R})} f(x) \, d\nu_2(x) = \lim_{m \to \infty} \int_{P^n(\mathbb{R})} f(x) \, d([g_m]_* \mu_2)(x)$$

$$= \lim_{m \to \infty} \int_{P^n(\mathbb{R})} f([g_m]x) \, d\mu_2(x)$$

$$= \lim_{m \to \infty} \int_{P^n(\mathbb{R}) - [N]} f([g_m]x) \, d\mu_2(x) \quad (\text{since } \mu_2([N]) = 0)$$

$$= 0,$$

where in the last step we used the dominated convergence theorem and that $\lim_{m \to \infty} f([g_m]x) = 0$ for $x \in P^n(\mathbb{R}) - [N]$. $\square$

We sketch now the proof that $PGL(n + 1, \mathbb{R})$-orbits in $\mathcal{M}(P^n(\mathbb{R}))_1$ are locally closed (hence the action is tame). Let $\mu \in \mathcal{M}(P^n(\mathbb{R}))_1$ and denote by $\mathcal{O}$ the $PGL(n + 1, \mathbb{R})$-orbit of μ. It will suffice to show that there is an open set $U(\mu) \subset \mathcal{M}(P^n(\mathbb{R}))_1$ such that $\bar{\mathcal{O}} \cap U(\mu) = \mathcal{O}$.

Let $\mathcal{C}$ be the set of all compact subsets of $P^n(\mathbb{R})$. $\mathcal{C}$ is a compact metric space with the *Hausdorff metric*

$$d(A, B) := \max \left\{ \sup_{y \in A} d(x, B), \sup_{y \in B} d(y, A) \right\}$$

If $A \in \mathcal{C}$, there is a natural identification of $\mathcal{M}(A)_1$ with the measures in $\mathcal{M}(P^n(\mathbb{R}))_1$ whose supports are contained in A.

In order to simplify the exposition, let us assume that the support of μ is not contained in the union of a finite number of proper projective subspaces. Define the subset $K(\mu) \subset \mathcal{C}$ that consists of sets of the form $[V] \cup [W]$, where V and W are proper complementary subspaces of $\mathbb{R}^{n+1}$, together with the sets A of the form $A = \bigcup_{i=1}^{l}[V_i]$ such that the sum of the dimensions of the V_i, as well as the dimension of the subspace spanned by the union of all the V_i, is at most n. Let $M_{K(\mu)}$ be the set of probability measures on $P^n(\mathbb{R})$ whose supports are contained in elements of $K(\mu)$. It can be shown that $M_{K(\mu)}$ is closed and $PGL(n + 1, \mathbb{R})$-invariant. Furstenberg's lemma 6.7.4 can now be used to conclude that the open set $U(\mu) = \mathcal{M}(P^n(\mathbb{R}))_1 - M_{K(\mu)}$ satisfies the property $\bar{\mathcal{O}} \cap U(\mu) = \mathcal{O}$.

The proof that the isotropy groups are algebraic also depends in an essential way on Furstenberg's lemma. As already mentioned, the reader will find all the details in [36].

We finish the chapter with the following variant of Furstenberg's lemma which will be needed in chapter 10. It is also due to Zimmer.

Proposition 6.7.5 Let $W \subset P^n(\mathbb{R})$ be a quasi-projective irreducible variety that is not contained in a proper projective subspace. Let X be a projective irreducible variety and $f : X \to \bar{W}$ a regular surjection. Suppose that h_j is a sequence in $PGL(n+1, \mathbb{R})$ such that $h_j(W) \subset W$ and $h_j f$ converges uniformly to a continuous function $\theta : X \to \bar{W}$. Then $\{h_j\}$ is bounded in $PGL(n+1, \mathbb{R})$.

Proof. Fix a metric on $P^n(\mathbb{R})$ and choose $\epsilon > 0$ such that $d(\bar{W}, Y) \geq \epsilon$ for every proper projective subspace $Y \subset P^n(\mathbb{R})$. Choose j_1 large enough so that $j \geq j_1$ implies

$$\sup_j d(h_j f(x), \theta(x)) \leq \frac{\epsilon}{2}.$$

Let $T_j = h_j / \|h_j\|$. Then if h_j is not bounded in $PGL(n+1, \mathbb{R})$, by passing to a subsequence we can assume that T_j converges in the space of $(n+1)$-by-$(n+1)$ real matrices to a matrix T with $\|T\| = 1$ and $\det T = 0$. Let $X_1 = f^{-1}(P^n(\mathbb{R}) - [\ker T])$. Then $X_1 \subset X$ is (Zariski-) open and dense. Observe that if $y \in P^n(\mathbb{R}) - [\ker T]$, then $h_j y = T_j y$ converges to Ty, in the image V of T in $P^n(\mathbb{R})$. Therefore, if $x \in Z_1$, then $h_j f(x)$ converges to $Tf(x) \in V$. Since $h_i f(x)$ converges to $\theta(x)$, we have $\theta(X_1) \subset V$. Therefore $\theta(X) \subset V$. However, this contradicts the fact that θ must be surjective. Namely, if $y \in \bar{W}$, choose $x_j \in X$ such that $f(x_j) = h_j^{-1}(y)$. By passing to a subsequence, we can assume that x_j converges to some $x \in X$. Then $h_j f(x_j)$ converges to $\theta(x)$ since X is compact and $h_j f$ converges uniformly to θ. Thus $y = \theta(x)$. $\square$

7

Semisimple Lie Groups

We discuss in this chapter the main general results concerning the structure of semisimple Lie groups. Some simplification is achieved, and not much generality is lost, by assuming that our Lie groups are linear groups, that is, subgroups of the complex general linear group. The definition of semisimple Lie algebras given below is a "shortcut," which will allow us to arrive more quickly at a few basic results. The more standard treatment as well as some of the important topics not covered here can be found in [17] or [31].

7.1 Reductive and Semisimple Lie Algebras

Even though our semisimple groups are defined as subgroups of $GL(n, \mathbb{C})$, we still think of them as *real* groups; as we have seen before, they can always be represented as subgroups of the real general linear group by means of the embedding of $GL(n, \mathbb{C})$ into $GL(2n, \mathbb{R})$ used in chapter 5.

We will also be considering at times algebraic groups. Recall that these are, for our purposes, subgroups of some $GL(n, \mathbb{C})$ whose elements satisfy polynomial equations in the matrix entries. By a real algebraic group we mean the set of real points of an algebraic group $G \subset GL(n, \mathbb{C})$ defined by polynomials with real coefficients, that is, the group $G \cap GL(n, \mathbb{R})$.

We denote by A^* the complex conjugate transpose of A. The *Cartan involution* of $GL(n, \mathbb{C})$ is the homomorphism

$$\Theta : A \mapsto (A^*)^{-1}.$$

The Cartan involution of $\mathfrak{gl}_n(\mathbb{C})$ is the Lie algebra isomorphism induced from Θ and is given by $\theta : X \mapsto -X^*$. Θ is indeed a group homomorphism and an involution, that is, Θ^2 is the identity map, as one can easily check.

Let G be a connected Lie subgroup of $GL(n, \mathbb{C})$. We say that G is a *reductive* group if it is conjugate to a subgroup that is stable under the Cartan involution

119

Θ. In other words, G is reductive if there is $g \in GL(n, \mathbb{C})$ such that gGg^{-1} is mapped into itself by Θ. A Lie algebra $\mathfrak{g} \subset \mathfrak{gl}_n(\mathbb{C})$ is reductive if it is conjugate by an element in $GL(n, \mathbb{C})$ to a θ-stable subalgebra. In particular, G is reductive if and only if $\mathfrak{g}$ is.

We recall that the center of a group G is the subgroup

$$Z(G) = \{a \in G \mid ag = ga \text{ for all } g \in G\}.$$

The center is clearly a normal subgroup of G. The center of a Lie algebra $\mathfrak{g}$ is the subalgebra

$$\mathfrak{z}(\mathfrak{g}) = \{X \in \mathfrak{g} \mid [X, Y] = 0 \text{ for all } Y \in \mathfrak{g}\}$$

and is an ideal of $\mathfrak{g}$. (A subalgebra $\mathfrak{n} \subset \mathfrak{g}$ is an ideal if $[X, Y] \in n$ for all $X \in n$ and all $Y \in \mathfrak{g}$.)

A Lie algebra $\mathfrak{g} \subset \mathfrak{gl}_n(\mathbb{C})$ is *semisimple* if it is reductive and has trivial center. $G \subset GL(n, \mathbb{C})$ is a *semisimple Lie group* if its Lie algebra is semisimple.

As an example, we show that $SU(p, q)$, $p + q \geq 2$, is a semisimple group. Recall that A belongs to $SU(p, q)$ if, by definition, $A^* I_{p,q} A = I_{p,q}$ and $\det A = 1$ (see chapter 5). It is immediate from the definition that if $A \in SU(p, q)$ then $\Theta(A) \in SU(p, q)$, so the group is reductive. We show by direct computation that its center is finite. First observe that the diagonal matrix $\text{diag}[\lambda_1, \ldots, \lambda_{p+q}]$ is in $SU(p, q)$ if and only if $|\lambda_k| = 1$ for each k and the product of the diagonal entries is 1. We can choose one such matrix, call it B, such that its diagonal entries are all mutually distinct. If $A = (a_{ij})$ is in the center, $AB = BA$ implies that $a_{ij}\lambda_j = \lambda_i a_{ij}$ (no summation involved), so all off-diagonal entries of A must be zero. Moreover, we clearly have that $AXA^{-1} = X$ for all $X \in \mathfrak{su}(p, q)$. Thus, we look for a diagonal element $\text{diag}[a_1, \ldots, a_{p+q}]$ of $SU(p, q)$ that commutes with all matrices $X = \begin{pmatrix} X_{11} & X_{12} \\ X_{12}^* & X_{22} \end{pmatrix}$ of trace 0 such that $X_{11}^* = -X_{11}$ (a p-by-p matrix) and $X_{22}^* = -X_{22}$ (a q-by-q matrix). By inspecting the form of X, it is clear that for every index (i, j) with $i \neq j$, we can always find a matrix $X = (x_{ij})$ in $\mathfrak{su}(p, q)$ such that $x_{ij} \neq 0$. Writing out the product we obtain $a_i x_{ij} = x_{ij} a_j$ so $a_i = a_j$ for all i and j. Consequently, $A = \lambda I$ for some complex number λ such that $\lambda^{p+q} = 1$. Therefore, the center is the finite subgroup of $U(1)$ of $(p+q)$th roots of 1, so the center of $\mathfrak{g}$ is trivial and $SU(p, q)$ is semisimple.

A similar argument shows that the classical groups of chapter 5 are all semisimple.

Exercise 7.1.1 Show that $SL(n, \mathbb{R})$ is semisimple.

Let $\mathfrak{g} \subset \mathfrak{gl}_n(\mathbb{C})$ be a reductive Lie algebra, and assume, possibly after having to conjugate $\mathfrak{g}$ by some matrix in $GL(n, \mathbb{C})$, that it is stable under the involution

θ. One defines an inner product on $\mathfrak{g}$ as follows. Given, $X, Y \in \mathfrak{g}$, set

$$\langle X, Y \rangle := \mathrm{Re}(\mathrm{Tr}(XY^*)) = -\mathrm{Re}(\mathrm{Tr}(X\theta(Y))).$$

Exercise 7.1.2 Show that $\langle \cdot, \cdot \rangle$ is a positive-definite symmetric bilinear form.

We denote by $A^\dagger$ the adjoint of a linear transformation A of $\mathfrak{g}$ with respect to the inner product $\langle \cdot, \cdot \rangle$ and by $\mathrm{ad}(Z)$ the linear transformation defined by $\mathrm{ad}(Z)(X) = [Z, X]$, $X \in \mathfrak{g}$. Observe that the Jacobi identity can be written in the following form: $\mathrm{ad}([X, Y]) = [\mathrm{ad}(X), \mathrm{ad}(Y)]$, for all $X, Y \in \mathfrak{g}$.

Exercise 7.1.3 Verify the following statements.

1. $\mathrm{ad}(\theta(Z)) = -\mathrm{ad}(Z)^\dagger$.
2. $\langle \theta(X), \theta(Y) \rangle = \langle X, Y \rangle$ for all $X, Y \in \mathfrak{g}$.
3. If $\mathfrak{a}$ is an ideal of $\mathfrak{g}$, then $\theta(\mathfrak{a})$ is also an ideal.

A direct sum decomposition of a Lie algebra $\mathfrak{h}$ is a direct sum (in the sense of vector spaces) $\mathfrak{h} = \mathfrak{h}_1 \oplus \cdots \oplus \mathfrak{h}_k$ such that $\mathfrak{h}_1, \ldots, \mathfrak{h}_k$ are ideals of $\mathfrak{h}$. In particular, $[X, Y] = 0$ whenever $X \in \mathfrak{h}_i$, $Y \in \mathfrak{h}_j$, $i \neq j$.

Exercise 7.1.4 Given an arbitrary ideal $\mathfrak{a}$ of the θ-stable Lie algebra $\mathfrak{g}$, define the subspace $\mathfrak{a}^\perp := \{X \in \mathfrak{g} \mid \langle X, Y \rangle = 0 \text{ for all } Y \in \mathfrak{a}\}$.

1. Show that $\mathfrak{a}^\perp$ is also an ideal of $\mathfrak{g}$ and $\mathfrak{g} = \mathfrak{a} \oplus \mathfrak{a}^\perp$.
2. Show that $\theta(\mathfrak{a}^\perp) = \theta(\mathfrak{a})^\perp$.

Exercise 7.1.5 For an arbitrary Lie algebra $\mathfrak{h}$, denote by $[\mathfrak{h}, \mathfrak{h}]$ the subspace of $\mathfrak{h}$ spanned by all elements $[X, Y]$ for $X, Y \in \mathfrak{h}$.

1. Show that $[\mathfrak{h}, \mathfrak{h}]$ is an ideal of $\mathfrak{h}$.
2. If $\mathfrak{g}$ is a θ-stable subalgebra of $\mathfrak{gl}_n(\mathbb{C})$, show that $[\mathfrak{g}, \mathfrak{g}]$ also is θ-stable.

Exercise 7.1.6 If $\mathfrak{g}$ is θ-stable, show that $[\mathfrak{g}, \mathfrak{g}]^\perp = \mathfrak{z}(\mathfrak{g})$. In particular,

$$\mathfrak{g} = [\mathfrak{g}, \mathfrak{g}] \oplus \mathfrak{z}(\mathfrak{g}).$$

Conclude that, if $\mathfrak{g}$ is a semisimple Lie algebra (not necessarily θ-stable), then $\mathfrak{g} = [\mathfrak{g}, \mathfrak{g}]$.

A Lie algebra is said to be *simple* if its only ideals are $\{0\}$ and itself.

Proposition 7.1.7 Let $\mathfrak{g}$ be a θ-stable semisimple Lie subalgebra of $\mathfrak{gl}_n(\mathbb{C})$. Then $\mathfrak{g}$ decomposes as an $\langle \cdot, \cdot \rangle$-orthogonal direct sum $\mathfrak{g}_1 \oplus \cdots \oplus \mathfrak{g}_k$ of θ-stable nonabelian simple ideals.

Proof. Let $\mathfrak{a}$ be a nonzero ideal of $\mathfrak{g}$ of minimal dimension. The Jacobi identity implies that $[\mathfrak{a}, \mathfrak{a}]$ is also an ideal of $\mathfrak{g}$. Moreover, $[\mathfrak{a}, \mathfrak{a}] \neq \{0\}$ since, otherwise, $\mathfrak{a}$ would be contained in the center of $\mathfrak{g}$, which is zero. Therefore, $[\mathfrak{a}, \mathfrak{a}] = \mathfrak{a}$.

We claim that $\mathfrak{a}$ is θ-stable. Once this is shown, the proposition follows by considering $\mathfrak{a}^{\perp}$ and proceeding by induction.

Observe that $[\theta(\mathfrak{a}), \mathfrak{a}]$ is an ideal of $\mathfrak{g}$ contained in $\mathfrak{a}$, so either it coincides with $\mathfrak{a}$ or it must be zero by the minimality of $\mathfrak{a}$. In the latter case, we would conclude that for any $Z = \sum_i [X_i, Y_i] \in \mathfrak{a} = [\mathfrak{a}, \mathfrak{a}]$,

$$\langle Z, Z \rangle = \sum_i \langle [X_i, Y_i], Z \rangle = -\sum_i \langle Y_i, [\theta(X_i), Z] \rangle \in \langle \mathfrak{a}, [\theta(\mathfrak{a}), \mathfrak{a}] \rangle,$$

so that Z, hence $\mathfrak{a}$, would be zero, which is not the case. Therefore, $[\theta(\mathfrak{a}), \mathfrak{a}] = \mathfrak{a}$. It follows from this that the ideal $\mathfrak{a} \cap \theta(\mathfrak{a})$, which contains $[\theta(\mathfrak{a}), \mathfrak{a}]$, must coincide with $\mathfrak{a}$, so $\mathfrak{a} = \theta(\mathfrak{a})$ as claimed. □

7.2 The Adjoint Group

If $\mathfrak{g}$ is a real Lie algebra of dimension N, the vector space $\mathfrak{g}_{\mathbb{C}} = \mathfrak{g} \otimes_{\mathbb{R}} \mathbb{C}$ has a natural structure of a Lie algebra over $\mathbb{C}$ such that

$$[X \otimes 1, Y \otimes 1] = [X, Y] \otimes 1.$$

It is called the *complexification* of $\mathfrak{g}$.

The group of Lie algebra automorphisms of $\mathfrak{g}$, defined as the subgroup of all $A \in GL(\mathfrak{g}_{\mathbb{C}})$ such that $A[u, v] = [Au, Av]$ for all $u, v \in \mathfrak{g}_{\mathbb{C}}$, is in a natural way an algebraic group defined over $\mathbb{R}$. Namely, let $\{e_1, \ldots, e_N\}$ be a basis for $\mathfrak{g}$ and define real numbers c_{ij}^k by $[e_i, e_j] = \sum_k c_{ij}^k e_k$. Then the group $\mathrm{Aut}(\mathfrak{g}_{\mathbb{C}})$ of $\mathbb{C}$-linear automorphisms of $\mathfrak{g}_{\mathbb{C}}$ becomes identified with the subgroup of $GL(N, \mathbb{C})$ consisting of all complex matrices $A = (a_{ij})$ that satisfy the polynomial equations with real coefficients

$$\sum_k c_{ij}^k a_{kl} - \sum_{r,s} c_{rs}^l a_{ir} a_{js} = 0 \quad (\Leftrightarrow A[e_i, e_j] - [Ae_i, Ae_j] = 0).$$

Let G be a connected Lie group and let $\mathfrak{g}$ be its Lie algebra. Then G has a canonical representation into the group of real automorphisms of $\mathfrak{g}$, the *adjoint representation*, defined as

$$\mathrm{Ad} : G \to GL(\mathfrak{g})$$

such that $\mathrm{Ad}(g)(X) = gXg^{-1}$. Although the adjoint representation is not in general faithful, its kernel is precisely the center of G, as one can easily check. If G is semisimple, its center Z is a closed subgroup with trivial Lie algebra, and therefore Z is discrete. Therefore, for a connected semisimple G, G/Z is isomorphic to $\mathrm{Ad}(G)$.

Exercise 7.2.1 If G is a connected semisimple Lie group, show that G/Z is a connected semisimple Lie group with trivial center. (Hint: The main point is to show that $\mathrm{Ad}(G)$ has trivial center. Show that for any A in the center of $\mathrm{Ad}(G)$ and all $X, Y \in \mathfrak{g}$, $A[X, Y] = [X, Y]$. But $\mathfrak{g}$ semisimple implies $[\mathfrak{g}, \mathfrak{g}] = \mathfrak{g}$ so A is the identity transformation.)

Exercise 7.2.2 Show that the Lie algebra of $\mathrm{Aut}(\mathfrak{g}_\mathbb{C})$ is the linear space of all $\mathbb{C}$-linear derivations of $\mathfrak{g}_\mathbb{C}$, that is, of all complex linear maps δ from $\mathfrak{g}_\mathbb{C}$ into $\mathfrak{g}_\mathbb{C}$ such that $\delta[X, Y] = [\delta X, Y] + [X, \delta Y]$ for all $X, Y \in \mathfrak{g}_\mathbb{C}$.

A derivation δ of $\mathfrak{g}_\mathbb{C}$ is called an *inner* derivation if it is of the form $\delta X = \mathrm{ad}(Z)X := [Z, X]$ for some $Z \in \mathfrak{g}_\mathbb{C}$.

The *Killing form* of $\mathfrak{g}$ is the symmetric bilinear form

$$\kappa(X, Y) := \mathrm{Tr}(\mathrm{ad}(X) \circ \mathrm{ad}(Y)),$$

for $X, Y \in \mathfrak{g}$. It is ad-invariant, in the following sense:

$$\kappa(\mathrm{ad}(H)X, Y) + \kappa(X, \mathrm{ad}(H)Y) = 0$$

for each $H, X, Y \in \mathfrak{g}$.

Exercise 7.2.3 Prove that κ is ad-invariant. (Use $\mathrm{ad}([X, Y]) = [\mathrm{ad}(X), \mathrm{ad}(Y)]$, $X, Y \in \mathfrak{g}$, and the fact that $\mathrm{Tr}(AB) = \mathrm{Tr}(BA)$ for linear transformations A, B.) Also show that $\kappa(\theta X, \theta Y) = \kappa(X, Y)$, $X, Y \in \mathfrak{g}$. (Express $\mathrm{Tr}(\mathrm{ad}(\theta X) \circ \mathrm{ad}(\theta Y))$ in terms of an $\langle \cdot, \cdot \rangle$-orthonormal basis and use that θ is an isometry for $\langle \cdot, \cdot \rangle$.)

The Killing form is clearly a symmetric bilinear form. If $\mathfrak{g}$ is semisimple, κ is also nondegenerate. To see this, let $\{e_i \mid i = 1, \dots, n\}$ be an orthonormal basis for $\mathfrak{g}$ with respect to the inner product $\langle \cdot, \cdot \rangle$ and recall that the adjoint of $\mathrm{ad}(H)$ with respect to this inner product is $\mathrm{ad}(H)^\dagger = -\mathrm{ad}(\theta H)$. Then, if $\kappa(X, \cdot) = 0$,

$$0 = \kappa(-\theta(X), X)$$
$$= \sum_{i=1}^{n} \langle (-\mathrm{ad}(\theta X) \circ \mathrm{ad}(X)e_i, e_i \rangle$$
$$= \sum_{i=1}^{n} \langle \mathrm{ad}(\theta X)^\dagger \circ \mathrm{ad}(X)e_i, e_i \rangle$$
$$= \sum_{i=1}^{n} \langle \mathrm{ad}(X)e_i, \mathrm{ad}(\theta X)e_i \rangle,$$

so $\text{ad}(X) = 0$. But if $\mathfrak{g}$ is semisimple, this also implies that $X = 0$ since the kernel of ad is the center of $\mathfrak{g}$, which is trivial.

It is useful to introduce another inner product on $\mathfrak{g}$, defined by the equation

$$\langle X, Y \rangle_\kappa := -\text{Re}(\kappa(X, \theta Y)).$$

Exercise 7.2.4 Using the facts already established for the Killing form, show that if $\mathfrak{g}$ is a θ-stable semisimple Lie algebra, then

1. $\langle \cdot, \cdot \rangle_\kappa$ is a symmetric nondegenerate bilinear form;
2. with respect to $\langle \cdot, \cdot \rangle_\kappa$, the adjoint of $\text{ad}(X)$ is $\text{ad}(X)^\dagger = -\text{ad}(\theta X)$;
3. $\langle \theta(X), \theta(Y) \rangle_\kappa = \langle X, Y \rangle_\kappa$ for all $X, Y \in \mathfrak{g}$.

Theorem 7.2.5 Let G be a connected semisimple Lie group and Z its center. Then the adjoint representation defines an isomorphism between G/Z and the connected component of the identity of the group of real points of a linear algebraic group defined over $\mathbb{R}$.

The algebraic group of the previous theorem is called the *adjoint group* of G. In particular, if G is a connected semisimple Lie group with trivial center, then G is naturally isomorphic to the identity component of a real algebraic group.

To prove the theorem it suffices to show that the Lie algebra of the group of real points of $\text{Aut}(\mathfrak{g}_\mathbb{C})$ is isomorphic to $\mathfrak{g}$. Since that Lie algebra is the space of derivations of $\mathfrak{g}$, the theorem will be proved once we show the next lemma.

Lemma 7.2.6 Any derivation of a semisimple Lie algebra is inner.

Proof. Let δ be a derivation of $\mathfrak{g}$ and define a linear functional f on $\mathfrak{g}$ such that $f(X) = \text{Tr}(\text{ad}(X) \circ \delta)$ for each $X \in \mathfrak{g}$. Since κ is nondegenerate, we can find $Y \in \mathfrak{g}$ such that $f(X) = \kappa(X, Y)$ for all $X \in \mathfrak{g}$. Defining the derivation $D := \delta - \text{ad}(Y)$, we show that $D = 0$, as follows. First notice that $\text{Tr}(D \circ \text{ad}(Z)) = 0$ for all Z, by definition. Therefore, for all $Z_1, Z_2 \in \mathfrak{g}$,

$$\begin{aligned}
0 &= \text{Tr}(D \circ \text{ad}[Z_1, Z_2]) \\
&= \text{Tr}(D \circ [\text{ad}(Z_1), \text{ad}(Z_2)]) \\
&= \text{Tr}(D \circ \text{ad}(Z_1) \circ \text{ad}(Z_2) - D \circ \text{ad}(Z_2) \circ \text{ad}(Z_1)) \\
&= \text{Tr}(D \circ \text{ad}(Z_1) \circ \text{ad}(Z_2) - \text{ad}(Z_1) \circ D \circ \text{ad}(Z_2)) \\
&= \text{Tr}([D, \text{ad}(Z_1)] \circ \text{ad}(Z_2)) \\
&= \text{Tr}(\text{ad}(DZ_1) \circ \text{ad}(Z_2)) \\
&= \kappa(DZ_1, Z_2),
\end{aligned}$$

which shows that $DZ_1 = 0$ for all Z_1. Therefore $D = 0$. $\quad\square$

Table 7.1.   *The maximal compact*
subgroups of the classical groups

G	K
$SL(n, \mathbb{C})$	$SU(n)$
$SO(n, \mathbb{C})$	$SO(n)$
$Sp(n, \mathbb{C})$	$Sp(n) = Sp(n, \mathbb{C}) \cap U(2n)$
$SL(n, \mathbb{R})$	$SO(n)$
$SL(n, \mathbb{H})$	$Sp(n)$
$SO(m, n)$	$S(O(m) \times O(n))$
$SU(m, n)$	$S(U(m) \times U(n))$
$Sp(m, n)$	$Sp(m) \times Sp(n)$
$Sp(n, \mathbb{R})$	$U(n)$
$SO^*(2n)$	$U(n).$

7.3   The Cartan Decomposition

Since θ is an involution, that is, $\theta^2 = \mathrm{id}$, its only eigenvalues are 1 and -1. We define subspaces $\mathfrak{k}$ and $\mathfrak{p}$ of the (as always, θ-stable) Lie algebra $\mathfrak{g}$ as follows:

$$\mathfrak{k} := \{X \in \mathfrak{g} \mid \theta(X) = X\},$$
$$\mathfrak{p} := \{X \in \mathfrak{g} \mid \theta(X) = -X\}.$$

Since θ is a Lie algebra automorphism, $\mathfrak{k}$ is a Lie subalgebra, but $\mathfrak{p}$ is only a subspace. As a vector space, $\mathfrak{g} = \mathfrak{k} \oplus \mathfrak{p}$.

Exercise 7.3.1   Verify the relations $[\mathfrak{k}, \mathfrak{k}] \subset \mathfrak{k}$, $[\mathfrak{k}, \mathfrak{p}] \subset \mathfrak{p}$, $[\mathfrak{p}, \mathfrak{p}] \subset \mathfrak{k}$. Also show that the decomposition $\mathfrak{g} = \mathfrak{k} \oplus \mathfrak{p}$ is orthogonal with respect to both $\langle \cdot, \cdot \rangle$ and $\langle \cdot, \cdot \rangle_\kappa$.

If G is the Lie subgroup of $GL(n, \mathbb{C})$ with Lie algebra $\mathfrak{g}$, then $\mathfrak{k}$ is the Lie algebra of the subgroup

$$K = \{g \in G \mid \Theta(g) = g\}.$$

Exercise 7.3.2   Show that K is a subgroup of the unitary group $U(n)$. Therefore, K is a compact group. Show, moreover, that K normalizes $\mathfrak{p}$, that is, $kXk^{-1} \in \mathfrak{p}$ for all $k \in K$ and $X \in \mathfrak{p}$.

Table 7.1 gives the compact group K for each of the classical groups.

Lemma 7.3.3   Let $P : \mathbb{R}^n \to \mathbb{R}$ be a polynomial map. Suppose $(a_1, \ldots, a_n)$ has the property that $P(e^{ka_1}, \ldots, e^{ka_n}) = 0$ for all nonnegative integers k. Then $P(e^{ta_1}, \ldots, e^{ta_n}) = 0$ for all $t \in \mathbb{R}$.

Proof. A monomial $x_1^{l_1} \cdots x_n^{l_n}$, when evaluated at $(e^{ta_1}, \ldots, e^{ta_n})$ becomes $e^{t(\sum_{i=1}^n a_i l_i)}$. Collecting terms with the same exponentials, we may assume that $P(e^{ta_1}, \ldots, e^{ta_n})$ takes the form $\sum_{j=1}^N c_j e^{tb_j}$. We may further assume that all c_j are nonzero and $b_1 < b_2 < \cdots < b_N$. We argue by contradiction and suppose that $N > 0$. Multiplying by e^{-tb_N} and changing notation, we may assume that $b_N = 0$. We pass to the limit in the expression $\sum_{j=1}^N c_j e^{tb_j}$ as t approaches $+\infty$ through integer values and find $c_N = 0$. But this is a contradiction.   □

Proposition 7.3.4 Set $K = U(n)$ and let $\mathfrak{p}$ be the space of all $n \times n$ Hermitian matrices. Then the map

$$F : K \times \mathfrak{p} \to GL(n, \mathbb{C}), \quad (k, X) \mapsto k \exp X$$

is a smooth diffeomorphism. F restricted to $\{e\} \times \mathfrak{p}$ is a smooth diffeomorphism onto the space S of positive Hermitian $n \times n$ matrices.

Proof. Given $g \in GL(n, \mathbb{C})$, g^*g is a positive Hermitian matrix, so there are $A \in U(n)$ and positive numbers λ_i such that $g^*g = A \operatorname{diag}[\lambda_1, \ldots, \lambda_n] A^{-1}$. Define $C = A \operatorname{diag}[\sqrt{\lambda_1}, \ldots, \sqrt{\lambda_n}] A^{-1}$. Then $C^2 = g^*g$ and $C^* = C$, so $C^*C = g^*g$. It follows that $(Cg^{-1})^{-1} = (Cg^{-1})^*$, so $Cg^{-1} \in K$. Therefore $g = kC$, for some $k \in K$, showing that the map is surjective.

We wish to show that the map F is also injective. First notice that if $kC = k'C'$ for $k, k' \in K$ and positive Hermitian matrices C, C', then $Ck^{-1} = (kC)^* = (k'C')^* = C'k'^{-1}$, which implies that

$$C^2 = (Ck^{-1})(kC) = (C'k'^{-1})(k'C') = C'^2.$$

The previous lemma implies that C belongs to the subgroup $H \subset GL(n, \mathbb{C})$ consisting of the Zariski closure of the set of all powers C^{2n} for $n \in \mathbb{Z}$. (Indeed, let Q be any polynomial that vanishes on H. Since

$$C^2 = A \operatorname{diag}[\lambda_1, \ldots, \lambda_n] A^{-1},$$

and $Q(C^{2m}) = 0$ for all $m \in \mathbb{N}$, we obtain a polynomial $P(x_1, \ldots, x_n)$ such that $P(\lambda_1^m, \ldots, \lambda_n^m) = 0$ for each m. The lemma now implies that $Q(C) = P(\lambda_1^{\frac{1}{2}}, \ldots, \lambda_n^{\frac{1}{2}}) = 0$.)

On the other hand, also by the previous lemma, C' must commute with each element of H since C' commutes with all C^{2m} (commuting with C' can, of course, be characterized by polynomial equations). Therefore C and C' commute. It follows that $D = C'C^{-1}$ is a positive Hermitian matrix such that $D^2 = I$; therefore $D = I$ and $C = C'$. This shows that F is injective, hence bijective.

To show that F is a diffeomorphism, it suffices to prove that it is a local diffeomorphism at each $(k, Z_0) \in K \times \mathfrak{p}$. In fact, since the dimensions of the target space and domain are the same, it suffices to show that $DF_{(k, Z_0)}$ is injective. Write $\gamma(t) := \exp(Z_0 + tZ)$, $Z \in \mathfrak{p}$. Then

$$DF_{(k, Z_0)}(X, Z) = \frac{d}{dt}\bigg|_{t=0} (k \exp(tX)\gamma(t)) = kX\gamma(0) + k\gamma'(0).$$

Suppose that this equation is 0, so that $X + \gamma'(0)\gamma(0)^{-1} = 0$. Our goal is to show that $X = Z = 0$.

Notice that

$$\theta(\gamma'(0)\gamma(0)^{-1}) = \frac{d}{dt}\bigg|_{t=0} \Theta(\gamma(t)\gamma(0)^{-1})$$

$$= \frac{d}{dt}\bigg|_{t=0} \gamma(t)^{-1}\gamma(0) \qquad \text{(since } \gamma(t) \text{ is Hermitian)}$$

$$= -\gamma(0)^{-1}\gamma'(0).$$

Therefore, recalling that $\theta(X) = X$, we obtain $X - \gamma(0)^{-1}\gamma'(0) = 0$. Therefore

$$\gamma'(0)\gamma(0)^2 = \gamma(0)^2\gamma'(0).$$

Since $\gamma(0)$ is Hermitian symmetric, we can apply the same argument used already to show commutativity of C and C' to conclude, here, that $\gamma(0)$ and $\gamma'(0)$ commute. We get at once $\gamma'(0) = 0$ and $X = 0$.

We still need to show $Z = 0$. Since $\gamma'(0)$ commutes with $\gamma(0) = \exp Z_0$, we also conclude that $\gamma'(0)$ commutes with Z_0. (Use the lemma once again to prove that $\gamma'(0)$ commutes with $(\exp(\frac{1}{m}Z_0) - I)/m$ for each m and then pass to the limit as $m \to \infty$ to conclude that $\gamma'(0)$ and Z_0 commute.) Expanding $\exp(Z_0 + tZ)$ in a Taylor series and taking the derivative term by term yields

$$\gamma'(0) = \sum_{n=0}^{\infty} \frac{1}{(n+1)!} \left(\sum_{k=0}^{n} Z_0^k Z Z_0^{n-k} \right).$$

Therefore

$$0 = Z_0\gamma'(0) - \gamma'(0)Z_0$$

$$= \sum_{n=0}^{\infty} \frac{1}{(n+1)!} \left(Z_0^{n+1} Z - Z Z_0^{n+1} \right)$$

$$= (\exp Z_0)Z - Z \exp Z_0.$$

It follows that Z_0 and Z commute. Hence $0 = \gamma'(0) = (\exp Z_0)Z$ and, finally, $Z = 0$.

To conclude the proof of the proposition, observe that

$$A\,\mathrm{diag}[\lambda_1, \ldots, \lambda_n]A^{-1} \in \mathfrak{p} \mapsto A\,\mathrm{diag}[\exp \lambda_1, \ldots, \exp \lambda_n]A^{-1} \in S$$

is a bijection.   □

The inverse of the map F is called the *polar decomposition*.

Theorem 7.3.5 (Cartan decomposition) Let G be a subgroup of $GL(n, \mathbb{C})$ defined by polynomial equations, with real coefficients, in the real and imaginary parts of the matrix entries. Let $\mathfrak{g}$ be the Lie algebra of G. Suppose that G is stable under the Cartan involution. Let $K = G \cap U(n)$ and $\mathfrak{p}$ be the subspace of Hermitian matrices in $\mathfrak{g}$. Then the map

$$T : K \times \mathfrak{p} \to G, \quad (k, X) \mapsto k \exp X$$

is a surjective diffeomorphism.

Proof. Since T is the restriction to $K \times \mathfrak{p}$ of the smooth diffeomorphism of the previous proposition, T will also be a diffeomorphism once we show it is surjective.

Let $g \in G$ and write $g = k \exp X$, the polar decomposition of g in $GL(n, \mathbb{R})$. By hypothesis,

$$\Theta(g)^{-1} = g^* = (\exp X)k^{-1}$$

lies in G, so $\Theta(g)^{-1}g = \exp(2X)$ also does. Since X is Hermitian, we can write $2X = A\,\mathrm{diag}[a_1, \ldots, a_n]A^{-1}$, for some $A \in U(n)$, where $a_1, \ldots, a_n$ are real. For each integer k, $(\exp(2X))^k = A\,\mathrm{diag}[e^{a_1 k}, \ldots, e^{a_n k}]A^{-1} \in G$, so an application of the previous lemma shows that $\exp X \in G$. Therefore both $\exp X$ and k lie in G, showing surjectivity of T and concluding the proof.   □

Corollary 7.3.6 K is a maximal compact subgroup of G.

Proof. If K_1 is another compact subgroup of G that properly contains K, then there is $k_1 = k \exp X \in K_1$ such that $X \neq 0$. But then $\exp X = k^{-1}k_1 \in K_1$ so $\exp(nX) \in K_1$ for all n. But this is impossible since the eigenvalues of $\exp(nX)$ are not bounded.   □

Proposition 7.3.7 Let G be a subgroup of $GL(n, \mathbb{C})$ defined by polynomial equations, with real coefficients, in the real and imaginary parts of the matrix entries. Then the center of G is finite.

Proof. Denote the center by Z. We know that Z is discrete in G since G is semisimple. In order to show that Z is finite, it suffices to prove it is contained in a maximal compact subgroup. We may assume that G is Θ-stable. Let K be as in the previous theorem. Let $z \in Z$ and write $z = k \exp X$. Since G is stable under Θ and Θ is a homomorphism from G into itself, it follows that Z is also Θ-stable. Therefore $\Theta(z)^{-1}z = (\exp X)k^{-1}k \exp X = \exp(2X)$ also belongs to Z. But Z is an algebraic subgroup of G, so another application of the previous lemma gives $\exp(tX) \in Z$ for all $t \in \mathbb{R}$. Therefore, if X were not 0, Z would have dimension greater than 0, a contradiction. Therefore $X = 0$ and $Z \subset K$. $\square$

It is also immediate from the theorem that the number of connected components of G is the same as the number of connected components of K, which is finite.

7.4 The Restricted Root Space Decomposition

For each $X \in \mathfrak{p}$, the operator $\mathrm{ad}(X)$ on $\mathfrak{g}$ is self-adjoint with respect to the inner product $\langle \cdot, \cdot \rangle_\kappa$. Therefore $\mathrm{ad}(X)$ is diagonalizable with real eigenvalues. Let $\mathfrak{a}$ be a maximal abelian algebra in $\mathfrak{p}$. More precisely, $\mathfrak{a}$ is abelian and is not properly contained in a subspace of $\mathfrak{p}$ consisting of commuting elements. The operators $\mathrm{ad}(X)$, $X \in \mathfrak{a}$, commute since

$$0 = \mathrm{ad}([X, Y]) = \mathrm{ad}(X) \circ \mathrm{ad}(Y) - \mathrm{ad}(Y) \circ \mathrm{ad}(X)$$

for $X, Y \in \mathfrak{a}$. Therefore, it is possible to find a basis for $\mathfrak{g}$ that simultaneously diagonalizes all the operators $\mathrm{ad}(X)$, $X \in \mathfrak{a}$.

Denote by $\mathfrak{a}^*$ the space of real linear functionals on $\mathfrak{a}$. We now define

$$\mathfrak{g}_\lambda := \{X \in \mathfrak{g} \mid [H, X] = \lambda(H)X, \text{ for all } H \in \mathfrak{a}\}.$$

If $\lambda \in \mathfrak{a}^*$ is nonzero and $\mathfrak{g}_\lambda$ is nonzero, we say that λ is a *root* of $(\mathfrak{a}, \mathfrak{g})$ (or a *restricted root* of $\mathfrak{g}$), with associated root space $\mathfrak{g}_\lambda$. The set of all such roots is denoted $\Phi(\mathfrak{a}, \mathfrak{g})$. We denote by $\mathfrak{g}_0$ the centralizer of $\mathfrak{a}$ in $\mathfrak{g}$, that is, $\mathfrak{g}_0$ is the subspace of all X in $\mathfrak{g}$ such that $[X, H] = 0$ for all $H \in \mathfrak{a}$. Therefore, we have the direct sum decomposition of vector spaces

$$\mathfrak{g} = \mathfrak{g}_0 \oplus \bigoplus_{\lambda \in \Phi(\mathfrak{a}, \mathfrak{g})} \mathfrak{g}_\lambda$$

called the *restricted root space decomposition* of $\mathfrak{g}$.

Exercise 7.4.1 If $\lambda, \mu \in \Phi(\mathfrak{a}, \mathfrak{g})$, show that $[\mathfrak{g}_\lambda, \mathfrak{g}_\mu] \subset \mathfrak{g}_{\lambda+\mu}$ if $\lambda + \mu \in \Phi(\mathfrak{a}, \mathfrak{g})$ and $[\mathfrak{g}_\lambda, \mathfrak{g}_\mu]$ is zero otherwise. Moreover, $[\mathfrak{g}_0, \mathfrak{g}_\lambda] \subset \mathfrak{g}_\lambda$.

Exercise 7.4.2 Show that $\lambda \in \Phi(\mathfrak{a}, \mathfrak{g})$ if and only if $-\lambda \in \Phi(\mathfrak{a}, \mathfrak{g})$ and that θ restricts to an isomorphism between $\mathfrak{g}_\lambda$ and $\mathfrak{g}_{-\lambda}$.

Exercise 7.4.3 If λ and μ are distinct roots, show that $\mathfrak{g}_\lambda$ and $\mathfrak{g}_\mu$, as well as $\mathfrak{g}_\lambda$ and $\mathfrak{g}_0$, are orthogonal subspaces with respect to $\langle \cdot, \cdot \rangle_\kappa$.

Exercise 7.4.4 Show that $\mathfrak{g}_0$ is stable under θ and that we have a $\langle \cdot, \cdot \rangle_\kappa$-orthogonal direct sum $\mathfrak{g}_0 = \mathfrak{a} \oplus \mathfrak{m}$, where $\mathfrak{m} = \{X \in \mathfrak{k} \mid [X, Y] = 0 \text{ for all } Y \in \mathfrak{a}\}$.

The subalgebra $\mathfrak{a}$ will be called an $\mathbb{R}$-*split Cartan subalgebra* of $\mathfrak{g}$. A more descriptive name is "maximal abelian $\mathbb{R}$-diagonalizable subalgebra."

We define an ordering on the set of roots $\Phi = \Phi(\mathfrak{a}, \mathfrak{g})$. Choose $H \in \mathfrak{a}$ in the complement of $\bigcup_{\lambda \in \Phi} \ker \lambda$. Since $\lambda(H)$ is nonzero for all $\lambda \in \Phi$, we can write Φ as the disjoint union of the sets Φ^+ and Φ^- of roots λ such that $\lambda(H) > 0$ and $\lambda(H) < 0$, respectively. Due to exercise 7.4.2, $\Phi^- = -\Phi^+$.

A Lie algebra $\mathfrak{n}$ is said to be *nilpotent* if the sequence

$$\mathfrak{n}_{(0)} = \mathfrak{n}, \ \mathfrak{n}_{(1)} = [\mathfrak{n}, \mathfrak{n}], \ldots, \mathfrak{n}_{(n+1)} = [\mathfrak{n}, \mathfrak{n}_{(n)}], \ldots$$

eventually terminates at 0, that is, $\mathfrak{n}_{(n)} = \{0\}$ for some n. The Lie algebra is said to be *solvable* if the sequence

$$\mathfrak{n}^{(0)} = \mathfrak{n}, \ \mathfrak{n}^{(1)} = [\mathfrak{n}, \mathfrak{n}], \ldots, \mathfrak{n}^{(n+1)} = [\mathfrak{n}^{(n)}, \mathfrak{n}^{(n)}], \ldots$$

eventually terminates at 0.

Exercise 7.4.5 Show that $\mathfrak{n} := \bigoplus_{\lambda \in \Phi^+} \mathfrak{g}_\lambda$ is a nilpotent algebra. Also show that $\mathfrak{m} \oplus \mathfrak{a} \oplus \mathfrak{n}$ is solvable.

Proposition 7.4.6 (Iwasawa decomposition for $\mathfrak{g}$) Let $\mathfrak{g} \subset \mathfrak{gl}_n(\mathbb{C})$ be a θ-stable semisimple Lie algebra. Then, as a vector space, $\mathfrak{g}$ has the direct sum decomposition $\mathfrak{g} = \mathfrak{k} \oplus \mathfrak{a} \oplus \mathfrak{n}$. Moreover, there exists a basis $X_1, \ldots, X_n$ of $\mathfrak{g}$ such that the matrix representing $\mathrm{ad}(X)$, $X \in \mathfrak{g}$, has the following properties:

1. It is skew-Hermitian if $X \in \mathfrak{k}$.

2. It is diagonal with real entries if $X \in \mathfrak{a}$.
3. It is upper triangular with 0s on the diagonal, if $X \in \mathfrak{n}$.

Proof. We first show that the sum is direct. It is clear that $\mathfrak{a} \cap \mathfrak{n} = \{0\}$, so we need to check that any $X \in \mathfrak{k} \cap (\mathfrak{a} \oplus \mathfrak{n})$ is 0. Such an X must be fixed by θ and must have the form $X = X_0 + \sum_{\lambda \in \Phi^+} X_\lambda$, $X_0 \in \mathfrak{g}_0$ and $X_\lambda \in \mathfrak{g}_\lambda$. Since $\theta(X) = X$, we can write $2X_0 + \sum_{\lambda \in \Phi^+} X_\lambda - \sum_{\lambda \in \Phi^+} \theta(X_\lambda) = 0$. Therefore, as the root space decomposition is a direct sum, we conclude that each component must be zero, so $X = 0$.

All that is left to do is show that $\bigoplus_{\lambda \in \Phi^-} \mathfrak{g}_\lambda$ is contained in $\mathfrak{k} \oplus \mathfrak{a} \oplus \mathfrak{n}$. But this is a consequence of the fact that any $X \in \mathfrak{g}_{-\lambda}$, $\lambda \in \Phi^+$, can be represented as $X = X + \theta(X) - \theta(X)$, where $X + \theta(X) \in \mathfrak{k}$, $\theta(X) \in \mathfrak{g}_\lambda$.

Take an $\langle \cdot, \cdot \rangle_\kappa$-orthonormal basis $X_1, \ldots, X_n$ of $\mathfrak{g}$ adapted to the restricted root space decomposition, having the property that if $X_i \in g_\lambda$ and $X_j \in g_\mu$ with $i < j$ then $\lambda \geq \mu$. For $X \in \mathfrak{k}$, we have by exercise 7.2.4

$$\mathrm{ad}(X)^\dagger = -\mathrm{ad}(\theta(X)) = -\mathrm{ad}(X),$$

which implies that the matrix representing $\mathrm{ad}(X)$ is skew-Hermitian. The remaining properties are immediate consequences of the restricted root space decomposition. □

Unless specified otherwise, we assume from now on that $\mathfrak{g}$ is semisimple and θ-stable, and that $\mathfrak{a}$ is an $\mathbb{R}$-split Cartan subalgebra of $\mathfrak{a}$. The dimension of $\mathfrak{a}$ is called the *real rank* of $\mathfrak{g}$. This definition seems to depend on the choice of $\mathfrak{a}$ in $\mathfrak{p}$. It turns out, however, that any two such subalgebras are conjugate by an element of K, so their dimensions are the same. This is proved by the next proposition.

Proposition 7.4.7 If $\mathfrak{a}_1$ is another choice of maximal abelian subspace of $\mathfrak{p}$, we can find $k \in K$ (the subgroup of G pointwise fixed by Θ) such that $\mathfrak{a}_1 = k\mathfrak{a}k^{-1}$. Since any $X \in \mathfrak{p}$ is contained in a maximal abelian subspace, it follows that

$$\mathfrak{p} = \bigcup_{k \in K} k\mathfrak{a}k^{-1}.$$

Proof. We choose $H \in \mathfrak{a}$ (resp., $H_1 \in \mathfrak{a}_1$) such that no root in $\Phi(\mathfrak{a}, \mathfrak{g})$ (resp., $\Phi(\mathfrak{a}_1, \mathfrak{g})$) vanishes on H (resp., H_1). We now define a smooth function on K by $f(k) := \langle \mathrm{Ad}(k)H, H_1 \rangle_\kappa$. Since K is compact, there must be some k_0 that minimizes f. We claim that $k_0^{-1}\mathfrak{a}_1 k_0 = \mathfrak{a}$. To see this, it suffices to show that

$k_0^{-1} \mathfrak{a}_1 k_0 \subset \mathfrak{g}_0 \cap \mathfrak{p} = \mathfrak{a}$, and equality will be a consequence of the maximality of $\mathfrak{a}_1$. Inclusion in $\mathfrak{p}$ is clear since K normalizes $\mathfrak{p}$.

We now show inclusion in $\mathfrak{g}_0$. Since k_0 is a critical point of f, for any $X \in \mathfrak{k}$ we have

$$
\begin{aligned}
0 = \left. \frac{d}{dt} \right|_{t=0} f(e^{tX} k_0) &= \left. \frac{d}{dt} \right|_{t=0} \langle \mathrm{Ad}(e^{tX} k_0) H, H_1 \rangle_\kappa \\
&= \left. \frac{d}{dt} \right|_{t=0} \langle \mathrm{Ad}(e^{tX}) \mathrm{Ad}(k_0) H, H_1 \rangle_\kappa \\
&= \langle \mathrm{ad}(X)(\mathrm{Ad}(k_0) H), H_1 \rangle_\kappa \\
&= \langle -\mathrm{ad}(\mathrm{Ad}(k_0) H) X, H_1 \rangle_\kappa \\
&= \langle X, \mathrm{ad}(\theta (\mathrm{Ad}(k_0) H)) H_1 \rangle_\kappa \\
&= \langle X, -\mathrm{ad}(\mathrm{Ad}(k_0) H) H_1 \rangle_\kappa \\
&= \langle X, [H_1, \mathrm{Ad}(k_0) H] \rangle_\kappa.
\end{aligned}
$$

Therefore, as $\mathfrak{k}$ and $\mathfrak{p}$ are orthogonal and $[H_1, \mathrm{Ad}(k_0) H] \subset [\mathfrak{p}, \mathfrak{p}] \subset \mathfrak{k}$, we have $\langle X, [H_1, \mathrm{Ad}(k_0) H] \rangle_\kappa = 0$ for all $X \in \mathfrak{g}$, so $[H_1, \mathrm{Ad}(k_0) H] = 0$. But no root of $(\mathfrak{a}_1, \mathfrak{g})$ vanishes on H_1, so $[\mathfrak{a}_1, \mathrm{Ad}(k_0) H] = 0$ or, equivalently, $[k_0^{-1} \mathfrak{a}_1 k_0, H] = 0$. By the same reason, as no root of $(\mathfrak{a}, \mathfrak{g})$ vanishes on H, we conclude that $k_0^{-1} \mathfrak{a}_1 k_0 \subset \mathfrak{g}_0$. □

Theorem 7.4.8 (KAK decomposition) Let G be a closed subgroup of $GL(n, \mathbb{C})$ and suppose that it is the common set of zeros of some set of polynomials, with real coefficients, in the real and imaginary parts of the matrix entries, and let $\mathfrak{g}$ be its Lie algebra. Suppose that G is semisimple and stable under the Cartan involution. Then $G = KAK$, where A is the connected abelian subgroup of G with Lie algebra $\mathfrak{a}$. More precisely, for each $g \in G$ we can find (not necessarily unique) $a \in A$ and $k_1, k_2 \in K$ such that $g = k_1 a k_2$.

Proof. The Cartan decomposition gives $g = k \exp X$ and by the previous proposition, $\exp X = k_2^{-1} \exp H k_2$, so $g = k k_2^{-1} \exp H k_2$. □

7.5 Root Spaces for the Classical Groups

We describe now the restricted root space decomposition in some particular cases. As a first class of examples let us take $\mathfrak{g} = \mathfrak{sl}(n, \mathbb{F})$, where $\mathbb{F} = \mathbb{R}, \mathbb{C}$, or $\mathbb{H}$ (recall that $\mathfrak{sl}(n, \mathbb{H})$ is isomorphic to $\mathfrak{su}^*(2n)$). In this case $\mathfrak{k}$ is the subalgebra of skew-Hermitian $n \times n$ matrices (skew-symmetric if $\mathbb{F} = \mathbb{R}$) and

$\mathfrak{p}$ is the subspace consisting of Hermitian $n \times n$ matrices of trace 0. Denote by $\mathfrak{a}$ the abelian algebra consisting of real diagonal matrices of trace 0. Then $\mathfrak{a} \subset \mathfrak{p}$, and by a simple computation we see that the subalgebra consisting of all matrices in $\mathfrak{p}$ that commute with each element of $\mathfrak{a}$ is $\mathfrak{a}$ itself. Therefore, $\mathfrak{a}$ is an $\mathbb{R}$-split Cartan subalgebra of $\mathfrak{sl}(n, \mathbb{F})$ and the real rank of $\mathfrak{sl}(n, \mathbb{F})$ is $n - 1$.

For each i, $1 \le i \le n$, define $f_i \in \mathfrak{a}^*$ by $f_i(\text{diag}[a_1, \ldots, a_n]) = a_i$ and set $\alpha_{ij} := f_j - f_i$. Define $\mathfrak{g}_{ij} := \mathbb{F} E_{ij}$, where E_{ij} is the matrix with 1 at the (i, j) entry and 0 at the other positions. Notice that $\dim \mathfrak{g}_{ij} = 1, 2, 4$ for $\mathbb{R}, \mathbb{C}, \mathbb{H}$. One can easily check that

$$\mathfrak{sl}(n, \mathbb{F}) = \mathfrak{g}_0 \oplus \bigoplus_{i \ne j} \mathfrak{g}_{ij},$$

where $\mathfrak{g}_0$ is the subalgebra of all diagonal matrices of trace 0. We can write $\mathfrak{g}_0 = \mathfrak{m} \oplus \mathfrak{a}$, where $\mathfrak{m}$ is trivial if $\mathbb{F} = \mathbb{R}$, $\mathfrak{m}$ is the subalgebra of all diagonal matrices with imaginary entries for $\mathbb{F} = \mathbb{C}$, and $\mathfrak{m}$ is the direct sum of n copies of $\mathfrak{su}(2)$ if $\mathbb{F} = \mathbb{H}$. (Notice that $\mathfrak{su}(2)$ is isomorphic to the algebra of imaginary quaternions.)

Before considering other examples, we make some general remarks about the way the root space decomposition of a subalgebra is placed inside a larger Lie algebra.

Proposition 7.5.1 Let $\mathfrak{g}_1$ be a semisimple subalgebra of a semisimple Lie algebra $\mathfrak{g}$, and let $\mathfrak{a}_1$ be a Cartan subalgebra of $\mathfrak{g}_1$ contained in a Cartan subalgebra $\mathfrak{a}$ of $\mathfrak{g}$. Then each root of $(\mathfrak{g}_1, \mathfrak{a}_1)$ is the restriction to $\mathfrak{a}_1$ of a root of $(\mathfrak{g}, \mathfrak{a})$. The root space $(\mathfrak{g}_1)_\alpha$ is the intersection of $\mathfrak{g}_1$ with the direct sum of the $\mathfrak{g}_\beta$ for all β such that $\beta|_{\mathfrak{a}_1} = \alpha$.

Proof. For each $\gamma \in \mathfrak{a}_1^*$ we denote by V_γ the direct sum of root spaces $\mathfrak{g}_\beta$ for $\mathfrak{a}$ such that $\beta|_{\mathfrak{a}_1} = \gamma$. Then $\mathfrak{g}$ is the direct sum of a finite number of V_γ. For each root α of $(\mathfrak{g}_1, \mathfrak{a}_1)$ and each X in the root space $(\mathfrak{g}_1)_\alpha$ we can uniquely write $X = \sum_\gamma X_\gamma$, $X_\gamma \subset V_\gamma$. Therefore, for all $H \in \mathfrak{a}_1$,

$$\sum_\gamma \alpha(H) X_\gamma = \alpha(H) X = \text{ad}(H) X = \sum_\gamma \text{ad}(H) X_\gamma = \sum_\gamma \gamma(H) X_\gamma,$$

and we conclude that whenever X_γ is not 0, $\gamma(H) = \alpha(H)$ for all $H \in \mathfrak{a}_1$. In other words, $(\mathfrak{g}_1)_\alpha$ is contained in V_α, so $(\mathfrak{g}_1)_\alpha = V_\alpha \cap \mathfrak{g}_1$. $\square$

We now use the previous lemma to compute the restricted root space decomposition for $\mathfrak{g}_\mathbb{F} = \mathfrak{su}(p, q)_\mathbb{F}$, $p \ge q$. Recall that $\mathfrak{g}_\mathbb{F} = \mathfrak{so}(p, q)$, $\mathfrak{su}(p, q)$, or

$\mathfrak{sp}(p, q)$ for $\mathbb{F} = \mathbb{R}, \mathbb{C}$, or $\mathbb{H}$, respectively. It was shown in an earlier exercise that

$$\mathfrak{g}_\mathbb{F} = \{X \in \mathfrak{sl}(p + q, \mathbb{F}) \mid X^*L + LX = 0\},$$

where

$$L := \begin{pmatrix} 0 & 0 & F \\ 0 & I & 0 \\ F & 0 & 0 \end{pmatrix}, \qquad F := \begin{pmatrix} 0 & \cdots & 1 \\ \vdots & & \vdots \\ 1 & \cdots & 0 \end{pmatrix}.$$

The middle block is the identity matrix of size $p - q$ and F is a square block of size q with 1s on the SW-NE diagonal and 0s everywhere else. The point of writing the algebra in this form, rather than taking the representation for which L is diagonal, is that we can now show that the intersection of $\mathfrak{g}_\mathbb{F}$ with the diagonal Cartan subalgebra of $\mathfrak{sl}(p + q, \mathbb{F})$ is a Cartan subalgebra for $\mathfrak{g}_\mathbb{F}$. In fact, a simple computation shows the following. The algebra $\mathfrak{g}_\mathbb{F}$ is still θ-stable, and elements of $\mathfrak{p} = \{X \in \mathfrak{g}_\mathbb{F} \mid X^* = X\}$ have the form

$$X = \begin{pmatrix} A & B & C \\ B^* & 0 & -B^*F \\ C^* & -FB & -FAF \end{pmatrix},$$

where $A^* = A$ and $C^* = -FCF$. If we set $\mathfrak{a} = \mathfrak{p} \cap \{\text{diagonal matrices}\}$, then $\mathfrak{a}$ is the set of diagonal matrices of the form

$$\mathrm{diag}[x_1, \ldots, x_q, 0, \ldots, 0, -x_q, \ldots, -x_1],$$

for $x_i \in \mathbb{R}$. Another simple computation shows that any element of $\mathfrak{g}_\mathbb{F}$ that commutes with all elements of $\mathfrak{a}$ must take the form

$$\mathrm{diag}[\lambda_1, \ldots, \lambda_q, E, -\bar{\lambda}_q, \ldots, -\bar{\lambda}_1]$$

where $\lambda_i \in \mathbb{F}$ and E is a square block of size q such that $E^* = -E$. Of all such matrices, the only ones that lie in $\mathfrak{p}$ are the ones already in $\mathfrak{a}$, so $\mathfrak{a}$ is indeed an $\mathbb{R}$-split Cartan subalgebra. In particular, the real rank is q.

The roots and root spaces can now be obtained with the help of the previous proposition. Define as before f_i to be the linear map on the space of diagonal matrices in $\mathfrak{sl}(p + q, \mathbb{F})$ that evaluates the ith diagonal entry. The roots of $\mathfrak{g}_\mathbb{F}$ are then the restrictions of $f_i - f_j$ to $\mathfrak{a}$. Notice that if α is the restriction of $f_i - f_j$ for $1 \leq i < j \leq p + q$, then $\mathfrak{g}_\alpha$ is contained in the subspace of upper-triangular matrices with 0s on the diagonal, and $\mathfrak{g}_{-\alpha} = \theta(\mathfrak{g}_\alpha)$ is contained in the subspace of lower-triangular matrices with 0s on the diagonal.

Any root of $\mathfrak{g}_{\mathbb{F}}$ is, therefore, one of the following:

$$\pm f_i, \pm 2 f_i, \pm(f_i + f_j),\ 1 \leq i, j \leq q, i \neq j,\ \text{and}\ \pm(f_i - f_j),\ 1 \leq i < j \leq q.$$

In order to get the corresponding roots spaces, it is convenient to observe that if C is a square matrix of size q, then $F C^t F$ is the matrix obtained from C by "flipping" it about the SW-NE diagonal. The root space $\mathfrak{g}_{f_i}$ consists of all matrices of the form

$$\begin{pmatrix} 0 & B & 0 \\ 0 & 0 & -B^* F \\ 0 & 0 & 0 \end{pmatrix},$$

where B has entries 0 in all places except the ith row, the entries of which are arbitrary elements of $\mathbb{F}$. The dimension of $\mathfrak{g}_{f_i}$ is $(p - q) \dim_{\mathbb{R}} \mathbb{F}$. The elements of $\mathfrak{g}_{2 f_i}$ have the form

$$\begin{pmatrix} 0 & 0 & C \\ 0 & 0 & 0 \\ 0 & 0 & 0 \end{pmatrix}, \quad C = F \operatorname{diag}[0, \ldots, 0, x_i, 0 \ldots, 0] F,$$

where x_i is real. Therefore $\mathfrak{g}_{2 f_i}$ has real dimension 1. Elements of $\mathfrak{g}_{f_i + f_j}$ take the form

$$\begin{pmatrix} 0 & 0 & M F \\ 0 & 0 & 0 \\ 0 & 0 & 0 \end{pmatrix},$$

where $M = (m_{rs})_{1 \leq r, s \leq q}$ has all entries 0 except for m_{ij} (which is an arbitrary element of $\mathbb{F}$) and $m_{ji} = -\bar{m}_{ij}$. The dimension of $\mathfrak{g}_{f_i + f_j}$ is $\dim_{\mathbb{R}} \mathbb{F}$. Elements of $\mathfrak{g}_{f_i - f_j}$ have the form

$$\begin{pmatrix} A & 0 & 0 \\ 0 & 0 & 0 \\ 0 & 0 & -F A^* F \end{pmatrix},$$

where the square block A of size q has all entries 0 except for a_{ij}, which can be any element in $\mathbb{F}$. Thus, $\dim_{\mathbb{R}} \mathfrak{g}_{f_i - f_j} = \dim_{\mathbb{R}} \mathbb{F}$.

Exercise 7.5.2 Obtain the real ranks and the restricted root space decompositions for $\mathfrak{sp}(2n, \mathbb{R})$ and $\mathfrak{sp}(2n, \mathbb{C})$. It may be convenient to use the representation $\mathfrak{sp}(2n, \mathbb{F}) = \{X \in \mathfrak{sl}(2n, \mathbb{F}) \mid X^t L + L X = 0\}$, where

$$L = \begin{pmatrix} 0 & F \\ -F & 0 \end{pmatrix}.$$

Table 7.2. *Real ranks of the classical*
noncompact real Lie algebras

$\mathfrak{g}$	Real rank
$\mathfrak{su}(p,q),\ p \geq q$	q
$\mathfrak{so}(p,q),\ p \geq q$	q
$\mathfrak{sp}(p,q),\ p \geq q$	q
$\mathfrak{sp}(2n,\mathbb{R})$	n
$\mathfrak{so}^*(2n)$	$\left[\frac{n}{2}\right]$
$\mathfrak{sl}(n,\mathbb{R})$	$n-1$
$\mathfrak{su}^*(2n)$	$n-1$

Table 7.2 shows the real ranks of the classical noncompact real Lie algebras.

7.6 The Iwasawa Decomposition

We have seen in the previous section that if $\mathfrak{g}$ is a real semisimple Lie algebra stable under the Cartan involution θ and $\mathfrak{k}$ is the eigenspace of θ associated to the eigenvalue 1, then $\mathfrak{g}$ decomposes as a direct sum of subspaces $\mathfrak{g} = \mathfrak{k} \oplus \mathfrak{a} \oplus \mathfrak{n}$, and each subspace is a subalgebra. Recall that $\mathfrak{a}$ is a maximal abelian subalgebra of $\mathfrak{p}$ (the eigenspace of θ associated to the eigenvalue -1) and $\mathfrak{n}$ is a nilpotent subalgebra.

We prove in this section a corresponding decomposition for a Lie group with Lie algebra $\mathfrak{g}$, called the *Iwasawa decomposition*.

Exercise 7.6.1 Set $G = SL(n,\mathbb{C})$, $K = SU(n)$. Let A be the abelian subgroup of positive diagonal matrices, and let N be the upper-triangular group with 1s on the diagonal. Use the Gram–Schmidt orthogonalization process to prove that the map

$$K \times A \times N \to G, \quad (k,a,n) \mapsto kan$$

is a bijection. (Note: If $\{e_1, \ldots, e_n\}$ denotes the standard basis of $\mathbb{C}^n$ and $g \in G$, apply the orthogonalization process to $\{ge_1, \ldots, ge_n\}$. You'll get an orthonormal basis $\{u_1, \ldots, u_n\}$ such that the linear span of $\{ge_1, \ldots, ge_i\}$ is equal to the linear span of $\{u_1, \ldots, u_i\}$ and $u_i \in \mathbb{R}^+(ge_i) + \text{span}\{v_1, \ldots, v_{i-1}\}$ for $i = 1, \ldots, n$. Define k by $k^{-1}u_i = e_i$, for each i.)

Lemma 7.6.2 Let L be a Lie group with Lie algebra $\mathfrak{l}$, and suppose that $\mathfrak{l}$ has a vector space direct sum decomposition $\mathfrak{l} = \mathfrak{u} \oplus \mathfrak{v}$, where $\mathfrak{u}$ and $\mathfrak{v}$ are Lie subalgebras. Let U and V denote subgroups of L with Lie algebras $\mathfrak{u}$ and $\mathfrak{v}$, respectively. Then the map

$$M : U \times V \to L, \quad (u,v) \mapsto uv$$

is everywhere regular (i.e., its differential is everywhere a surjective linear map).

Proof. Let $(X, Y) \in T_{(u,v)}(U \times V)$ be such that $X = \frac{d}{dt}|_{t=0} \exp(t X_0)u$ and $Y = \frac{d}{dt}|_{t=0} \exp(t Y_0)v$, where $X_0 \in \mathfrak{u}$ and $Y_0 \in \mathfrak{v}$. Then

$$DM_{(u,v)}(X, Y) = \frac{d}{dt}\bigg|_{t=0} \exp(t X_0)u \exp(t Y_0)v = (X_0 + \mathrm{Ad}(u)(Y_0))uv.$$

If $DM_{(u,v)}(X, Y) = 0$, then $X_0 + \mathrm{Ad}(u)(Y_0) = 0$, and hence $Y_0 = -\mathrm{Ad}(u^{-1})$ (X_0) lies in $\mathfrak{u}$. Therefore $Y_0 = 0$, $X_0 = 0$, proving that M is everywhere an immersion. Since the dimensions of the domain and range are the same, M is everywhere regular.   □

Exercise 7.6.3 Show that the exponential map is a diffeomorphism from the subalgebra of $\mathfrak{gl}(n, \mathbb{R})$ of upper-triangular matrices with 0s on the diagonal (resp., the real diagonal matrices) onto the subgroup of $GL(n, \mathbb{R})$ of upper-triangular matrices with 1s on the diagonal (resp., the positive diagonal matrices). (Hint: For $n = 1$ the result is clear. For $n \geq 2$, write an arbitrary upper-triangular matrix with 1s on the diagonal in the form

$$U = \begin{pmatrix} 1 & x \\ 0 & U_1 \end{pmatrix},$$

where U_1 is an upper-triangular square block of size $n - 1$ with 1s on the diagonal and x is a row vector of dimension $n - 1$. Use induction to argue that $U_1 = \exp N_1$, for some nilpotent upper-triangular N_1 of size $n - 1$, and show that $U = \exp N$, where

$$N = \begin{pmatrix} 0 & x P^{-1} \\ 0 & N_1 \end{pmatrix}$$

and $P = I + \frac{1}{2}N_1 + \cdots + \frac{1}{n!}N_1^{n-1}$.)

Theorem 7.6.4 (Iwasawa decomposition) Let $G \subset GL(n, \mathbb{C})$ be a Θ-stable semisimple Lie group, let $\mathfrak{g} = \mathfrak{k} \oplus \mathfrak{a} \oplus \mathfrak{n}$ be the Iwasawa decomposition of the Lie algebra $\mathfrak{g}$ of G, and let A and N be the connected subgroups of G with Lie algebras $\mathfrak{a}$ and $\mathfrak{n}$, respectively. Then the multiplication map

$$K \times A \times N \to G, \quad (k, a, n) \mapsto kan$$

is a diffeomorphism onto G. The groups A and N are simply connected.

Proof. We first show the result for $\check{G} = \mathrm{Ad}(G)$, regarded as the closed subgroup $\mathrm{Aut}(\mathfrak{g})^{\circ} \subset GL(\mathfrak{g})$, and then lift the decomposition to G.

Let $\check{K} = \mathrm{Ad}(K)$, $\check{A} = \mathrm{Ad}(A)$, and $\check{N} = \mathrm{Ad}(N)$. Proposition 7.4.6 implies that for some basis of $\mathfrak{g}$, the matrices representing elements of $\check{K}$ are rotation matrices, those representing elements of $\check{A}$ are positive diagonal matrices, and those for $\check{N}$ are upper triangular with 1s on the diagonal. The group K, hence also $\check{K}$, is compact.

The groups A and N are closed subgroups of G and $\check{A}$ and $\check{N}$ are closed subgroups of $\check{G}$. In fact, let $\bar{A}$ and $\bar{N}$ be the closures in G, which are connected Lie groups with Lie algebras denoted $\bar{\mathfrak{a}}$ and $\bar{\mathfrak{n}}$. Since $\mathrm{Ad}(A)$ (resp., $\mathrm{Ad}(N)$) is contained in the space of diagonal (resp., upper-triangular, with 1s on the diagonal) matrices, so is $\mathrm{Ad}(\bar{A})$ (resp., $\mathrm{Ad}(\bar{N})$). It follows that $\mathrm{ad}(\bar{\mathfrak{a}})$ (resp., $\mathrm{ad}(\bar{\mathfrak{n}})$) is contained in the space of diagonal (resp., upper-triangular, with 0s on the diagonal) matrices. Therefore, $\mathrm{ad}(\bar{\mathfrak{a}}) \cap \mathrm{ad}(\bar{\mathfrak{n}}) = 0$, whence $\bar{\mathfrak{a}} \cap \bar{\mathfrak{n}} = 0$. Moreover, $\mathfrak{k} \cap (\bar{\mathfrak{a}} \oplus \bar{\mathfrak{n}}) = 0$, since for any X in the intersection, $\mathrm{ad}(X)$ is upper triangular with real diagonal entries such that $\mathrm{ad}(X)^{\dagger} = -\mathrm{ad}(X)$. Therefore, $\mathfrak{g}$ is the direct sum of $\mathfrak{k}$, $\bar{\mathfrak{a}}$, and $\bar{\mathfrak{n}}$. It follows that $\mathfrak{a} = \bar{\mathfrak{a}}$ and $\mathfrak{n} = \bar{\mathfrak{n}}$ and $A = \bar{A}$, $N = \bar{N}$. Since G modulo the finite center Z is diffeomorphic to its (closed) image in $GL(\mathfrak{g})$, the groups $\check{A}$, $\check{N}$, $\check{K}$ are also closed.

The map from $\check{A} \times \check{N}$ into $\check{G}$ given by $(\check{a}, \check{n}) \mapsto \check{a}\check{n}$ is clearly one-to-one since $\check{a}$ can be recovered from the diagonal entries of the product. Moreover, $\check{A}\check{N}$ is a group, since $\check{N}$ is normalized by $\check{A}$. It is, in fact, a closed subgroup of $\check{G}$: if $\check{a}_m\check{n}_m \to \check{x}$ and $\check{a}$ is the diagonal matrix with the same diagonal entries as $\check{x}$, then $\check{a}_m$ must converge to $\check{a} \in \check{A}$ and $\check{n}_m$ must converge to $\check{n} = \check{a}^{-1}\check{x} \in \check{N}$, so $\check{x}$ lies in $\check{A}\check{N}$.

Note that the Lie algebra of $\check{A}\check{N}$ is $\check{\mathfrak{a}} \oplus \check{\mathfrak{n}}$ and that the multiplication map $\check{A} \times \check{N} \to \check{A}\check{N}$ is a bijection, so by lemma 7.6.2, $\check{A} \times \check{N}$ and $\check{A}\check{N}$ are diffeomorphic.

The image of the multiplication map $\check{K} \times \check{A}\check{N} \to \check{G}$ is the product of a compact set and a closed set, so it is closed. It is also open since by the previous lemma the map is everywhere regular. But $\check{G}$ is connected, so the multiplication map

$$\check{K} \times \check{A} \times \check{N} \to \check{G}$$

is surjective. The map is also injective since any rotation matrix with positive eigenvalues must be the identity. But a smooth everywhere regular bijection must be a diffeomorphism.

The map $\mathrm{Ad} : G \to \check{G}$ is a finite covering sending A onto $\check{A}$ and N onto $\check{N}$, and the kernel of Ad is contained in $K . \check{A}$ is simply connected by the Cartan decomposition, since it is diffeomorphic to its Lie algebra via the exponential map. $\check{N}$ is also simply connected, due to exercise 7.6.3 and the fact that it is a subgroup of the group of upper-triangular matrices with 0s on the diagonal. Therefore, $\mathrm{Ad} : A \to \check{A}$ and $\mathrm{Ad} : N \to \check{N}$ are diffeomorphisms.

The multiplication map $K \times A \times N \to G$ is a smooth regular map, due to the previous lemma. It is also surjective: Given $g \in G$, write $\check{g} = \check{k}\check{a}\check{n}$ and let a and n be the unique elements in A and N, respectively, such that $\mathrm{Ad}(a) = \check{a}$

and $\mathrm{Ad}(n) = \check{n}$. Then $\mathrm{Ad}(g(kan)^{-1}) = I$, so $g(kan)^{-1}$ belongs to the center of G. But the center is contained in K, so $g = k_1 an$ for some $k_1 \in K$.

The proof will be complete if we show that the foregoing multiplication map is injective. If $kan = k_1 a_1 n_1$, then $k' := (k_1)^{-1} k = a_1 a^{-1}(a n_1 n^{-1} a^{-1}) \in AN$, so $\mathrm{Ad}(k')$ is upper triangular with positive diagonal entries. Since it is also a rotation matrix we must have k' in the kernel of Ad. But Ad is injective on AN, so $k' = e$. Therefore $k = k_1$. It also follows that $a = a_1$ and $n = n_1$, concluding the proof. $\square$

7.7 Representations of $\mathfrak{sl}(2, \mathbb{C})$

Before we can derive further information about the structure of semisimple Lie algebras, it is necessary to describe the finite-dimensional linear representations of $\mathfrak{sl}(2, \mathbb{C})$.

We begin with a couple of results about linear representations of general semisimple Lie algebras. Recall from section 3.8 that a representation is said to be completely reducible if it decomposes as a direct sum of irreducible subrepresentations.

Theorem 7.7.1 (Complete reducibility) Let $\mathfrak{g}$ be a semisimple Lie algebra, and let $\rho : \mathfrak{g} \to \mathfrak{gl}(W)$ be a linear representation on a finite-dimensional complex vector space W. Then ρ is completely reducible.

Proof. We give the proof only for $\mathfrak{sl}(2, \mathbb{R})$. It will be apparent, however, that the argument is much more general. The main argument that will be used is known as Weyl's *unitary trick*.

Before specializing to $\mathfrak{sl}(2, \mathbb{R})$ assume more generally that $\mathfrak{g}$ is a real, θ-stable subalgebra of the general linear algebra. Denote by $\mathfrak{g}_{\mathbb{C}} = \mathfrak{g} \otimes \mathbb{C}$ the complexification of $\mathfrak{g}$ (for example, if $\mathfrak{g} = \mathfrak{sl}(2, \mathbb{R})$, then $\mathfrak{g}_{\mathbb{C}} = \mathfrak{sl}(2, \mathbb{C})$) and extend ρ to $\mathfrak{g}_{\mathbb{C}}$ by linearity.

Let $\mathfrak{k} \oplus \mathfrak{p}$ be the Cartan decomposition of $\mathfrak{g}$. Then $\mathfrak{u} := \mathfrak{k} \oplus i\mathfrak{p}$ is also a Lie subalgebra of $\mathfrak{g}_{\mathbb{C}}$, as one easily sees by using the bracket relations

$$[\mathfrak{k}, \mathfrak{k}] \subset \mathfrak{k}, \quad [\mathfrak{k}, \mathfrak{p}] \subset \mathfrak{p}, \quad [\mathfrak{p}, \mathfrak{p}] \subset \mathfrak{k}.$$

(If $\mathfrak{g} = \mathfrak{sl}(2, \mathbb{R})$, then $\mathfrak{u} = \mathfrak{su}(2)$.) Notice that each vector in $\mathfrak{u}$ is fixed by θ, so $\mathfrak{u}$ is the Lie algebra of a subgroup U of the unitary group. In the general case, it can be shown that U is a compact group; in our special case, $U = SU(2)$, which is seen to be compact as follows. Notice that

$$SU(2) = \left\{ \begin{pmatrix} \alpha & \beta \\ -\bar{\beta} & \bar{\alpha} \end{pmatrix} \,\middle|\, \alpha, \beta \in \mathbb{C}, \ |\alpha|^2 + |\beta|^2 = 1 \right\}$$

is homeomorphic to the 3-sphere. It is also simply connected, so the restriction of ρ to $\mathfrak{su}(2)$ exponentiates to a representation of $SU(2)$ on W. (We are using here corollary 3.9.9.)

Suppose now that W has a subspace S that is $\mathfrak{sl}(2, \mathbb{R})$-invariant. By linearity, it is also $\mathfrak{sl}(2, \mathbb{C})$-invariant, and by restriction, $\mathfrak{su}(2)$-invariant. By exponentiating the representation of the latter algebra, we obtain that S is $SU(2)$-stable. Applying proposition 3.8.2, we obtain an $SU(2)$-invariant complement S'. Reversing the argument yields that S' is now $\mathfrak{su}(2)$-invariant, since we can differentiate the group representation to obtain a representation of the algebra. By linearity, S' is invariant under the complexification of $\mathfrak{su}(2)$, which is $\mathfrak{sl}(2, \mathbb{C}) = \mathfrak{su}(2, \mathbb{R}) \oplus i\,\mathfrak{su}(2, \mathbb{R})$, whence it is also invariant under $\mathfrak{sl}(2, \mathbb{R})$, by restriction.   □

Corollary 7.7.2 Let $R : G \to GL(W)$ be a continuous representation of a connected semisimple Lie group on a finite-dimensional complex vector space W. Then R is completely reducible.

Proof. This is immediate from the fact that $\rho = DR_e : \mathfrak{g} \to \mathfrak{gl}(W)$ is completely reducible.   □

We choose a basis of $\mathfrak{sl}(2, \mathbb{R})$ consisting of

$$h = \begin{pmatrix} 1 & 0 \\ 0 & -1 \end{pmatrix}, \qquad x = \begin{pmatrix} 0 & 1 \\ 0 & 0 \end{pmatrix}, \qquad y = \begin{pmatrix} 0 & 0 \\ 1 & 0 \end{pmatrix}.$$

Their bracket relations are: $[h, x] = 2x$, $[h, y] = -2y$, $[x, y] = h$.

Lemma 7.7.3 Let $\rho : \mathfrak{sl}(2, \mathbb{R}) \to \mathfrak{gl}(W)$ be a linear representation on a finite-dimensional complex vector space W. Then $\rho(h)$ is diagonalizable.

Proof. Let $\mathfrak{k} \oplus \mathfrak{p}$ denote the Cartan decomposition of $\mathfrak{sl}(2, \mathbb{R})$. Then h spans the one-dimensional Cartan subalgebra $\mathfrak{a} \subset \mathfrak{p}$. We now apply the unitary trick once again. Notice that $i\mathfrak{a} \subset \mathfrak{su}(2)$ is the Lie algebra of a compact abelian subgroup T of $SL(2, \mathbb{C})$. By corollary 3.8.4, the representation of T obtained from the original representation of $\mathfrak{sl}(2, \mathbb{R})$, using the same method applied in the proof of the previous theorem, decomposes as a direct sum of one-dimensional representations. Each one-dimensional subrepresentation is also $i\mathfrak{a}$-invariant, therefore also $\mathfrak{a}$-invariant.   □

Let $\rho : \mathfrak{sl}(2, \mathbb{C}) \to \mathfrak{gl}(W)$ be a linear representation, where W is a finite-dimensional complex vector space. By the previous lemma, $\rho(h)$ is diagonalizable.

Therefore, W decomposes as a direct sum of eigenspaces

$$W_\lambda = \{w \in W \mid \rho(h)w = \lambda w\}.$$

Lemma 7.7.4 If $w \in W_\lambda$, then $\rho(x)w \in W_{\lambda+2}$ and $\rho(y)w \in W_{\lambda-2}$.

Proof. By the bracket relations, we have

$$\rho(h)\rho(x)w = \rho([h, x])w + \rho(x)\rho(h)w$$
$$= 2\rho(x)w + \lambda\rho(x)w = (\lambda + 2)\rho(x)w.$$

The claim for y is shown similarly. $\qquad\square$

Theorem 7.7.5 Let ρ be an irreducible complex-linear representation of $\mathfrak{sl}(2, \mathbb{C})$ on a complex vector space W of dimension m. Then there is in W a basis $\{w_0, \ldots, w_{m-1}\}$ such that, using $n = m - 1$, we have

1. $\rho(h)w_i = (n - 2i)w_i$;
2. $\rho(x)w_0 = 0$;
3. $\rho(y)w_i = w_{i+1}$, with $w_{n+1} = 0$;
4. $\rho(x)w_i = i(n - i + 1)w_{i-1}$, with $w_{-1} = 0$.

Proof. Since $\dim W < \infty$, there must exist $W_\lambda \neq 0$ such that $W_{\lambda+2} = 0$. By the lemma, $\rho(x)w = 0$ for each $w \in W_\lambda$. Let w_0 be a nonzero vector in that W_λ.

Define $w_i = \rho(y)^i w_0$. Then $\rho(h)w_i = (\lambda - 2i)w_i$, by the lemma, so there is a minimum integer n with $\rho(y)^{n+1} w_0 = 0$. Then $w_0, \ldots, w_n$ are linearly independent and the first three of the following properties hold:

1. $\rho(h)w_i = (\lambda - 2i)w_i$;
2. $\rho(x)w_0 = 0$;
3. $\rho(y)w_i = w_{i+1}$, with $w_{n+1} = 0$;
4. $\rho(x)w_i = i(\lambda - i + 1)w_{i-1}$, with $w_{-1} = 0$.

The last equation can be shown by induction, the case $i = 0$ being equation 2. To prove the case $i + 1$ assuming that the equation is true for i, we write

$$\rho(x)w_i = \rho(x)\rho(y)w_i$$
$$= \rho([x, y])w_i + \rho(y)\rho(x)w_i$$
$$= \rho(h)w_i + \rho(y)\rho(x)w_i$$
$$= (\lambda - 2i)w_i + \rho(y)(i(\lambda - i + 1))w_{i-1}$$
$$= (i + 1)(\lambda - i)w_i.$$

By the previous equations, the subspace spanned by the vectors w_i is stable under ρ, and hence must be all of W by irreducibility.

It remains to show that $\lambda = n$. We write

$$\mathrm{Tr}\,\rho(h) = \mathrm{Tr}(\rho(x)\rho(y) - \rho(y)\rho(x)) = 0.$$

Therefore $\sum_{i=0}^{n}(\lambda - 2i) = 0$, whence $\lambda = n$ as claimed.   □

Corollary 7.7.6 Let $\rho : \mathfrak{sl}(2, \mathbb{R}) \to \mathfrak{gl}(V)$ be a finite-dimensional linear representation, where V is now a real vector space. Then V has a basis consisting of eigenvectors of $\rho(h)$, and each eigenvalue is an integer.

Proof. Let $W = V \otimes \mathbb{C}$ and decompose W into irreducible subspaces. By the theorem, the restriction of $\rho(h)$ to each irreducible subspace is diagonalizable with integral eigenvalues. Let $\{w_1, \ldots, w_l\}$ be a basis of W consisting of eigenvectors of $\rho(h)$. Since the eigenvalues are real and $\rho(h)$ maps V into itself, it follows that the real and imaginary parts of each $w_j = u_j + i v_j$ are eigenvectors associated to the same eigenvalue of w_j. Clearly, $\{u_j, v_j \mid j = 1, \ldots, l\}$ spans V, since $\{u_j + i v_j \mid j = 1, \ldots, l\}$ spans $W = V \otimes \mathbb{C}$. Pick now a basis for V from among the vectors u_j, v_j.   □

Exercise 7.7.7 Use the corollary and the proof of the theorem to show that a finite-dimensional real representation $\rho : \mathfrak{sl}(2, \mathbb{R}) \to \mathfrak{gl}(V)$ decomposes as a direct sum of irreducible representations, and in each irreducible subspace there is a basis with respect to which $\rho(h)$, $\rho(x)$, and $\rho(y)$ take the form described in the theorem.

It turns out that the preceding theorem imposes severe constraints on the structure of general semisimple Lie algebras and is a key ingredient used in their classification. We finish this section by indicating how representations of $\mathfrak{sl}(2, \mathbb{C})$ bear on the problem of understanding the structure of a general semisimple Lie algebra. The key remark is that to each restricted root of $(\mathfrak{a}, \mathfrak{g})$ and each nonzero $X \in \mathfrak{g}_\lambda$, there is a copy of $\mathfrak{sl}(2, \mathbb{R})$ in $\mathfrak{g}$ containing X, hence a representation of $\mathfrak{sl}(2, \mathbb{R})$ by restriction of the adjoint representation of $\mathfrak{g}$ to that copy of $\mathfrak{sl}(2, \mathbb{R})$.

In what follows, $\mathfrak{g}$ will denote a θ-stable semisimple Lie subalgebra of $\mathfrak{gl}(n, \mathbb{C})$. Let $\mathfrak{k} \oplus \mathfrak{p}$ be the Cartan decomposition associated to θ and let $\mathfrak{a} \subset \mathfrak{p}$ be an $\mathbb{R}$-split Cartan subalgebra in $\mathfrak{p}$. Recall that $\mathfrak{a} = \mathfrak{p} \cap \mathfrak{g}_0$.

Given any $\mu \in \mathfrak{a}^*$, let $H_\mu \in \mathfrak{a}$ be the element such that $\mu(H) = \langle H, H_\mu \rangle_\kappa$, for each $H \in \mathfrak{a}$. By duality, $\mathfrak{a}^*$ acquires an inner product defined by

$$\langle \mu, \lambda \rangle_\kappa := \langle H_\mu, H_\lambda \rangle_\kappa = \mu(H_\lambda) = \lambda(H_\mu)$$

for all $\mu, \lambda \in \mathfrak{a}^*$. The norm associated to $\langle \cdot, \cdot \rangle_\kappa$ will be denoted simply by $\| \cdot \|$.

Proposition 7.7.8 Let $\lambda \in \Phi(\mathfrak{a}, \mathfrak{g})$ be any root and choose $x_\lambda \in \mathfrak{g}_\lambda$ such that $\|x_\lambda\| = \sqrt{2}/\|\lambda\|$. Define $y_\lambda := -\theta x_\lambda \in \mathfrak{g}_{-\lambda}$ and $h_\lambda := 2H_\lambda/\|\lambda\|^2 \in \mathfrak{a}$. Then

$$[x_\lambda, y_\lambda] = h_\lambda, \ [h_\lambda, x_\lambda] = 2x_\lambda, \ [h_\lambda, y_\lambda] = -2y_\lambda.$$

Therefore, the linear span of $\{x_\lambda, \ y_\lambda, \ h_\lambda\}$ is a Lie subalgebra isomorphic to $\mathfrak{sl}(2, \mathbb{R})$.

Proof. First notice that

$$\theta[x_\lambda, y_\lambda] = [\theta x_\lambda, \theta y_\lambda] = [-y_\lambda, -x_\lambda] = -[x_\lambda, y_\lambda];$$

therefore $[x_\lambda, y_\lambda] \in [\mathfrak{g}_\lambda, \mathfrak{g}_{-\lambda}] \cap \mathfrak{p} \subset \mathfrak{g}_0 \cap \mathfrak{p} = \mathfrak{a}$. Moreover, for each $H \in \mathfrak{a}$,

$$\langle [x_\lambda, y_\lambda], H \rangle_\kappa = \langle \mathrm{ad}(\theta x_\lambda) x_\lambda, H \rangle_\kappa = -\langle \mathrm{ad}(x_\lambda)^\dagger x_\lambda, H \rangle_\kappa = -\langle x_\lambda, \mathrm{ad}(x_\lambda) H \rangle_\kappa$$
$$= \langle x_\lambda, [H, x_\lambda] \rangle_\kappa = \lambda(H) \|x_\lambda\|^2 = \langle 2H_\lambda/\|\lambda\|^2, H \rangle_\kappa;$$

therefore $[x_\lambda, y_\lambda] = h_\lambda$. We also have

$$[h_\lambda, x_\lambda] = \lambda(h_\lambda) x_\lambda = 2(\langle \lambda, \lambda \rangle / \|\lambda\|^2) x_\lambda = 2x_\lambda.$$

A similar computation gives the bracket relation involving y_λ and h_λ. $\square$

It follows from the definition of h_λ that if X is any element of $\mathfrak{g}_\mu$, then

$$[h_\lambda, X] = 2 \frac{\langle \lambda, \mu \rangle_\kappa}{\langle \lambda, \lambda \rangle_\kappa} X.$$

By theorem 7.7.5 and exercise 7.7.7 we conclude, in particular, that $2\langle \lambda, \mu \rangle_\kappa /$ $\langle \lambda, \lambda \rangle_\kappa \in \mathbb{Z}$ and it can be shown that only a small number of integer values can occur. This observation points to the fact that there are severe constraints on the way the set of roots can lie inside $\mathfrak{a}^*$. It will be seen in the next section, in fact, that the set of roots has a high degree of symmetry. The classification of semisimple Lie algebras hinges on the study of the geometry of the set of roots relative to the inner product $\langle \cdot, \cdot \rangle_\kappa$. For details, we refer the reader to [17].

Exercise 7.7.9 Draw a diagram describing the vectors H_α for each root α of $\mathfrak{sl}(3, \mathbb{R})$ inside the subalgebra $\mathfrak{a}$ of diagonal matrices with trace 0. (Identify $\mathfrak{a}$ with $\mathbb{R}^2$ and $\langle \cdot, \cdot \rangle_\kappa$ with the Euclidean metric. There are six roots, which are the vertices of an hexagon.)

7.8 The Weyl Group

We continue to use the notation of the previous section. Let $\alpha \in \mathfrak{a}^*$ be a root of $(\mathfrak{a}, \mathfrak{g})$ and let $H_\alpha \in \mathfrak{a}$ be the dual vector, so that $\alpha = \langle H_\alpha, \cdot \rangle_\kappa$. Denote by $r_\alpha : \mathfrak{a} \to \mathfrak{a}$ the orthogonal reflection in the hyperplane perpendicular to H_α, that is, r_α is the identity map on $H_\alpha^\perp$ and sends H_α to $-H_\alpha$. It is easy to check that

$$r_\alpha(H) := H - 2\frac{\langle H_\alpha, H \rangle_\kappa}{\langle H_\alpha, H_\alpha \rangle_\kappa} H_\alpha$$

for each $H \in \mathfrak{a}$.

The set of reflections $r_\alpha, \alpha \in \Phi(\mathfrak{a}, \mathfrak{g})$, generates a group of orthogonal transformations of $\mathfrak{a}$, which we denote W'.

Proposition 7.8.1 Let G be a semisimple θ-stable subgroup of $GL(n, \mathbb{C})$, and let K be the compact group consisting of fixed points of Θ. Denote by $N_K(\mathfrak{a})$ the subgroup of K that stabilizes $\mathfrak{a}$. Then for each root $\alpha \in \Phi(\mathfrak{a}, \mathfrak{g})$, there exists $k_\alpha \in N_K(\mathfrak{a})$ such that $r_\alpha = \text{Ad}(k_\alpha)|_{\mathfrak{a}}$.

Proof. Define $z_\alpha := \frac{\pi}{2}(x_\alpha - y_\alpha)$ and notice that $\theta z_\alpha = z_\alpha$, so that $z_\alpha \in \mathfrak{k}$. Set $k_\alpha := \exp(z_\alpha) \in K$. We claim that r_α coincides with conjugating by k_α. In order to prove the claim it is sufficient to check that conjugation by k_α sends H_α to $-H_\alpha$ and fixes each H in the orthogonal complement of H_α. It will follow, in particular, that k_α is in the normalizer of $\mathfrak{a}$ in K.

If $H \in H_\alpha^\perp$, then $\lambda(H) = 0$ and $\text{ad}(z_\alpha)^m H = 0$ for each positive integer m. Therefore, $\text{Ad}(k_\alpha)H = \exp(\text{ad}(z_\alpha))H = H$. Set $t_\alpha = \frac{\pi}{2}(x_\alpha + y_\alpha)$. Then,

$$\text{Ad}(k_\alpha)H_\alpha = \exp(\text{ad}(z_\alpha))H_\alpha$$

$$= \sum_{n=0}^{\infty} \frac{1}{n!}\text{ad}(z_\alpha)^n H_\alpha$$

$$= \left(1 - \frac{\pi^2}{2!} + \frac{\pi^4}{4!} - \cdots\right) H_\alpha - \frac{\|\alpha\|^2}{\pi}\left(\pi - \frac{\pi^3}{3!} + \cdots\right) t_\alpha$$

$$= \cos \pi H_\alpha - \|\alpha\|^2 \frac{\sin \pi}{\pi} t_\alpha$$

$$= -H_\alpha. \qquad \square$$

By duality, the reflection r_α is also defined on $\mathfrak{a}^*$, and takes the form

$$r_\alpha \mu = \mu - 2\frac{\langle \mu, \alpha \rangle_\kappa}{\langle \alpha, \alpha \rangle_\kappa} \alpha.$$

Using the same symbol for the reflections on $\mathfrak{a}$ and on $\mathfrak{a}^*$, we have immediately that $r_\alpha H_\mu = H_{r_\alpha(\mu)}$ and $\mu \circ r_\alpha = r_\alpha(\mu)$ for each $\mu \in \mathfrak{a}^*$.

Lemma 7.8.2 Given any root α, r_α permutes the elements of $\Phi(\mathfrak{a}, \mathfrak{g})$. Moreover, $\mathrm{Ad}(k_\alpha^{-1}) : \mathfrak{g} \to \mathfrak{g}$ maps the root space $\mathfrak{g}_\beta$ isomorphically onto $\mathfrak{g}_{r_\alpha(\beta)}$, for each $\beta \in \Phi(\mathfrak{a}, \mathfrak{g})$.

Proof. For each $H \in \mathfrak{a}$ and each $X \in \mathfrak{g}_\beta$ we have

$$
\begin{aligned}
\left[H, \mathrm{Ad}(k_\alpha^{-1})X\right] &= \mathrm{Ad}(k_\alpha^{-1})[\mathrm{Ad}(k_\alpha)H, X] \\
&= \beta(\mathrm{Ad}(k_\alpha)H)\mathrm{Ad}(k_\alpha^{-1})X \\
&= \beta(r_\alpha H)\mathrm{Ad}(k_\alpha^{-1})X \\
&= r_\alpha(\beta)\mathrm{Ad}(k_\alpha^{-1})X.
\end{aligned}
$$

Therefore, $r_\alpha(\beta)$ is also a root and $\mathrm{Ad}(k_\alpha^{-1})$ maps $\mathfrak{g}_\beta$ onto $\mathfrak{g}_{r_\alpha(\beta)}$ as claimed. $\quad\square$

We denote $W = W(\mathfrak{a}, \mathfrak{g}) = N_K(\mathfrak{a})/Z_K(\mathfrak{a})$, where $Z_K(\mathfrak{a})$ is the subgroup of $N_K(\mathfrak{a})$ that centralizes $\mathfrak{a}$. By the proposition, W contains the group W' of orthogonal transformations of $\mathfrak{a}$ generated by the reflections r_α. It turns out that $W = W'$, although this fact won't be needed later. Notice that the proof of the previous lemma shows that W permutes the roots. We call W the *Weyl group* of $(\mathfrak{a}, \mathfrak{g})$. It is shown next that W is a finite group.

Lemma 7.8.3 $N_K(\mathfrak{a})/Z_K(\mathfrak{a})$ is a finite group.

Proof. Since $N_K(\mathfrak{a})$ is a closed subgroup of K, it is a compact group. In order to show that the quotient is a finite group, it suffices to prove that the Lie algebra of $N_K(\mathfrak{a})$ coincides with the Lie algebra of $Z_K(\mathfrak{a})$, since this will imply that the dimension of the quotient is 0. The latter Lie algebra is $\mathfrak{m} = \mathfrak{g}_0 \cap \mathfrak{k}$. The former is the normalizer of $\mathfrak{a}$ in $\mathfrak{k}$, which we denote $N_\mathfrak{k}(\mathfrak{a})$.

To see that $N_\mathfrak{k}(\mathfrak{a}) = \mathfrak{m}$, let $X \in N_\mathfrak{k}(a)$ and write

$$
X = Z + H + \sum_{\alpha \in \Phi(\mathfrak{a}, \mathfrak{g})} X_\alpha,
$$

where $Z \in \mathfrak{m}$, $H \in \mathfrak{a}$, and $X_\alpha \in \mathfrak{g}_\alpha$. Since $X \in \mathfrak{k}$, it must be fixed by θ, so $X = Z + \sum_{\alpha \in \Phi^+}(X_\alpha + \theta X_\alpha)$, where Φ^+ is the set of positive roots. Acting on X with $\mathrm{ad}(H)$, for $H \in \mathfrak{a}$, we obtain $[H, X] = \sum_{\alpha \in \Phi^+} \alpha(H)(X_\alpha - \theta X_\alpha)$, which is also in $\mathfrak{a}$ since X normalizes $\mathfrak{a}$. It follows that $[H, X] = 0$ for each $H \in \mathfrak{a}$, so $X_\alpha = 0$, whence $X = Z$. $\quad\square$

We summarize the main results of this section in the next theorem.

Theorem 7.8.4 The Weyl group $W = W(\mathfrak{a}, \mathfrak{g})$ of a θ-stable semisimple linear Lie algebra $\mathfrak{g}$ is a finite group of orthogonal transformations of $\mathfrak{a}^*$, or $\mathfrak{a}$, permuting the roots. Each element of W is uniquely determined by the permutation it induces on $\Phi(\mathfrak{a}, \mathfrak{g})$. In particular, W acts on $\mathfrak{a}^*$ and on $\mathfrak{a}$ with no nonzero fixed points. Moreover, there are elements $k_\alpha \in N_K(\mathfrak{a})$, $\alpha \in \Phi(\mathfrak{a}, \mathfrak{g})$, such that W is generated by (the duals of) the maps $\mathrm{Ad}(k_\alpha)|_\mathfrak{a}$.

Proof. The only claim that remains to be checked is that each element of W is uniquely determined by the permutation it induces on $\Phi(\mathfrak{a}, \mathfrak{g})$. As pointed out already, W actually permutes the roots, so it suffices to check that $\mathfrak{a}^*$ is the linear span of the set of roots.

Let V denote the linear span of $\Phi(\mathfrak{a}, \mathfrak{g})$ in $\mathfrak{a}^*$. Then the orthogonal complement $V^\perp$ of V in $\mathfrak{a}^*$ is fixed by all the reflections r_α. Any vector $H \in \mathfrak{a}$ that is dual to some $\mu \in V$ must then be in the kernel of α. Any such vector must commute with all the root spaces, so it is in the center of $\mathfrak{g}$, which is trivial since $\mathfrak{g}$ is semisimple. Therefore, $V = 0$, proving the claim.   □

Exercise 7.8.5 Let $\{e_1, \dots, e_n\}$ be the standard basis of $\mathbb{R}^n$ and denote by $F_{i,j}$ the linear automorphism of $\mathbb{R}^n$ that sends e_i to e_j and e_j to $-e_i$ and fixes all the other e_k. Show that $F_{i,j}$ is an element of $N_{SO(n)}(\mathfrak{a})$, where $\mathfrak{a}$ is the subalgebra of diagonal matrices of trace 0. Therefore, each $F_{i,j}$ produces an element of $W(\mathfrak{a}, \mathfrak{sl}(n, \mathbb{R}))$. Describe how $\mathrm{Ad}(F_{i,j}) : \mathfrak{sl}(n, \mathbb{R}) \to \mathfrak{sl}(n, \mathbb{R})$ acts on the roots of $\mathfrak{sl}(n, \mathbb{R})$. (Recall that each root can be written as $\alpha_{ij} := f_j - f_i$, where $f_i(\mathrm{diag}[a_1, \dots, a_n]) = a_i$.)

7.9 Generation by Centralizers

We prove in this section a technical but important result about semisimple Lie groups of real rank at least 2, which will be needed in chapter 10. The result is that the group is generated by the centralizers of elements in $\mathfrak{a}$.

The following assumptions and notation will be in place throughout the section. Let $G \subset GL(n, \mathbb{C})$ be a connected, semisimple, Θ-stable linear group with finite center. Let $\mathfrak{g}$ be the Lie algebra of G and let $\mathfrak{g} = \mathfrak{k} \oplus \mathfrak{p}$ be the Cartan decomposition defined by θ. Let $\mathfrak{a}$ be a maximal abelian subalgebra of $\mathfrak{p}$.

We order the set of roots $\Phi(\mathfrak{a}, \mathfrak{g})$ as before, and denote by Φ^+ (resp., Φ^-) the set of positive (resp., negative) roots. Recall that the ordering can be defined as follows. Choose $X \in \mathfrak{a}$ such that $\alpha(X) \neq 0$ for each root α and define $\alpha \geq \beta$ if and only if $\alpha(X) \geq \beta(X)$. We label the elements of $\Phi^+ = \{\alpha_1, \dots, \alpha_s\}$ according to this ordering, so that

$$\alpha_1 \geq \alpha_2 \geq \cdots \geq \alpha_s.$$

Let $\mathfrak{n} = \bigoplus_{\lambda \in \Phi^+} \mathfrak{g}_\lambda$ and $\mathfrak{n}^- = \theta\mathfrak{n}$. Let K be, as before, the group consisting of all Θ-fixed points of G, and let $M = Z_K(\mathfrak{a})$ the subgroup of K centralizing

$\mathfrak{a}$. We have seen before that $\mathfrak{g} = \mathfrak{n}^- \oplus \mathfrak{m} \oplus \mathfrak{a} \oplus \mathfrak{n}$. Let N and A be the connected subgroups of G with Lie algebras $\mathfrak{n}$ and $\mathfrak{a}$, respectively.

Lemma 7.9.1 Let $\mathcal{W}$ be the image of the multiplication map

$$m : N^- \times MAN \to G.$$

Then $G = \bigcup_{i=1}^{\infty} \mathcal{W}^i$.

Proof. By lemma 7.6.2, $\mathcal{W}$ is an open set containing e. Therefore it generates G by lemma 3.7.3.   □

A much stronger statement than the previous lemma holds, namely, that $G = \mathcal{W}^2$. This is due to the fact that $\mathcal{W}$ is actually dense in G. (Notice that if $\mathcal{W}$ is open and dense, then $\mathcal{W} \cap g\mathcal{W}^{-1}$ is nonempty for each $g \in G$, so each g is the product of two elements in $\mathcal{W}$.) The weaker statement given in the lemma will suffice for us.

Recall that the exponential map of G restricts to a diffeomorphism between n and N. In particular, for each $\alpha \in \Phi^+$ (resp., Φ^-), there is a closed subgroup $U_\alpha \subset N$ (resp., $U_\alpha \subset N^-$) with Lie algebra $\mathfrak{g}_\alpha$ such that $\exp : \mathfrak{g}_\alpha \to U_\alpha$ is a diffeomorphism. It is clear that $\Theta U_\alpha = U_{-\alpha}$.

Lemma 7.9.2 With the foregoing notation, the multiplication map

$$m : U_{\alpha_1} \times \cdots \times U_{\alpha_i} \to W_i := U_{\alpha_1} \cdots U_{\alpha_i}$$

is a diffeomorphism onto a subgroup W_i of N, for each $i = 1, \ldots, s$. W_i is a normal subgroup of W_{i+1} and $W_s = N$.

Proof. Define $\mathfrak{w}_i := \mathfrak{g}_{\alpha_1} \oplus \cdots \oplus \mathfrak{g}_{\alpha_i}$. Then $\mathfrak{w}_i$ is a Lie subalgebra of $\mathfrak{n}$ and $\mathfrak{w}_i$ is an ideal of $\mathfrak{w}_{i+1}$ for each i, $1 \le i \le s - 1$. This is due to the fact that if $[\mathfrak{g}_{\alpha_k}, \mathfrak{g}_{\alpha_l}]$ is not 0, then $\alpha_k + \alpha_l$ is a positive root strictly greater than α_k and α_l, and the fact that $[\mathfrak{g}_{\alpha_k}, \mathfrak{g}_{\alpha_l}] \subset \mathfrak{g}_{\alpha_k + \alpha_l}$.

Let $\bar{W}_i$ be the connected subgroup of N with Lie algebra $\mathfrak{w}_i$. Since $U_{\alpha_{i+1}}$ normalizes $\bar{W}_i$, the set $\bar{W}_i U_{\alpha_{i+1}}$ is a subgroup of $\bar{W}_{i+1}$, which is open in $\bar{W}_{i+1}$ due to lemma 7.6.2, hence a Lie subgroup. Therefore, as $\bar{W}_{i+1}$ is connected, $\bar{W}_{i+1} = \bar{W}_i U_{\alpha_{i+1}}$. Notice that $\bar{W}_1 = W_1 = U_{\alpha_1}$. Therefore, we obtain by induction that $\bar{W}_i = W_i$ for each i and $W_s = N$. That the multiplication map is a diffeomorphism is also easily shown by induction and by the fact that $\exp : \mathfrak{g}_\alpha \to U_\alpha$ is a diffeomorphism.   □

We now introduce the assumption that $\dim a \ge 2$; in other words, G has real rank at least 2. For each $\alpha \in \Phi(\mathfrak{a}, \mathfrak{g})$, let $H_\alpha \in \mathfrak{a}$ be the dual vector to α. The

orthogonal complement in $\mathfrak{a}$ (relative to the inner product $\langle \cdot, \cdot \rangle_\kappa$) of the line $\mathbb{R}H_\alpha$ is the hyperplane denoted $H_\alpha^\perp$. Notice that $H_\alpha^\perp$ is nonzero by the rank assumption. Since $\alpha(H) = \langle H_\alpha, H \rangle_\kappa$, we have $H_\alpha^\perp = \ker(\alpha)$.
The centralizer of $H_\alpha^\perp$ in G will be denoted

$$Z_\alpha = \left\{ g \in G \mid \mathrm{Ad}(g)H = H \text{ for all } H \in H_\alpha^\perp \right\}.$$

Notice that Z_α is Θ-stable (hence, it is reductive). This is a consequence of the identity $\theta \circ \mathrm{Ad}(g) = \mathrm{Ad}(\Theta(g)) \circ \theta$ and of the fact that $\theta H = -H$ for each $H \in \mathfrak{a}$.

The Lie algebra of Z_α is $\mathfrak{z}_\alpha = Z_\mathfrak{g}(H_\alpha^\perp)$. Notice that $\mathfrak{g}_\alpha$ is contained in $\mathfrak{z}_\alpha$ since $[H, X] = \alpha(H)X = 0$ for each $X \in \mathfrak{g}_\alpha$ and $H \in H_\alpha^\perp = \ker(\alpha)$. Since $\mathfrak{z}_\alpha$ is θ-stable, we also have $\mathfrak{g}_{-\alpha} \subset \mathfrak{z}_\alpha$. Therefore, by proposition 7.7.8, z_α contains a θ-stable copy of $\mathfrak{sl}(2, \mathbb{R})$ whose (one-dimensional) Cartan subalgebra is spanned by H_α.

Each Z_α contains U_α and $U_{-\alpha}$, as well as the group MA. The following proposition is now an immediate consequence of the previous two lemmas.

Proposition 7.9.3 Each $g \in G$ can be written as a product $g = g_1 g_2 \cdots g_l$, where, for each i, $1 \leq i \leq l$, $g_i \in Z_\alpha$ for some $\alpha \in \Phi(\mathfrak{a}, \mathfrak{g})$.

8

Ergodic Theory – Part II

This chapter is mainly concerned with proving that certain measure-preserving actions are ergodic. Section 1 contains a general discussion of invariant measures on Lie groups and homogeneous spaces. We then introduce, in section 2, a characterization of ergodicity in terms of unitary representations, which is used in section 3 to prove the central result of the chapter, namely, Moore's ergodicity theorem. Moore's theorem asserts that if a G-space X is ergodic with respect to a finite G-invariant measure, where G is a simple noncompact Lie group with finite center, then for any closed noncompact subgroup H of G, the H-action on X is also ergodic.

After that, we derive Birkhoff's ergodic theorem and use it in section 5 to indicate, by means of a simple homogeneous example, a method for proving ergodicity of Anosov systems.

Finally, we introduce the notions of amenability and Kazhdan's property T.

8.1 Invariant Measures on Coset Spaces

Most of the examples of ergodic transformations given in chapter 2 were actions defined on the n-torus $\mathbb{T}^n$. The simple nature of the space allowed us in those examples to prove ergodicity (with respect to the Lebesgue measure) using the elementary theory of Fourier series.

A more general class of actions can be defined as follows. Let H be a topological group and let G and Λ be closed subgroups of H. We form the coset space H/Λ, which we regard as a G-space with G-action given by left translations. For example, if $H = \mathbb{R}^n$, $\Lambda = \mathbb{Z}^n$, and G is the group $\mathbb{Z}$ corresponding to translations in $\mathbb{R}^n$ by the vector $(u_1, \ldots, u_n) \in \mathbb{R}^n$, then the Lebesgue measure on $\mathbb{R}^n$ clearly induces a finite invariant measure on the compact space $\mathbb{T}^n = \mathbb{R}^n/\mathbb{Z}^n$. If the numbers $1, u_1, \ldots, u_n$ are rationally independent, then the proof of proposition 1.2.4 essentially shows that the $\mathbb{Z}$-action is ergodic. See also exercise 2.3.3.

Not all actions admit invariant measures. One example is given in the next exercise.

Exercise 8.1.1 The linear action of $SL(2, \mathbb{R})$ on $\mathbb{R}^2$ induces a transitive action on the circle S^1, regarded as the space of rays in $\mathbb{R}^2$ issuing from the origin. Show that this action does not preserve any Borel probability measure. (Hint: The $\mathbb{Z}$-action generated by $\begin{pmatrix} 1 & 1 \\ 0 & 1 \end{pmatrix}$ "squishes" all points toward the two horizontal rays. This forces a measure invariant under such an element to be supported in the union of these two points. It is possible to choose other elements for which invariant measures would be supported at other points. Therefore, there cannot exist a probability measure invariant under the entire group.)

Before discussing more general actions of G, we consider the action of G on itself by right or left translations, and invariant measures associated to these actions. Denote by L_g (resp., R_g) the diffeomorphism of G such that $L_g g' = gg'$ (resp., $R_g g' = g'g$). Haar's theorem asserts that a locally compact group G always admits left-invariant regular Borel (resp., right-invariant) measures and that these measures are unique up to a multiplicative constant. They are called left (resp., right) *Haar measures*.

Exercise 8.1.2 Let G be an n-dimensional Lie group and fix a nonzero alternating n-form ω at $T_e G$ and denote by Ω the unique left- (resp., right-) invariant n-form on G that agrees with ω at e. (The existence of Ω shows, in particular, that G is orientable. We take the orientation compatible with Ω as the positive orientation on G.) Define a measure μ on G by setting, for any relatively compact open set $A \subset G$, $\mu(A) := \int_A \omega$. Show that μ is a left (resp., right) Haar measure on G.

Exercise 8.1.3 Show that for any two nonzero left (or right) Haar measures μ_1 and μ_2 on G there is a constant c such that $\mu_1 = c\mu_2$. (Hint: First note that μ_i is absolutely continuous with respect to $\mu_1 + \mu_2$, and that the Radon–Nikodym derivative h_i is a left-invariant nonnegative measurable function. Use Fubini's theorem and the fact that G acts on itself transitively to conclude that h_i is constant $(\mu_1 + \mu_2)$-almost everywhere.)

Exercise 8.1.4 Show that any Haar measure is positive on nonempty open sets

Since right and left multiplication commute, given any $g \in G$ and left Haar measure μ on G, the measure $(R_g)_* \mu$ is also a left Haar measure; therefore, there exists a positive function of g, $\Delta_G(g)$, such that

$$(R_g)_* \mu = \Delta_G(g)\mu.$$

Δ_G is independent of μ. The function Δ_G is called the *modular function* of the locally compact group G. If $\Delta_G \equiv 1$, G is called a *unimodular* group.

Exercise 8.1.5 Show that Δ_G is a continuous homomorphism of G into the multiplicative group $\mathbb{R}^+$ of positive real numbers. As a consequence, if G is a connected semisimple

Lie group, show that G is unimodular. Show that compact, discrete, and abelian locally compact groups are also unimodular groups. (Hint: For the continuity of Δ, use that for every compactly supported function f on G and all $g_0 \in G$ we have $\Delta(g_0) \int_G f(g) \, d\mu(g) = \int_G f(gg_0) \, d\mu(g)$.)

Exercise 8.1.6 If G is a Lie group, show that its modular function is

$$\Delta(g) = |\det(\mathrm{Ad}(g))|,$$

where $\mathrm{Ad} : G \to \mathfrak{g}$ is the adjoint representation of G. (Hint: Show that if X is the left-invariant vector field on G that corresponds to $v \in \mathfrak{g}$, then $(R_g)_* X$ is the left-invariant vector field that corresponds to $\mathrm{Ad}(g)v$. Use this to show that if ω is a left-invariant n-form on G, with $n = \dim G$, then $R_g^* \omega = (\det \mathrm{Ad}(g))\omega$, for each $g \in G$.)

Lemma 8.1.7 If μ_G is a left Haar measure on G and $i : G \to G$ is the map that associates to each $g \in G$ its inverse, then

$$\frac{d(i_* \mu_G)}{d\mu_G} = \Delta.$$

In other words, for each $f \in C_c(G)$, $\int f(g) d(i_* \mu_G)(g) = \int_G f(g) \Delta(g) \, d\mu_G(g)$.

Proof. If h is the Radon–Nikodym derivative of a measure ν with respect to another measure μ, we write $\nu = h \cdot \mu$. We first claim that $i_* \mu_G$ and the measure $\Delta \cdot \mu_G$ are both right-invariant. This is clear for $i_* \mu_G$ since $i \circ L_g = R_{g^{-1}} \circ i$. To show it for $\Delta \cdot \mu_G$, let f be any continuous function on G with compact support. Then for each $g_0 \in G$,

$$\int_G f(gg_0) d(\Delta \cdot \mu_G)(g) := \int_G f(gg_0)\Delta(g) \, d\mu_G(g)$$

$$= \int_G f(gg_0)\Delta(gg_0)\Delta(g_0)^{-1} \, d\mu_G(g)$$

$$= \int_G f(g)\Delta(g)\Delta(g_0)^{-1} d((R_{g_0})_* \mu_G)(g)$$

$$= \int_G f(g)\Delta(g)\Delta(g_0)^{-1}\Delta(g_0) \, d\mu_G(g)$$

$$= \int_G f(g)\Delta(g) \, d\mu_G(g)$$

$$= \int_G f(g)d(\Delta \cdot \mu_G)(g).$$

The Radon–Nikodym derivative of one measure with respect to the other must be a positive constant, by exercise 8.1.3. Therefore, $\frac{d(i_* \mu_G)}{d(\Delta \cdot \mu_G)} = c > 0$. To see that $c = 1$ observe that

$$\mu_G = (i^2)_* \mu_G = i_*(c\Delta \cdot \mu_G) = c(\Delta \circ i) \cdot (i_* \mu_G)$$
$$= c^2(\Delta^{-1}) \cdot (\Delta \cdot \mu_G) = c^2 \mu_G. \qquad \square$$

Theorem 8.1.8 (A. Weil) Let G be a locally compact group and H a closed subgroup. If $\Delta_G(h) = \Delta_H(h)$ for all $h \in H$, then G/H admits a nonzero G-invariant measure. The measure, if it exists, is unique up to a scalar multiple. In particular, if G is a connected semisimple Lie group and H is a discrete subgroup, G/H admits a nonzero G-invariant measure.

Proof. (From [28].) To obtain a nontrivial invariant measure on G/H it will suffice to obtain a nonzero continuous linear functional $\mathcal{F}$ on $C_c(G/H)$, taking nonnegative functions into $[0, \infty)$, such that $\mathcal{F}[f \circ L_g] = \mathcal{F}[f]$ for all $f \in C_c(G/H)$. In fact, by the Riesz representation theorem, there is then an (invariant) measure μ on G/H such that $\mathcal{F}[f] = \int_{G/H} f \, d\mu$.

Define a map

$$I : \phi \in C_c(G) \mapsto I(\phi) \in C_c(G/H),$$

such that $I(\phi)(\pi(g)) := \int_H \phi(gh) \, d\mu_H(h)$. $I(\phi)(x)$, $x \in G/H$, is indeed well defined and $I(\phi)$ has compact support as one easily checks. Moreover, given $f \in C_c(G/H)$ and $\phi \in C_c(G)$, we have

$$I(\phi f \circ \pi) = I(\phi)f.$$

We first note that I is a surjective map. Namely, given $f \in C_c(G/H)$, we can find a relatively compact open set $U \subset G$ such that its image $\pi(U)$ contains the support of f. Choose a nonnegative (real-valued) function $\lambda \in C_c(G)$ such that $\lambda|_U \equiv 1$. It is clear that $I(\lambda)(x) > 0$ for all $x \in \pi(U)$. Define a function $h \in C_c(G/H)$ as follows:

$$h(x) := \begin{cases} I(\lambda)(x)^{-1} f(x), & \text{if } x \in \pi(U), \\ 0, & \text{if } x \notin \pi(U). \end{cases}$$

Therefore, $f = I(\lambda)h = I(\lambda h \circ \pi)$, where $\lambda h \circ \pi \in C_c(G)$.

We claim that for any $\phi \in C_c(G)$,

$$I(\phi) \equiv 0 \Rightarrow \int_G \phi(g) \, d\mu_G(g) = 0.$$

Once the claim is proved we will be able to define $\mathcal{F}$ by

$$\mathcal{F}[f] = \int_G \phi(g) \, d\mu_G(g),$$

where ϕ is any element in $C_c(G)$ such that $I(\phi) = f$.

To prove the claim, choose $\lambda \in C_c(G)$ such that $I(\lambda)(\pi(g)) = 1$ for all g contained in the support of ϕ, and suppose that $I(\phi)(x) = 0$ for all $x \in G/H$. Therefore,

$$
\begin{aligned}
0 &= \int_G \lambda(g) I(\phi)(\pi(g)) \, d\mu_G(g) \\
&= \int_G \lambda(g) \int_H \phi(gh) \, d\mu_H(h) \, d\mu_G(g) \\
&= \int_G \lambda(g) \int_H \phi(gh^{-1}) \Delta_H(h) \, d\mu_H(h) \, d\mu_G(g) \quad \text{(since } i_* \mu_H = \Delta \cdot \mu_H) \\
&= \int_G \int_H \lambda(g) \phi(gh^{-1}) \Delta_G(h) \, d\mu_H(h) \, d\mu_G(g) \quad \text{(since } \Delta_H = \Delta_G|_H) \\
&= \int_H \int_G \lambda(g) \phi(gh^{-1}) \Delta_G(h) \, d\mu_G(g) \, d\mu_H(h) \quad \text{(by Fubini's theorem)} \\
&= \int_H \int_G \lambda(gh) \phi(g) \, d\mu_G(g) \, d\mu_H(h) \quad \text{(by the definition of } \Delta_G) \\
&= \int_G \int_H \lambda(gh) \phi(g) \, d\mu_H(h) \, d\mu_G(g) \quad \text{(by Fubini's theorem)} \\
&= \int_G I(\lambda)(\pi(g)) \phi(g) \, d\mu_G(g) \\
&= \int_G \phi(\phi) \, d\mu_G(g) \quad \text{(since } I(\lambda) \circ \pi \equiv 1 \\
&\qquad\qquad\qquad\qquad \text{on supp } \phi).
\end{aligned}
$$

Therefore the linear functional $\mathcal{F}$ can indeed be defined and we obtain the desired measure. The measure on G/H is G-invariant and is unique up to a multiplicative constant, since that is true for μ_G. □

It follows from the preceding proof that the invariant measure μ on G/H, when it exists, can be chosen so that

$$
\int_G f(g) \, d\mu_G(g) = \int_{G/H} \int_H f(gh) \, d\mu_H(h) \, d\mu(gH)
$$

for all $f \in C_c(G)$.

A corollary of the previous theorem is that for any connected semisimple Lie group G and discrete subgroup Γ of G, the space G/Γ always admits nontrivial G-invariant measures.

We will later show that the invariant measures of $SL(n, \mathbb{R})/SL(n, \mathbb{Z})$ are finite. If a discrete subgroup Γ of a Lie group G has the property that G/Γ admits a finite nonzero G-invariant measure we say that Γ is a *lattice* of G.

The fact that $SL(n, \mathbb{Z})$ is a lattice of $SL(n, \mathbb{R})$ is a special case of the following hard theorem.

Theorem 8.1.9 (Borel–Harish-Chandra [6]) Let $G \subset GL(n, \mathbb{C})$ be a semisimple algebraic group defined over $\mathbb{Q}$. Then $G(\mathbb{Z}) = G \cap GL(n, \mathbb{Z})$ is a lattice subgroup of $G(\mathbb{R}) = G \cap GL(n, \mathbb{R})$.

8.2 Ergodicity and Unitary Representations

Ergodicity can be characterized in terms of the unitary representations of G. We assume that X is a G-space with a finite G-invariant measure μ and consider $H := L^2(X, \mu)$, with inner product

$$\langle f_1, f_2 \rangle := \int_X f_1(x) \overline{f_2(x)} \, d\mu(x).$$

Then $(H, \langle \cdot, \cdot \rangle)$ is a separable Hilbert space. (Recall that a Hilbert space is said to be *separable* if it has a countable orthonormal basis.)

For each $g \in G$, denote by $\pi(g)$ the linear operator on H defined as follows. To each element in H, represented by a square-integrable function f, the (class determined by the) function $\pi(g)f$ satisfies $(\pi(g)f)(x) := f(g^{-1}x)$. As the measure μ is preserved by G, each $\pi(g)$ satisfies the identity

$$\langle \pi(g)f_1, \pi(g)f_2 \rangle = \langle f_1, f_2 \rangle.$$

Therefore, $\pi(g)$ belongs to the group of unitary operators $\mathbf{U}(H)$ and we have a homomorphism $\pi : G \to \mathbf{U}(H)$.

More generally, let H be a separable Hilbert space and $\mathbf{U}(H)$ the group of all unitary operators on H. We give $\mathbf{U}(H)$ the *strong operator topology*, that is, the smallest topology that makes all the maps $U \mapsto Uf$, $f \in H$, $U \in \mathbf{U}(H)$, continuous. With respect to the strong operator topology, if $\{U_n\}$ is a sequence of elements in $\mathbf{U}(H)$, we have $\lim_{n \to \infty} U_n = U \in \mathbf{U}(H)$ if and only if $\| U_n f - Uf \| \to 0$ for all f in some set D whose linear span is dense in H.

$\mathbf{U}(H)$ can also be given the *weak operator topology*. A sequence $\{U_n\}$ converges to U in the weak operator topology if each matrix coefficient converges, that is, $\lim_{n \to \infty} \langle U_n e, f \rangle = \langle U e, f \rangle$ for all $e, f \in H$.

Exercise 8.2.1 Show that a sequence $U_n \in \mathbf{U}(H)$ converges to $U \in \mathbf{U}(H)$ in the strong operator topology if and only if it converges to U in the weak operator topology. Also show that, with respect to either topology, $\mathbf{U}(H)$ is a metrizable second countable topological space.

Exercise 8.2.2 Use Fubini's theorem to show that the homomorphism

$$\pi : G \to \mathbf{U}(L^2(X, \mu))$$

is measurable.

We saw earlier that any continuous homomorphism between Lie groups must be smooth. The next theorem, due to Mackey, shows that it is sufficient that the homomorphism be measurable.

Theorem 8.2.3 Suppose that L is a second countable topological group, G is a second countable locally compact group, and $\rho : G \to L$ is a measurable homomorphism, that is, a measurable map that is also a homomorphism of groups. Then ρ is continuous.

Lemma 8.2.4 If $A \subset G$ is a compact set with positive Haar measure, where G is as in the theorem, then AA^{-1} contains a neighborhood of $e \in G$.

Proof. We denote a Haar measure on G by μ. Since A is a compact set of positive measure, we can find an open set W containing A such that $\mu(W) < 2\mu(A)$. Compactness of A also implies that there is a neighborhood N of e such that N is symmetric (i.e., if $g \in N$ then $g^{-1} \in N$) and $gA \subset W$ for all $g \in N$. Since $\mu(gA) = \mu(A)$ and $\mu(W) < 2\mu(A)$, we have that $(gA) \cap A$ is nonempty for each $g \in N$, so N is contained in AA^{-1}. $\square$

To prove the theorem, we may assume without loss of generality that ρ is surjective. Let U be an open neighborhood of the identity in L and $V \subset U$ a symmetric open neighborhood with $V^2 \subset U$. Let $\{l_n\}$ be a countable dense set in L and $g_n \in G$ such that $\rho(g_n) = l_n$. Then L is the union of the $l_n V$, so G is the union of the $g_n \rho^{-1}(V)$. It follows that for some n, $\mu(g_n \rho^{-1}(V)) > 0$ so $\mu(\rho^{-1}(V)) > 0$. Since $\rho^{-1}(U)$ contains $\rho^{-1}(V)\rho^{-1}(V)$, it contains in particular a subset KK^{-1}, where K is compact with positive Haar measure. The theorem now follows from the previous lemma.

It follows from the theorem that the representation $\pi : G \to \mathbf{U}(L^2(X, \mu))$ is continuous. Thus, we have a unitary representation of G into $\mathbf{U}(L^2(X, \mu))$. (A unitary representation of G is defined in general as a continuous homomorphism of G into the unitary group of a separable Hilbert space.) Let H_0 be the (G-invariant) orthogonal complement in $L^2(X, \mu)$ to the subspace $\mathbb{C}$ of constant functions. If the G-space X is not ergodic, then the characteristic function of a G-invariant measurable set that is neither null nor conull defines a nonzero vector in H_0 fixed by G. The converse is given in the next proposition.

Proposition 8.2.5 If X is a G-space with finite invariant measure μ, and π is the unitary representation of G on the orthogonal complement H_0 of $\mathbb{C}$ in $L^2(X, \mu)$, then X is ergodic if and only if there are no nonzero $\pi(G)$-invariant vectors in H_0.

Proof. We need to show that if the G-space is ergodic then its unitary representation does not have nonzero fixed vectors in H_0. Let f be a fixed vector in H_0. It then corresponds to a (square-integrable) measurable G-invariant function on X relative to μ. By proposition 2.3.1, there is a strictly G-invariant measurable function $\tilde{f}$ that is μ-almost everywhere equal to f. Therefore $\tilde{f}$ is constant μ-almost everywhere, by the definition of ergodicity. It follows that f represents an element in $H_0 \cap \mathbb{C} = \{0\}$. $\square$

8.3 Moore's Ergodicity Theorem

We discuss now a result due to C. C. Moore concerning ergodicity for subgroups of semisimple Lie groups. The proof of Moore's theorem given here is due to R. Ellis and M. Nerurkar [10].

Let X be a G-space with a finite invariant measure. X is called *irreducible* if every normal subgroup of G not contained in the center acts ergodically on X.

Theorem 8.3.1 (Moore's ergodicity theorem [36]) Suppose that G is a semisimple Lie group with finite center and no compact simple factors, and that X is an irreducible G-space with finite G-invariant measure. If H is a closed noncompact subgroup of G, then H also acts ergodically on X.

In view of the characterization of ergodicity in terms of the unitary representation of G on $L^2(X, \mu)$, in order to prove Moore's theorem it suffices to show the following: For any connected noncompact simple Lie group G with finite center, and unitary representation π of G with no nonzero invariant vectors, a closed subgroup L of G such that $\pi|_L$ has nonzero invariant vectors must be compact. Observe that given a nontrivial L-invariant vector $v \in H$, the function $f(g) := \langle \pi(g)v, w \rangle$ is constant on L. Therefore, the proof will be complete once we show that for all $v, w \in H$, $\langle \pi(g)v, w \rangle$ approaches 0 as $g \to \infty$ in G. This is the content of the next theorem.

Theorem 8.3.2 [36, 2.2.20] Let G_i be a connected noncompact simple Lie group with finite center, and let $G = \prod_i G_i$ be a finite direct product. Let H be a separable Hilbert space with inner product $\langle \cdot, \cdot \rangle$, and $\pi : G \to \mathrm{U}(H)$ a unitary representation such that for each i, $\pi|_{G_i}$ has no invariant vectors. Then for all $u, v \in H$,

$$\lim_{g \to \infty} \langle \pi(g)u, v \rangle = 0.$$

In other words, $\pi(g)$ tends to 0 in the weak operator topology as g tends to infinity in G.

Recall (theorem 7.4.8) that a connected semisimple Lie group G with finite center admits a KAK decomposition, where K is a maximal compact subgroup, A is abelian, and $\text{Ad}(A)$ is diagonalizable over $\mathbb{R}$. (In that theorem, we assumed that G was linear algebraic. We have seen in theorem 7.2.5, however, that G modulo its center is in a natural way the component of the identity of a real linear algebraic group. Since the finite center is contained in the maximal compact subgroup K, it is easy to see that the KAK decomposition also holds here.)

Lemma 8.3.3 Let π be as in the previous theorem. If

$$\lim_{A \ni a \to \infty} \langle \pi(a)u, v \rangle = 0$$

for all $u, v \in H$, then the conclusion of the theorem holds.

Proof. By the KAK decomposition each element g of G can be written as $g = kak'$, where $k, k' \in K$ and $a \in A$.

Let $g_n = k_n a_n k'_n$ be a sequence in G such that $g_n \to \infty$ in G, that is, g_n eventually leaves any compact subset in G. Since K is compact, the condition is equivalent to $a_n \to \infty$ in A. Let u, v be arbitrary elements of H and suppose for a contradiction that for some subsequence, still denoted g_n, there exists $\epsilon > 0$ such that $|\langle \pi(g_n)u, v \rangle| \geq \epsilon$ for all n. Passing to yet another subsequence, we may assume that k_n and k'_n converge to k and k', respectively. Then, from the fact that π is continuous in the operator norm, it follows that $|\langle \pi(k)\pi(a_n)\pi(k')u, v \rangle| \geq \epsilon/2$ for all sufficiently large n. Therefore, $|\langle \pi(a_n) \pi(k')u, \pi(k^{-1})v \rangle| \geq \epsilon/2$ for all sufficiently large n, a contradiction. $\quad\square$

For any topological group G and $a \in G$, define the *stable group* of a, G_a^{s}, as the closure of $\{g \in G \mid \lim_{m \to +\infty} a^m g a^{-m} = e\}$. Also define the *unstable group* of a as $G_a^{\text{u}} = G_{a^{-1}}^{\text{s}}$.

Lemma 8.3.4 Let H be a separable Hilbert space and G a locally compact, second countable group. Let $\pi : G \to \mathbf{U}(H)$ be a (strongly continuous) unitary representation of G. Let v be a vector in H that is fixed by an element $a \in G$. Then v is also fixed by all elements in G_a^{s} and G_a^{u}.

Proof. Suppose that $a^m g a^{-m}$ converges to the identity element and let v be any vector fixed by $\pi(a)$. Hence v is also fixed by $\pi(a)^{-1}$ and, since $\pi(a)$ is unitary, we have for all m

$$\|\pi(g)v - v\| = \|\pi(a^m g a^{-m})v - v\| \to 0.$$

Therefore, by continuity, $\pi(g)v = v$ for all $g \in G_a^s$. A similar argument applies to G_a^u. □

Lemma 8.3.5 Let G be a connected noncompact semisimple Lie group with finite center, and let A be the abelian subgroup defined in the KAK decomposition. Let H be a separable Hilbert space and $\pi : G \to \mathrm{U}(H)$ a unitary representation such that no normal subgroup of G not contained in the center leaves invariant a nonzero vector in H. Fix an element $a \in A$, $a \neq e$, and define

$$W = \{v \in H \mid \pi(a)v = v\}.$$

Then $W = 0$.

Proof. It suffices to show that W is stable under G since, if this is the case, a would be in the kernel of $g \in G \mapsto \pi(g)|_W$, whence the kernel would be a noncentral normal subgroup of G fixing W pointwise. Therefore $W = 0$.

We now show that W is stable under G. Notice that the subgroup $\check{G}$ of G that stabilizes W is closed, hence a Lie subgroup. If $\check{\mathfrak{g}}$ denotes the Lie algebra of $\check{G}$, then it suffices to show that $\mathfrak{g} = \check{\mathfrak{g}}$.

The group A is easily seen to be diffeomorphic to its Lie algebra $\mathfrak{a}$ via the exponential map. (Recall that we can identify $\mathfrak{a}$ and A with their images under ad and Ad, for the adjoint representations of $\mathfrak{g}$ and G, respectively, and that $\mathrm{ad}(\mathfrak{a})$ may be assumed to lie in the subalgebra of diagonal matrices of $\mathfrak{gl}(n, \mathbb{R})$, where n is the dimension of G, whereas $\mathrm{ad}(A)$ will then lie in the subgroup of diagonal matrices in $GL(n, \mathbb{R})$ with positive entries.) Therefore, we can write $a = \exp X$ for some $X \in \mathfrak{a}$. Let $\Phi(\mathfrak{a}, \mathfrak{g})$ be the set of roots of $(\mathfrak{a}, \mathfrak{g})$.

The Lie algebra $\mathfrak{g}$ decomposes as a direct sum

$$\mathfrak{g} = \mathfrak{u}^- \oplus \mathfrak{z} \oplus \mathfrak{u}^+,$$

where $\mathfrak{z}$ is the Lie algebra of the centralizer of a, $\mathfrak{u}^-$ is the subspace of $\mathfrak{g}$ spanned by the root spaces $\mathfrak{g}_\alpha$ such that $\alpha(X) < 0$, and $\mathfrak{u}^+$ is the subspace of $\mathfrak{g}$ spanned by the root spaces $\mathfrak{g}_\alpha$ such that $\alpha(X) > 0$. Clearly $\mathfrak{z} \subset \check{\mathfrak{g}}$ since any element that centralizes a must stabilize W. On the other hand, $\mathfrak{u}^-$ (resp., $\mathfrak{u}^+$) must be contained in the Lie algebra of the stable (resp., unstable) group of a, since for each $Y \in \mathfrak{g}_\alpha, \alpha \in \Phi(\mathfrak{a}, \mathfrak{g})$, and $c = \exp U \in A$, we have

$$c(\exp Y)c^{-1} = \exp(\mathrm{Ad}(c)Y)$$
$$= \exp(e^{\mathrm{ad}(U)}Y)$$
$$= \exp(e^{\alpha(U)}Y).$$

It is now a consequence of the previous lemma that $\mathfrak{u}^+, \mathfrak{u}^- \subset \check{\mathfrak{g}}$. Therefore $\mathfrak{g} = \check{\mathfrak{g}}$. $\qquad\qquad\qquad\qquad\qquad\qquad\qquad\qquad\qquad\qquad\qquad\square$

We recall the following basic fact, known as Alaoglu's theorem. Denote by $\mathbf{B}(H)$ the space of bounded operators on a separable Hilbert space H and let $\mathcal{B}$ denote the unit ball in $\mathbf{B}(H)$. Then $\mathcal{B}$ is compact in the weak operator topology.

Lemma 8.3.6 Denote by $\overline{\pi(A)}$ the closure of the subset $\pi(A) \subset \mathbf{U}(H) \subset \mathcal{B}$ in the weak operator topology of $\mathbf{B}(H)$. Then

$$\overline{\pi(A)}\,\overline{\pi(A)} \subset \overline{\pi(A)}.$$

Proof. For each $a \in A$, it is immediate to verify that the map

$$U \in \mathbf{B}(H) \mapsto \pi(a)U \in \mathbf{B}(H)$$

is continuous in the weak operator topology. Therefore, the preimage of $\overline{\pi(A)}$ under this map is closed. That preimage contains $\pi(A)$ so it also contains $\overline{\pi(A)}$. Since a is arbitrary, we have

$$\pi(A)\,\overline{\pi(A)} \subset \overline{\pi(A)}.$$

Apply now the same argument to the map

$$U \in \mathbf{B}(H) \mapsto UT \in \mathbf{B}(H),$$

where T is an arbitrary element of $\overline{\pi(A)}$. We conclude that $\overline{\pi(A)}\,T \subset \overline{\pi(A)}$ for all $T \in \overline{\pi(A)}$, which is the claim. $\qquad\qquad\qquad\qquad\square$

Exercise 8.3.7 Let T_m be a sequence of unitary operators on a Hilbert space H converging weakly to a bounded operator T. Let U_m be another sequence of unitary operators on H converging strongly to a bounded invertible operator U whose inverse U^{-1} is also bounded. Show that the sequences $T_m U_m$ and $U_m T_m$ converge weakly to UT and TU, respectively.

If T is a bounded operator on a Hilbert space H, its *adjoint operator* T^* is defined by the identity

$$\langle T^*u, v \rangle = \langle u, Tv \rangle$$

for all $u, v \in H$. It is a basic fact in functional analysis that the identity indeed defines a bounded operator on H and that $\|T^*\| = \|T\|$. Moreover, the map $T \in \mathbf{B}(H) \mapsto T^* \in \mathbf{B}(H)$ is easily seen to be continuous in the weak operator topology.

Although not necessary, we make now the simplifying assumption that our semisimple Lie algebra is linear and θ-stable, where θ is the standard Cartan involution introduced in chapter 7. Recall (proposition 7.7.8) that if Y is a nonzero root vector in $\mathfrak{g}_\alpha$, then Y and $\theta(Y)$ generate a three-dimensional subalgebra of $\mathfrak{g}$, which is isomorphic to $\mathfrak{sl}(2, \mathbb{R})$, and that $[Y, \theta(Y)] \in \mathfrak{a}$.

We now begin the proof of theorem 8.3.2 (keep in mind lemma 8.3.3). Let $a_m \in A$ be a sequence tending to infinity such that $\pi(a_m)$ converges in the weak operator topology to an operator $T \in \mathcal{B}$. We want to show that $T = 0$.

We write $a_m = \exp(X_m)$. By choosing a subsequence, we may assume without loss of generality that the line $\mathbb{R}X_m$ converges to $\mathbb{R}X \subset \mathfrak{a}$, where X is a nonzero element of $\mathfrak{a}$. Clearly there must exist $\alpha \in \Phi(\mathfrak{a}, \mathfrak{g})$ such that $\alpha(X) \neq 0$, since otherwise (having in mind the root space decomposition of $\mathfrak{g}$) X would commute with every element of $\mathfrak{g}$. But a semisimple Lie algebra has trivial center, which would be a contradiction.

Therefore, we may select a root α such that $\alpha(X_m)$ tends to $-\infty$ as m tends to $+\infty$. (Recall that $-\alpha$ is a root if α is a root.)

Choose a nonzero $Y \in \mathfrak{g}_\alpha$ and denote

$$u_\alpha(t) := \exp(tY).$$

Then, for all $t \in \mathbb{R}$, we have (by using the identities that appear at the end of the proof of lemma 8.3.5)

$$a_m u_\alpha(t) = u_\alpha\left(e^{\alpha(X_m)}t\right)a_m.$$

Recall that $\pi(a_m)$ converges weakly to T and observe that $\pi(u_\alpha(e^{\alpha(X_m)}t))$ as well as $\pi(u_{-\alpha}(e^{\alpha(X_m)}t))$ converge strongly to the identity operator as m tends to ∞. Therefore, we can apply exercise 8.3.7 to conclude

$$T\pi(u_\alpha(t)) = T,$$

for all $t \in \mathbb{R}$. Of course we also have $T^*T\pi(u_\alpha(t)) = T^*T$. These equations show that T^*T and T cannot be invertible; if they were, the kernel of π in G would contain the infinite group generated by the elements $u_\alpha(t)$, whence it could not be contained in the finite center of G. (Recall that we are assuming that no noncentral normal subgroup of G has invariant vectors under the representation π.)

Denote $S = T^*T \in \overline{\pi(A)}\,\overline{\pi(A)} \subset \overline{\pi(A)}$. Since S is not invertible, it must belong to the boundary $\overline{\pi(A)} - \pi(A)$. Therefore, S is a weak limit for a sequence $\pi(a'_m)$, with $a'_m \in A$ tending to infinity.

We now repeat for S the same argument that was used for T. Namely, write $a'_m = \exp(X'_m)$, and assume after passing to a subsequence that the line $\mathbb{R}X'_m$ converges to $\mathbb{R}X'$ for some nonzero $X' \in \mathfrak{a}$. Fix a root α' such that $\alpha'(X') < 0$. Choose a nonzero $Y' \in \mathfrak{g}_{\alpha'}$ and denote now

$$u_{\alpha'}(t) := \exp(tY') \quad \text{and} \quad u_{-\alpha'}(t) := \exp(t\theta Y').$$

Then, for all $t \in \mathbb{R}$, we have as before

$$a'_m u_{\alpha'}(t) = u_{\alpha'}\left(e^{\alpha'(X'_m)}t\right)a'_m \quad \text{and} \quad u_{-\alpha'}(t)a'_m = a'_m u_{\alpha'}\left(e^{\alpha'(X'_m)}t\right).$$

Applying π and passing to the limit $m \to \infty$, we conclude, as we did for T, that

$$S\pi(u_{\alpha'}(t)) = S \quad \text{and} \quad \pi(u_{-\alpha'}(t))S = S.$$

Notice that

$$(S\pi(u_{\alpha'}(t)))^* = \pi(u_{\alpha'}(t))^*S^* = \pi(u_{\alpha'}(-t))S.$$

Therefore, we have

$$\pi(u_{\alpha'}(t))S = S \quad \text{and} \quad \pi(u_{-\alpha'}(t))S = S$$

for each $t \in \mathbb{R}$.

The closure in G of the group generated by $u_{\alpha'}(t)$ and $u_{-\alpha'}(t)$, $t \in \mathbb{R}$, is a Lie group locally isomorphic to $SL(2, \mathbb{R})$, and its Lie algebra contains $0 \neq [Y', \theta Y'] \in \mathfrak{a}$ (proposition 7.7.8). By strong continuity of π, we have $\pi(a))S = S$, where $a := \exp([Y', \theta Y'] \in A$.

We claim that $S = 0$. Notice that $\pi(a)$ fixes each element in the image of S in H, so $W := \{v \in H \mid \pi(a)v = v\}$ contains that image. By lemma 8.3.5, $W = 0$, so S is indeed 0.

On the other hand, $S = T^*T$, so for each $v \in H$

$$0 = \|Sv\| = \langle T^*Tv, v\rangle = \langle Tv, Tv\rangle = \|Tv\|,$$

whence T itself must be zero. This concludes the proof of theorem 8.3.2, and with it the proof of Moore's theorem.

Corollary 8.3.8 Let G be a simple noncompact Lie group with finite center, and let Γ be a lattice in G. Then any closed noncompact subgroup L of G acts ergodically on G/Γ by left translations.

Proof. G clearly acts ergodically on G/Γ, since the action is transitive. By Moore's theorem, the L-action must also be ergodic.   □

8.4 Birkhoff's Ergodic Theorem

The next theorem is a fundamental result in ergodic theory. We state the theorem for $\mathbb{R}$-actions; a similar statement holds for $\mathbb{Z}$-actions, for which the integrals $\frac{1}{T}\int_0^T (\cdots)\, dt$ are replaced with summations $\frac{1}{n}\sum_{i=0}^{n-1} (\cdots)$.

Theorem 8.4.1 (Ergodic theorems) Let $(X, \mathcal{A}, \mu)$ be a finite measure space. Let $\Phi_t : X \to X$ be a flow on X that preserves μ. For each measurable function $f : X \to \mathbb{C}$ define

$$f_T^+(x) = \frac{1}{T}\int_0^T f(\Phi_t x)\, dt \quad \text{and} \quad f_T^-(x) = \frac{1}{T}\int_0^T f(\Phi_{-t} x)\, dt.$$

1. (von Neumann) If $f \in L^2(X, \mu)$, then f_T^+ and f_T^- converge in $L^2(X, \mu)$ to $\mathbb{R}$-invariant square-integrable functions f^+ and f^-.
2. (Birkhoff) If $f \in L^1(X, \mu)$, the limits

$$f^+(x) := \lim_{T\to\infty} f_T^+(x) \quad \text{and} \quad f^-(x) := \lim_{T\to\infty} f_T^-(x)$$

exist and are equal for μ-almost every $x \in X$. Moreover, f^+ and f^- belong to $L^1(X, \mu)$ and are $\mathbb{R}$-invariant.

If the action is ergodic, we have in both cases that f^+ and f^- are constant μ-a.e. and are μ-a.e. equal to $\int_X f(x)\, d\mu(x)$.

Part 1 is shown (for $\mathbb{Z}$-actions) in exercise 8.4.3. We will give later a proof of part 2 of the theorem for $\mathbb{Z}$-actions rather than $\mathbb{R}$-actions.

Theorem 8.4.1 can be regarded as an abstract formulation of the law of large numbers in probability theory. The next exercise illustrates this point.

Exercise 8.4.2 Consider again the urn problem discussed in section 2.2. Recall that the shift map σ is a measure-preserving transformation of the probability space $(\Omega, \mathcal{B}, \mu)$ and that it is ergodic. Use Birkhoff's theorem to show that, in the long run, the average number of balls in each urn is half the total number of balls. (Hint: Define the function $F : \Omega \to \{0, 1, \ldots, n\}$ such that $F(\omega) = \omega(0)$. Then for each performance ω of the urn experiment the average number of balls in urn 1 (from the present up to positive eternity) is the limit of $\frac{1}{n} \sum_{i=0}^{n} F(\sigma^i(\omega))$. Therefore, for μ-a.e. ω, the limit is $\int_\Omega F(\omega)\,d\mu(\omega) = \sum_{i=0}^{n} kp_k$.)

The next exercise gives von Neumann's ergodic theorem for $\mathbb{Z}$-actions.

Exercise 8.4.3   Consider a $\mathbb{Z}$-action generated by a measure-preserving transformation T of the probability space $(X, \mathcal{B}, \mu)$. Let $H = L^2(X, \mathcal{B}, \mu)$, a Hilbert space, and define the unitary operator U on H by $Uf := f \circ T$. Define the subspaces $I = \{f \in H \mid Uf = f\}$ and $B = \{f \in H \mid f = g - Ug \text{ for some } g \in H\}$ and show:

1. I is closed and H is the direct sum of I and the closure of B.
2. For each f in the closure of B, $\frac{1}{n}\sum_{i=0}^{n-1} U^i f$ converges to 0 in H.
3. For each $f \in H$, the limit of $\frac{1}{n}\sum_{i=0}^{n-1} U^i f$ exists and is the orthogonal projection of f into I. (Therefore, f^+ and f^- represent the same element $\bar{f}$ of $L^2(X, \mathcal{B}, \mu)$ and $\bar{f}$ is the projection of f onto the subspace of $\mathbb{R}$-invariant square-integrable functions.)

We give now a proof of Birkhoff's theorem for $\mathbb{Z}$-actions, taken from [16]. The key point is to show that the limits exist. We write $F = \Phi_1$, so that the $\mathbb{Z}$-action is generated by F. Let $\mathcal{I} := \{A \in \mathcal{B} \mid F^{-1}(A) = A\}$ be the σ-algebra of F-invariant sets.

For each $f_1 \in L^1(X, \mathcal{B}, \mu)$, define

$$H_n := \max\left\{ \left. \sum_{i=0}^{k-1} f_1 \circ F^i \;\right|\; k \leq n \right\}.$$

It follows from the definition that H_n is a monotonically increasing sequence of functions and that

$$H_{n+1}(x) = f_1(x) + \max\{0, H_n(F(x))\}$$

for each $x \in X$.

Define $A := \{x \in X \mid \sup_n \sum_{i=0}^{n} f_1(F^i(x)) = \infty\} \in \mathcal{I}$ and notice that the difference

$$H_{n+1}(x) - H_n(F(x)) = f_1(x) - \min\{0, H_n(F(x))\} \geq f_1(x)$$

decreases to $f_1(x)$ for each $x \in A$. We write

$$0 \leq \int_A (H_{n+1} - H_n)\,d\mu \text{ (since } H_n \text{ is monotonically increasing)}$$

$$= \int_A (H_{n+1} - H_n \circ F)\,d\mu \text{ (since } \mu \text{ is } F\text{-invariant)}$$

$$\to \int_A f_1\,d\mu,$$

where the limit is due to the dominated convergence theorem and to the fact
that the integrand of the second integral decreases to f_1 on the set A. Therefore,
$\int_A f_1 \, d\mu \geq 0$.

Moreover, $(1/n) \sum_{k=0}^{n-1} f_1 \circ F^k \leq H_n/n$ by the definition of H_n, so for each
x in the complement of A we have

$$\limsup_{n \to \infty} \frac{1}{n} \sum_{k=0}^{n-1} f_1(F^k(x)) \leq 0.$$

Now let $f \in L^1(X, \mathcal{B}, \mu)$ and denote by $f_{\mathcal{I}}$ the Radon–Nikodym derivative of
the signed measure $\nu(C) := \int_C f \, d\mu$ with respect to μ, restricted to the σ-
algebra $\mathcal{I}$. In particular, $\int_C f \, d\mu = \int_C f_{\mathcal{I}} \, d\mu$ for all $C \in \mathcal{I}$, and $f_{\mathcal{I}}$ is $\mathcal{I}$-
measurable, hence F-invariant.

Fix $\epsilon > 0$ and define $f_1 := f - f_{\mathcal{I}} - \epsilon$. Recall that $\int_A f_1 \, d\mu \geq 0$. Notice
on the other hand that

$$\int_A f_1 \, d\mu = \int_A (f - f_{\mathcal{I}} - \epsilon) \, d\mu = -\epsilon \mu(A) \leq 0;$$

therefore $\mu(A) = 0$ and the inequality $\limsup_{n \to \infty} \frac{1}{n} \sum_{k=0}^{n-1} f_1(F^k(x)) \leq 0$
(which was seen before to hold on the complement of A) holds μ-almost
everywhere.

Since $f_{\mathcal{I}}$ is F-invariant we have

$$\frac{1}{n} \sum_{k=0}^{n-1} f_1 \circ F^k = \left(\frac{1}{n} \sum_{k=0}^{n-1} f \circ F^k \right) - f_{\mathcal{I}} - \epsilon.$$

Therefore,

$$L := \limsup_{n \to \infty} \frac{1}{n} \sum_{k=0}^{n-1} f \circ F^k \leq f_{\mathcal{I}} + \epsilon$$

μ-almost everywhere. Replacing f with $-f$ and changing sign gives

$$l := \liminf_{n \to \infty} \frac{1}{n} \sum_{k=0}^{n-1} f \circ F^k \geq f_{\mathcal{I}} - \epsilon$$

μ-almost everywhere. Therefore,

$$f_{\mathcal{I}} - \epsilon \leq l \leq L \leq f_{\mathcal{I}} + \epsilon.$$

Since $\epsilon > 0$ is arbitrary, $f^+(x) = f_{\mathcal{I}}(x)$ for μ-a.e. $x \in X$. This concludes the
proof of Birkhoff's theorem.

The obtained function $f_{\mathcal{I}}$ is the *conditional expectation* of f given the σ-algebra $\mathcal{I}$ of F-invariant Borel sets of X. In general, the conditional expectation is defined as follows. Let $\mathcal{I}$ be a σ-subalgebra of $\mathcal{B}$. Then any $f \in L^1(X, \mathcal{B}, \mu)$ defines a complex or real signed measure ν on $\mathcal{I}$ by

$$\nu(A) := \int_A f \, d\mu$$

for $A \in \mathcal{I}$, which is absolutely continuous with respect to μ (restricted to $\mathcal{I}$). By the Radon–Nikodym theorem there exists a unique (modulo sets of measure zero) function $f_{\mathcal{I}} \in L^1(X, \mathcal{I}, \mu)$ such that $\nu(A) = \int_A f_{\mathcal{I}} \, d\mu$ for all $A \in \mathcal{I}$. The conditional expectation $f_{\mathcal{I}}$ of f given $\mathcal{I}$ is characterized (almost everywhere) by the properties

1. $\int_A f_{\mathcal{I}} \, d\mu = \int_A f \, d\mu$ for all $A \in \mathcal{I}$;
2. $f_{\mathcal{I}} \in L^1(X, \mathcal{I}, \mu)$.

8.5   Ergodicity of Anosov Systems

We discuss now another idea that can be used to prove ergodicity of certain systems without the use of unitary representations. For the next proposition we assume that X is a compact metric space with metric d and μ is a finite measure, positive on open sets. Let a^t be a continuous one-parameter group of transformations (a continuous $\mathbb{R}$-action) on X. (The same discussion holds for a $\mathbb{Z}$-action.) For $x \in X$, define the *stable* and *unstable sets* $V^s(x)$ and $V^u(x)$ according to the formulas

$$V^s(x) = \left\{ y \in X \,\middle|\, \lim_{t \to +\infty} d(a^t x, a^t y) = 0 \right\},$$

$$V^u(x) = \left\{ y \in X \,\middle|\, \lim_{t \to -\infty} d(a^t x, a^t y) = 0 \right\}.$$

Proposition 8.5.1 [1, 2.6] With the definitions just given, let $f : X \to \mathbb{C}$ be an $\mathbb{R}$-invariant measurable function, where the $\mathbb{R}$-action is given by the continuous flow a^t and X is compact. Then f is almost everywhere constant on stable and unstable sets. More precisely, we can find null sets N^s and N^u such that $f(x) = f(y)$ for any $x \in X \backslash N^s$, $y \in V^s(x) \cap (X \backslash N^s)$ and $f(x) = f(z)$ for any $x \in X \backslash N^u$, $z \in V^u(x) \cap (X \backslash N^u)$.

Proof. We show the proposition for the stable set. The unstable set is treated similarly by noting that it is the stable set for the action $t \mapsto a^{-t}$.

For $C \in \mathbb{R}$ define the $\mathbb{R}$-invariant function $f_C(x) := \min\{f(x), C\}$. It suffices to prove the statement for f_C. Given $\epsilon > 0$, let $h_\epsilon : X \to \mathbb{C}$ be a continuous function such that $\int_X |f_C - h_\epsilon| \, d\mu \leq \epsilon$. The Birkhoff ergodic theorem implies that

$$h_\epsilon^+(x) = \lim_{T \to \infty} \frac{1}{T} \int_0^T h_\epsilon(a^t x) \, dt$$

exists for a.e. x. Using the invariance of μ and f_C, we obtain

$$\int_X \left| f_C(y) - \frac{1}{T} \int_0^T h_\epsilon(a^t y) \, dt \right| d\mu(y) \leq \int_X |f_C(y) - h_\epsilon(y)| \, d\mu(y).$$

Since h_ϵ is uniformly continuous, $h_\epsilon^+(y) = h_\epsilon^+(x)$ whenever $y \in V^s(x)$ and $h_\epsilon^+(x)$ is defined. Hence there is a null set N_ϵ such that h_ϵ^+ exists and is constant on the stable sets in $X \backslash N_\epsilon$. Therefore, $f_C^+(y) = \lim_{m \to \infty} h_\epsilon^+(x)$ is constant on the stable sets in $X \backslash (\bigcup_n N_{\frac{1}{n}})$. But f_C and f_C^+ agree almost everywhere.   □

The following exercise illustrates how one can use the proposition to prove ergodicity in one example. Notice that the claim in the exercise is an immediate corollary of Moore's theorem, although the suggested argument gives a more elementary proof.

Exercise 8.5.2 Let $M = SL(n, \mathbb{R})/\Gamma$, where Γ is a uniform lattice, and let A be the diagonal subgroup of $SL(n, \mathbb{R})$. Show that the action of A on M by left translations is ergodic with respect to the finite invariant measure. Hint: Choose an element X in the Lie algebra $\mathfrak{a}$ of A such that $\alpha(X) \neq 0$ for each root α. Show that the Lie algebra of $G = SL(n, \mathbb{R})$ decomposes as $\mathfrak{g} = V^- \oplus \mathfrak{a} \oplus V^+$, where V^+ (resp., V^-) is the direct sum of root spaces $\mathfrak{g}_\alpha$ such that $\alpha(X) > 0$ (resp., $\alpha(X) < 0$). The right-invariant subbundle of TG obtained by right translating $V^\pm$ defines a subbundle $E^\pm$ of M. Use the Frobenius theorem to show that $E^\pm$ is the tangent bundle of a smooth foliation $\mathcal{E}^\pm$. Show that each leaf of $\mathcal{E}^+$ (resp., $\mathcal{E}^-$) is an unstable (resp., stable) set for the action of $a^t = \exp(tX)$ by left multiplication. Show that a measurable a^t-invariant function f on M is (a.e.) constant on the leaves of $\mathcal{E}^\pm$. Using Fubini's theorem and the general facts about smooth foliations discussed in section 3.2, conclude that f is a.e. constant on M.

When $n = 2$, and the uniform lattice Γ acts without fixed points on $SL(2, \mathbb{R})/SO(2)$, then $M = \Gamma \backslash SL(2, \mathbb{R})/SO(2)$ has a natural Riemannian metric that makes it a compact surface of constant negative curvature. (We are taking the quotient by Γ on the left side now in order to conform to the more standard notation.) $V = \Gamma \backslash SL(2, \mathbb{R})$ corresponds to the circle bundle of unit vectors on M, and the flow given by right translations by the one-parameter (diagonal) subgroup A corresponds to the *geodesic flow* of M. The geodesic flow φ_t has the following geometric definition: For each $v = \Gamma g \in V$, $\varphi_t(v) = va^t$ is the unique unit vector tangent to the geodesic determined by v, at a distance t

from the base point of v, pointing in the same direction along the geodesic as v. The previous exercise shows, therefore, that the geodesic flow of a compact surface of constant negative curvature is ergodic. D. V. Anosov proved that the geodesic flow of any compact Riemannian manifold whose sectional curvatures are all negative is also ergodic.

For Anosov diffeomorphisms, we state without proof the following fact (see [18]).

Theorem 8.5.3 Let f be an Anosov diffeomorphism of a compact manifold M, and suppose that f preserves a probability measure μ on M induced from an f-invariant volume density. Then f is ergodic with respect to μ.

We saw in chapter 1 that M is "foliated" by the stable and unstable manifolds of the Anosov diffeomorphism f, giving M a local product structure (theorem 1.5.2). It may seem at first that theorem 8.5.3 is an immediate consequence of proposition 8.5.1 and the Fubini-type argument used in exercise 8.5.2. This is indeed the case if $W^s(x)$ and $W^u(x)$ depend differentiably on x. The key technical difficulty in proving the theorem (or its counterpart for Anosov flows) is that the Anosov foliations are only continuous, so if we try to compare the measures of subsets in different stable manifolds that are mapped bijectively onto each other by "sliding along" unstable manifolds, it is possible in principle for a set of positive measure to be sent to a set of measure 0. In other words, the foliation might fail to be *absolutely continuous*, and Fubini's theorem could fail to hold.

It turns out that W^s and W^u are indeed absolutely continuous. An elementary account of this fact can be found in [1].

8.6   Amenability and Kazhdan's Property T

Moore's theorem is the first result in ergodic theory that we have encountered so far in this book that is specific to semisimple Lie groups. We discuss now another property that is special to semisimple groups of "higher rank," called the *Kazhdan property* or *property T*.

Suppose for now that G is only a locally compact second countable group and $\rho : G \to U(H)$ a (strongly continuous) unitary representation of G on a separable Hilbert space H. We say that ρ *almost has invariant vectors* if for each $\epsilon > 0$ and compact subset $K \subset G$, there exists a unit vector $v \in H$ such that $\|\rho(g)v - v\| < \epsilon$ for all $g \in K$.

Take, for example, $G = \mathbb{R}$ and define a unitary representation ρ of $\mathbb{R}$ into $H = L^2(\mathbb{R}, \lambda)$, where λ is Lebesgue measure, by $(\rho(t)f)(x) = f(x-t)$. Then

ρ almost has invariant vectors. In fact, given any $\epsilon > 0$ and compact $K \subset \mathbb{R}$, choose an interval I of length l, and let f_I be the characteristic function of I normalized so as to have L^2 norm equal to 1. It is immediate that we can make $\|\rho(t)f_I - f_I\| < \epsilon$ for all $t \in K$ by taking l to be sufficiently large. Notice, on the other hand, that ρ cannot have nonzero invariant vectors, since the only $\mathbb{R}$-invariant functions on $\mathbb{R}$ are constant, and nonzero constant functions are not in $L^2(\mathbb{R}, \lambda)$.

Returning to a general G, we say that G has the *Kazhdan property* if any unitary representation ρ of G that almost has invariant vectors actually has a unit invariant vector. By the previous paragraph, $\mathbb{R}$ does not have the Kazhdan property.

Exercise 8.6.1 Show that if G is a compact group, then G has the Kazhdan property. (For each $\epsilon > 0$, take a unit vector v such that $\|\rho(g)v - v\| < \epsilon$. Now average v over G, that is, take $\bar{v} := \int_G \rho(g)v \, d\lambda(g)$, where λ is a left-invariant Haar measure on G, normalized so that $\lambda(G) = 1$, and show that $\bar{v}$ is invariant.)

Exercise 8.6.2 For K a compact Lie group and λ a Haar measure on $G = \mathbb{R}^n \times \mathbb{Z}^m \times K$, show that a unitary representation of G on $L^2(G, \lambda)$ defined by $(\rho(g)f)(x) = f(g^{-1}x)$ almost has invariant vectors. Conclude that G does not have the Kazhdan property, unless $G = K$.

Suppose λ is a left Haar measure on G. The unitary representation ρ of G on $L^2(G, \lambda)$ defined by $(\rho(g)f)(x) := f(g^{-1}x)$ is called the *regular representation* of G. We call G an *amenable group* if its regular representation almost has invariant vectors. The previous exercises show that abelian and compact Lie groups are amenable. In fact, the following holds (see [36, 4.1.7]).

Proposition 8.6.3 Suppose that a Lie group G contains a normal solvable subgroup R such that G/R is compact. Then G is amenable.

A fundamental fact about the Kazhdan property is contained in the next theorem.

Theorem 8.6.4 (Kazhdan) Let G be a connected semisimple Lie group with finite center, each of whose factors has $\mathbb{R}$-rank at least 2. Then G, as well as any lattice in G, has the Kazhdan property.

A proof of this theorem would require a long detour into the general theory of unitary group representations and will not be pursued here. We refer the reader to [36].

It was noted in the preceding exercises that compact groups both have the Kazhdan property and are amenable. The converse is also true, that is, if G is amenable and has the Kazhdan property, then G is compact. Namely, by amenability, the regular representation of G almost has invariant vectors, whence it has a nonzero invariant vector by the Kazhdan property. But the only invariant functions on G are the constant functions, and these are integrable if and only if G has finite Haar measure. Having finite measure implies that G is compact. (Note: It is easy to check that the action of G on itself by left translation is a proper action, that is, given any compact subset $K \subset G$, the set $\{g \in G \mid gK \cap K \neq \emptyset\}$ is relatively compact. Therefore, if G is not compact and K is any compact subset of positive measure, we can find a sequence $g_n \in G$ such that all the translates $g_n K$ are disjoint and have the same measure. By σ-additivity of the measure we conclude that the measure of G is infinite.)

Notice that if G has the Kazhdan property and $\rho : G \to A$ is a surjective homomorphism, then A also has the Kazhdan property. In fact, if π is a unitary representation of A on a separable Hilbert space H such that π almost has invariant vectors, then $\pi \circ \rho$ also almost has invariant vectors, and hence there exists a unit vector $v \in H$ fixed by $\pi(\rho(G)) = \pi(H)$. An immediate consequence of this remark is that the quotient $G/(G, G)$ of G by its commutator is compact, since it is an abelian group with the Kazhdan property. In particular, if G is a reductive group with the Kazhdan property, its center must be a compact group.

The remark of the previous paragraph, that a homomorphic image of a Kazhdan group is also Kazhdan, has a very useful generalization for actions of Kazhdan groups by automorphisms of principal bundles. (Notice that it makes sense to think of a homomorphism of G into a group A as defining an action of G on the principal A-bundle A, whose base space is a single point.) Before stating the theorem, we make some general remarks about vector bundles with Hilbert spaces as fibers.

Let A be a locally compact second countable group and Λ a left Haar measure on A. Denote by $(H_0, \langle \cdot, \cdot \rangle_0)$ the Hilbert space $L^2(A, \lambda)$ with its usual inner product. Let $p : P \to M$ be a principal A-bundle over a measure space M. (This is really just the product $M \times A$, as far as the measurable structure is concerned, but it will be convenient for our later needs to use the principal-bundle language here.) Let μ be a probability measure on M and suppose that G acts on P by bundle automorphisms such that the action on M leaves μ invariant.

Given any ξ in the fiber P_x of $x \in M$, we can identify P_x with A via the obvious map $\xi : A \to P_x, \xi(a) = \xi a$. Notice that $\lambda_x := \xi_* \lambda$ is an A-invariant measure on P_x and is independent of the choice of $\xi \in P_x$, due to the fact that λ is a left-invariant measure on A. Define a Hilbert space $(E(x), \langle \cdot, \cdot \rangle_x)$ by $E(x) = L^2(P_x, \lambda_x)$, with the usual L^2 inner product. Then each $\xi \in P_x$ induces an isometry between H_0 and $E(x)$, by sending $f \in H_0$ to $f \circ \xi^{-1} \in E(x)$.

Denote by $\pi : A \to H_0$ the regular representation of A. Each vector in $E(x)$ will be written as ξv, where $\xi \in P_x$ and $v \in H_0$. Notice that $(\xi a)v = \xi(\pi(a)v)$, for each $a \in A$, and that $\langle \xi v, \xi w \rangle_x = \langle v, w \rangle_0$, for each $\xi \in P_x$.

We define a (measurable) vector bundle E over M, as in chapter 6, by taking the associated bundle induced by π. In other words, $E = (P \times H_0)/A$, where A acts on the product by

$$(\xi, v) \cdot a := (\xi a, \pi(a)^{-1} v).$$

(We remind the reader again that E is simply the product $M \times E$, once a measurable trivialization of P has been chosen.)

Let $(H, \langle \cdot, \cdot \rangle)$ denote the Hilbert space of L^2 sections of E, with inner product defined as follows:

$$\langle \phi, \psi \rangle := \int_M \langle \phi(x), \psi(x) \rangle_x \, d\mu(x)$$

for $\phi, \psi \in H$.

Suppose now that a group G acts on P via bundle automorphisms such that the action on M is measure preserving. Then G also acts on H, as follows. If $g \in G$, and $\phi \in H$ is represented by a section of E, also denoted ϕ, we define another section $g_*\phi$ of E by

$$(g_*\phi)(x) = g\phi(g^{-1}x).$$

This action indeed preserves the inner product of H. In fact, first notice that

$$\langle \xi v, \xi w \rangle_x = \langle v, w \rangle_0 = \langle g\xi v, g\xi w \rangle_{gx}.$$

Therefore, since μ is G-invariant,

$$\langle g_*\phi, g_*\psi \rangle = \int_M \langle g\phi(g^{-1}x), g\psi(g^{-1}x) \rangle_x \, d\mu(x)$$

$$= \int_M \langle \phi(g^{-1}x), \psi(g^{-1}x) \rangle_{g^{-1}x} \, d\mu(x)$$

$$= \int_M \langle \phi(x), \psi(x) \rangle_x \, d\mu(x)$$

$$= \langle \phi, \psi \rangle.$$

If σ is a section of P and $g \in G$, we also denote by $g_*\sigma$ the new section such that $(g_*\sigma)(x) = g\sigma(g^{-1}x)$.

Lemma 8.6.5 Suppose the setup and notation are those of the previous paragraph. If A is an amenable group and G is a second countable locally compact group acting on P by bundle automorphisms, then the unitary representation of G on H almost has invariant vectors.

Proof. (This argument is contained in the proof of [36, 9.1.1].) Fix a measurable section σ of P, and define for each subset $A_0 \subset A$ and $g \in G$ the set

$$S(g, A_0) := \{x \in M \mid (g_* \sigma)(x), (g_*^{-1} \sigma)(x) \in \sigma(x)A_0\}.$$

For each $g \in G$, the sets $S(g, A_i)$ increase to M as the A_i increase to A.

Let K be any compact symmetric subset of G, and let η denote the left Haar measure on G. Since A is σ-compact, we can find a sequence of compact subsets $A_i \subset A$, increasing to A, such that $(\eta \times \mu)(\{(g, x) \in K \times M \mid x \in S(g, A_i)\})$ increases to $\eta(K)$. Therefore, for each $\epsilon > 0$, we can find a compact subset $A_0 \subset A$ such that

$$(\eta \times \mu)(\{(g, x) \in K \times M \mid x \in S(g, A_0)\}) \geq 1 - \frac{\epsilon}{3}\eta(K).$$

By an application of Fubini's theorem, we have $\eta(K_0) > 0$, where

$$K_0 := \left\{ g \in K \mid \mu(S(g, A_0)) \geq 1 - \frac{\epsilon}{3} \right\}.$$

K_0 is clearly symmetric, due to the way $S(g, A_0)$ was defined. Recall (lemma 8.2.4) that $K_0^2 = K_0 K_0^{-1}$ contains an open neighborhood of the identity element of G. We denote one such neighborhood by W.

It is easy to check that

$$S(g_1 g_2, A_1 A_2) \supset S(g_1, A_1) \cap g_1 S(g_2, A_2)$$

for any $g_1, g_2 \in G$ and subsets $A_1, A_2 \subset A$. Thus, for each $g = g_1 g_2 \in K_0^2$,

$$\mu\left(S\left(g, A_0^2\right)\right) \geq \mu(S(g_1, A_0) \cap g_1 S(g_2, A_0)) \geq 1 - \frac{2}{3}\epsilon,$$

where we have used the fact that μ is G-invariant.

Since K is compact, we can find $g_1, \ldots, g_l$ in G such that $K \subset \bigcup_{i=1}^n g_i W$. We can also choose a sufficiently large compact subset $B \subset A$ such that

$$\mu(S(g_i, B)) > 1 - \frac{\epsilon}{3}$$

for each i. The argument used to show $\mu(S(g, A_0^2)) \geq 1 - 2\epsilon/3$ now implies

$$\mu\big(S\big(g, BA_0^2\big)\big) > 1 - \epsilon$$

for each $g \in K$.

Since A is amenable, we can find $v \in H_0$ such that

$$\|\pi(a)v - v\|_0 < \epsilon$$

for all $a \in BA_0^2$. Define a section φ_v of E by $\varphi_v(x) := \sigma(x)v$. Notice that, for each $x \in S(g, BA_0^2)$, we have $(g_*\varphi_v)(x) = (g_*\sigma)(x)v \in \sigma(x)\pi(BA_0^2)v$, so $\|(g_*\varphi_v)(x) - \varphi_v(x)\|_x < \epsilon$ for each $x \in S(g, BA_0^2)$. Therefore, for each $g \in K$, we have

$$\mu(\{x \in M \mid \|(g_*\varphi_v)(x) - \varphi_v(x)\| < \epsilon\}) \geq \mu\big(S\big(g, BA_0^2\big)\big) > 1 - \epsilon.$$

Since the function $\|(g_*\varphi_v)(x) - \varphi_v(x)\|^2$ is bounded above by $4\|v\|_0$, it follows that

$$\int_M \|(g_*\varphi_v)(x) - \varphi_v(x)\|^2 \, d\mu(x) \leq C^2\epsilon$$

for some positive constant C. The conclusion is that $\|g_*\varphi_v - \varphi_v\| \leq C\epsilon^{\frac{1}{2}}$. Since ϵ is arbitrary, the lemma follows.   □

If G has the Kazhdan property and the conditions of the previous lemma are in place, then there exists a nontrivial G-invariant L^2-integrable section of E. Denoting such a section by φ, invariance means that $g_*\varphi = \varphi$ for each $g \in G$. Notice that $\|\varphi(x)\|_x = \|(g_*\varphi)\|_x = \|g\varphi(g^{-1}x)\|_x = \|\varphi(g^{-1}x)\|_{g^{-1}x}$, so $\|\varphi(x)\|$ is a G-invariant function. By ergodicity, this is an essentially constant function. Therefore, after multiplying φ by a suitable constant, we may assume without loss of generality that $\|\varphi(x)\|_x$ is equal to 1 for μ-a.e. x.

The section φ can also be described by means of an A-equivariant G-invariant measurable map $\mathcal{G} : P \to L^2(A, \lambda)_1$, where the latter set is the space of unit vectors in H_0 (recall proposition 6.2.1). Namely,

$$\mathcal{G}(\xi) := \varphi(p(\xi)) \circ \xi$$

(each $\xi \in P$ being identified with the map $\xi : A \to P_x$ that sends a into ξa; thus the notation). Equivariance means that $\mathcal{G}(\xi a) = \pi(a)^{-1}\mathcal{G}(\xi)$, where π is the regular representation of A.

We are ready to prove the main result of the section.

Theorem 8.6.6 (Zimmer) Let G be a group with the Kazhdan property. Suppose that G acts by automorphisms of a principal A-bundle P over a measure space M, leaving invariant an ergodic probability measure μ on M. Suppose that A is an amenable group. Then there exists a G-invariant measurable K-reduction of P, where K is a compact subgroup of A.

Proof. For each f, $f' \in H_0 = L^2(A, \lambda)$, we claim that $\langle \pi(a)f, f' \rangle$ goes to 0 as a goes to infinity in A. This can be seen as follows. Choose sufficiently large compact sets K_1 and K_2 in A such that $\|\chi_{K_1} f - f\|$ and $\|\chi_{K_2} f' - f'\|$ are small, where χ_{K_i} is the characteristic function of K_i. For all a outside a sufficiently large compact set, we have $aK_1 \cap K_2 = \emptyset$. It follows from this and a simple computation that $\langle \pi(a)f, f' \rangle$ can be made as small as we like.

By the previous paragraph, each A-orbit in H_0 is closed in norm. To see this, suppose that $a_n \in A$ is a sequence such that $\|\pi(a_n)f - f'\| \to 0$ as $n \to \infty$. We claim that f' lies in the A-orbit of f. If f and f' are both 0, there is nothing to prove. Suppose that one of the two is not the zero element of H_0 and write

$$\|\pi(a_n)f - f'\|^2 = \|f\|^2 + \|f'\|^2 - \langle \pi(a_n)f, f' \rangle - \langle f', \pi(a_n)f \rangle$$

(where we have used $\|\pi(a_n)f\| = \|f\|$). Therefore, the set $\{a_n\}$ is relatively compact in A and the claim holds. Theorem 1.2.2 (and exercise 8.2.1) can now be used to conclude that the A-action on $L^2(A, \lambda)_1$ is tame.

We already know that there is a measurable G-invariant A-equivariant function $\mathcal{G} : P \to L^2(A, \lambda)_1$. Thus, we may apply proposition 6.4.2. In that proposition it was assumed that V, whose role is played here by $L^2(A, \lambda)_1$, is an algebraic A-space. A quick glance at the short proof shows, however, that all that was needed there is the fact that algebraic actions are tame. Therefore, the same proof allows us to conclude that $\mathcal{G}$ takes values in a single A-orbit $\pi(A)f_0$, for some f_0 in $L^2(A, \lambda)_1$. We obtain in this way a measurable A_0-reduction of P, where A_0 is the isotropy subgroup of f_0.

It remains to show that the isotropy subgroup of a nonzero $f_0 \in L^2(A, \lambda)$ is a compact subgroup of A. But this is clear from the remark made at the start of this proof that $\pi(a)f_0$ goes to 0 as a goes to infinity. $\square$

Exercise 8.6.7 Suppose that a group G with the Kazhdan property acts smoothly on a manifold M and that the action preserves a Borel probability measure and a *conformal structure*, that is, there exists a Riemannian metric $\eta(x) = \langle \cdot, \cdot \rangle_x$ on TM such that $(g_*\eta)(x) = \lambda(g, x)\eta(x)$, where $\lambda(g, x)$ is a positive measurable function and $(g_*\eta)_x(u, v) := \eta_{g^{-1}x}(Dg_x^{-1}u, Dg_x^{-1}v)$. (This means that the action preserves angles, as defined by η, but not necessarily lengths. We say that η and $g_*\eta$ are in the same conformal class.) Show that there exists a G-invariant measurable Riemannian metric on TM in the same conformal class as η.

9

Oseledec's Theorem

We saw in chapter 8 that the existence of a finite invariant measure for a $\mathbb{Z}$- or $\mathbb{R}$-action forces almost all orbits of the system to behave in the long run in the statistically regular manner described by Birkhoff's ergodic theorem. The present chapter shows another kind of regularity in the asymptotic behavior of measure-preserving actions, now having to do with the linearization of the action along orbits.

We begin by characterizing the exponential rates at which "infinitesimally close" orbits diverge from or approach one another. Those rates are called the *Lyapunov exponents* and are important initial data from which much information about the global properties of the system can be derived.

The main goal of the present chapter is to prove that those exponential rates actually exist, for almost every orbit, for smooth $\mathbb{Z}$-actions on a compact manifold preserving a probability measure. This is essentially the content of Oseledec's multiplicative ergodic theorem.

We end the chapter with a brief discussion of Pesin's theory. For more details and proofs on this subject, we refer the reader to [23], [16], [27], and the references cited therein.

9.1 Lyapunov Exponents

Let $(M, \mathcal{B})$ be a measurable space and $\pi : E \to M$ a "measurable vector bundle" over M. By that we mean simply that E is a measurable space isomorphic to the product $M \times V$, where V is a finite-dimensional vector space over $\mathbb{R}$, and the isomorphism maps fibers to fibers linearly.

Even though our measurable bundles are trivial, the more geometric terminology will be useful, since E will often arise as a topological bundle for which only the Borel structure is of immediate interest, yet we may not wish to select any particular trivialization. Notice, however, that any topological vector

174

bundle over a manifold M is trivial when regarded as a measurable (Borel) vector bundle.

The basic definitions concerning vector and principal bundles were introduced at the beginning of chapter 6. We fix on each fiber E_x a norm $\| \cdot \|_x$ coming from an inner product $\langle \cdot, \cdot \rangle_x$ that depends measurably on x. Measurability of $x \mapsto \langle \cdot, \cdot \rangle_x$ means that if X and Y are measurable sections of E (i.e., sections that are also measurable maps from M into E), then the function $x \mapsto \langle X(x), Y(x) \rangle_x$ is measurable. By a measurable *automorphism* of E we mean a measurable bijection with measurable inverse, $F : E \to E$, such that $\pi \circ F = f \circ \pi$, where $f : M \to M$ is an isomorphism of the measurable space and, for each $x \in M$, $F_x := F|_{E(x)} : E(x) \to E(f(x))$ is a linear isomorphism.

If F is a measurable automorphism of E and $v \in E(x)$ is a nonzero element, we define the *upper Lyapunov exponent* of v by

$$\bar{\chi}^+(v) := \limsup_{m \to +\infty} \frac{1}{m} \ln \| F_x^m v \|.$$

If the limit, rather than lim sup, exists, we call it the *Lyapunov exponent of v*. We define $\bar{\chi}^+(0) = -\infty$.

Exercise 9.1.1 Let $v, w \in E(x)$ and $\lambda \neq 0$, $\lambda \in \mathbb{R}$. Show:

1. $\bar{\chi}^+(\lambda v) = \bar{\chi}^+(v)$.
2. $\bar{\chi}^+(v + w) \leq \max\{\bar{\chi}^+(v), \bar{\chi}^+(w)\}$. Equality holds if $\bar{\chi}^+(v) \neq \bar{\chi}^+(w)$.

Conclude that

$$\{0\} \cup \{v \in E(x) - \{0\} \mid \bar{\chi}^+(v) \leq a\}$$

is a vector subspace of $E(x)$. Hint: For the last claim in item 2, notice that if $\bar{\chi}^+(v) < \bar{\chi}^+(w)$, say, then

$$\bar{\chi}^+(v + w) \leq \bar{\chi}^+(w) = \bar{\chi}^+(w - v + v)$$
$$\leq \max\{\bar{\chi}^+(v + w), \bar{\chi}^+(-v)\} = \bar{\chi}^+(v + w).$$

Fix a real number $\chi > 0$ and $x \in M$. We define

$$E_\chi(x) := \{0\} \cup \{v \in E(x) - \{0\} \mid \bar{\chi}^+(v) \leq \chi\}.$$

By the previous exercise, $E_\chi(x)$ is a vector space for each $x \in M$. Moreover, if $\chi_1 \leq \chi_2$, then $E_{\chi_1}(x)$ is a subspace of $E_{\chi_2}(x)$. If n is the (finite) dimension of E, we can find for each $x \in M$ an integer $k(x) \leq n$ and real numbers $\chi_1(x) < \chi_2(x) < \cdots < \chi_{k(x)}(x)$ such that

$$\{0\} \subset E_{\chi_1(x)}(x) \subset \cdots \subset E_{\chi_{k(x)}(x)}(x) = E(x)$$

and for each $v \in E_{\chi_{i+1}}(x) - E_{\chi_i}(x)$, $\bar{\chi}^+(v) = \chi_{i+1}(x)$. (The inclusions are proper.)

The *multiplicity* of $\chi_i(x)$ is defined by

$$l_i(x) := \dim E_{\chi_i}(x) - \dim E_{\chi_{i-1}}(x).$$

Being the lim sup of a sequence of measurable functions on E, $\bar{\chi}^+$ is a measurable function on E. Therefore, for each χ, E_χ, as well as k and l_i, depends measurably on $x \in M$. The measurable subbundle E_χ, $\chi \in \mathbb{R}$, is F-invariant, that is,

$$F_x E_\chi(x) = E_\chi(f(x)), \qquad \chi(Fv) = \chi(v),$$

as a simple calculation shows. Consequently k, χ_i, and l_i are all f-invariant functions.

Exercise 9.1.2 Denote by $\bar{\chi}^+(F, v)$ the upper exponent of v, making explicit the automorphism F, and define

$$\underline{\chi}^-(F, v) := \liminf_{m \to -\infty} \frac{1}{m} \ln \|F_x^m v\|.$$

Show that $\underline{\chi}^-(F, v) = \bar{\chi}^+(F^{-1}, v)$. Conclude that

$$\{0\} \cup \{v \in E(x) - \{0\} \mid \underline{\chi}^-(F, v) \geq a\}$$

is a vector subspace of $E(x)$.

Two natural questions are whether the filtration of measurable subbundles E_{χ_i} arises from a (measurable) direct sum decomposition of E into F-invariant subbundles and whether the lim sup involved in the definition of exponents can be replaced with lim. When this is the case, we can think of the decomposition as a kind of "asymptotic spectral decomposition" for the action of F on E (or, more precisely, the $\mathbb{Z}$-action on E generated by F). Instead of eigenvalues, we have well-defined exponential rates at which $\|F_x^m v\|_{f^m(x)}$ grows as $m \to \pm\infty$. Points x for which such decomposition exists will be called regular. More precisely, we say that $x \in M$ is a *regular point* for the action of F on E if we can find a positive integer $k(x)$, $1 \leq k(x) \leq n$, where n is the fiber dimension of E, and numbers $\chi_1(x) < \cdots < \chi_{k(x)}(x)$, $\chi_i(x) \in \mathbb{R}$, as well as subspaces $S_i(x)$, $1 \leq i \leq k(x)$, such that the following properties are satisfied:

1. $E(x) = S_1(x) \oplus \cdots \oplus S_{k(x)}(x)$
2. The sequence

$$\frac{1}{m} \ln \left(\frac{\|F_x^m v\|}{\|v\|} \right)$$

converges to $\chi_i(x)$ uniformly in $v \in S_i(x) - \{0\}$ as $m \to \pm\infty$.

3. For each subset $\mathcal{I} \subset \{1, 2, \ldots, k(x)\}$, set

$$E_{\mathcal{I}}(x) = \bigoplus_{i \in \mathcal{I}} S_i(x), \qquad E_{\mathcal{I}^c}(x) = \bigoplus_{i \in \mathcal{I}^c} S_i(x)$$

and let $0 \leq \alpha(m, x) \leq \pi$ be the angle between $F_x^m E_{\mathcal{I}}(x)$ and $F_x^m E_{\mathcal{I}^c}(x)$. Then

$$\lim_{m \to \pm\infty} \frac{1}{m} \ln \sin \alpha(m, x) = 0.$$

4. For each $v \in E(x)$, the limits

$$\chi^+(v) := \lim_{m \to +\infty} \frac{1}{m} \ln \| F_x^m v \|, \qquad \chi^-(v) := \lim_{m \to -\infty} \frac{1}{m} \ln \| F_x^m v \|$$

exist and for each $a \in \mathbb{R}$

$$\bigoplus_{\chi_i \leq a} S_i(x) = \{0\} \cup \{v \in E(x) - \{0\} \mid \chi^+(v) \leq a\}, \qquad (9.1)$$

$$\bigoplus_{\chi_i \geq a} S_i(x) = \{0\} \cup \{v \in E(x) - \{0\} \mid \chi^-(v) \geq a\}. \qquad (9.2)$$

By (9.1), the numbers $\chi_i(x)$ and $k(x)$ as well as the filtration

$$E_{\chi_1}(x) \subset \cdots \subset E_{\chi_{k(x)}}(x)$$

are uniquely determined. Moreover, (9.1) and (9.2) imply that each $L_i(x)$ is also uniquely determined. It follows, in particular, that x is regular if and only if $f(x)$ is regular and that $k(f(x)) = k(x)$, $\chi_i(f(x)) = \chi_i(x)$, and $F_x S_i(x) = S_i(f(x))$.

Exercise 9.1.3 Let M be a compact smooth manifold, and let E be a subbundle of TM with one-dimensional fibers. Let μ be a probability measure on M, and let $f : M \to M$ be a diffeomorphism under which μ and E are invariant. Show that there exists a measurable f-invariant function $\chi : M \to \mathbb{R}$ such that for μ-a.e. $x \in M$ and each nonzero $v \in E(x)$,

$$\lim_{m \to \pm\infty} \frac{1}{m} \ln \| Df_x^m v \| = \chi(x)$$

(in particular, the limit exists). Hint: Use Birkhoff's ergodic theorem. More precisely, let X be a measurable vector field of length 1 with respect to some Riemannian metric on TM and define a function $g : \mathbb{Z} \times M \to \mathbb{R}$ by the equation $Df_x^m X(x) = g(m, x) X(f^m(x))$. Then notice that, for positive m,

$$\ln |g(m, x)| = \sum_{i=0}^{m-1} \ln |g(1, f^i(x))|.$$

Exercise 9.1.4 Let x be a regular point for the action of F on E, and let $\chi_i(x)$ be the Lyapunov exponents. Show that x is a regular point for the natural action of F on the dual bundle E^*, having exponents $-\chi_i(x)$. (Given $\alpha \in E^*(x)$, we define $F_x\alpha = \alpha \circ F_{f(x)}^{-1}$.) If F_1 and F_2 are automorphisms of the vector bundles E_1 and E_2 over M, above the same map $f : M \to M$, and x is a common regular point with Lyapunov exponents $\chi_i^{(1)}(x)$ and $\chi_j^{(2)}(x)$, then x is a regular point for the action of $F_1 \otimes F_2$ on $E_1 \otimes E_2$, with exponents $\chi_i^{(1)}(x) + \chi_j^{(2)}(x)$.

Show that if F leaves invariant a measurable volume form on E (i.e., a nonvanishing alternating n-form on E, where n is the fiber dimension of E) then the sum of the exponents $\chi_i(x)$ is 0 for each regular point x. (The same is true if F leaves invariant a volume density.) How can this fact be generalized if instead of a volume form F leaves invariant some measurable tensor field on E?

The following fundamental theorem will be proven later in the chapter. It is also referred to as the *multiplicative ergodic theorem*. Notice that Birkhoff's theorem is a special case of Oseledec's theorem, when E has 1-dimensional fibers (exercise 9.1.4).

Theorem 9.1.5 (Oseledec) Let $(M, \mathcal{B}, \mu)$ be a measure space with finite measure μ, let $\pi : E \to M$ be a measurable vector bundle over M with finite-dimensional fibers, and fix a measurable Riemannian metric on the fibers with norms denoted $\| \cdot \|_x$. (We will often omit the subscript x.) Let $F : E \to E$ be a measurable bundle automorphism such that $\pi \circ F = f \circ \pi$, where f is a measure-preserving transformation of M. Suppose, moreover, that

$$\int_M \max\{\ln\|F_x\|, \ln\|F_x^{-1}\|, 0\}\, d\mu(x) < \infty.$$

Then μ-a.e. $x \in M$ is a regular point.

Taking a cue from exercise 9.1.4, we reduce the proof of Oseledec's theorem to studying the asymptotic properties of products of matrices, as follows. Choose a measurable section σ of the bundle of orthonormal frames of E. Thus, for each $x \in M$, $\sigma(x)$ is a linear isomorphism that sends an orthonormal basis relative to the Euclidean metric on $\mathbb{R}^n$ to an orthonormal basis relative to the Riemannian metric on $E(x)$. The automorphisms $F_x : E(x) \to E(f(x))$ can now be expressed, by means of σ, as a matrix $L(x) \in GL(n, \mathbb{R})$, and F_x^m (for, say, m positive) corresponds to the matrix $A(m, x) = L(f^{m-1}(x)) \cdots L(f(x))L(x)$.

In the next section, we consider the case in which L takes values in the group of lower-triangular matrices. The proof of the theorem will later be reduced to this special case.

9.2 Products of Triangular Matrices

We first recall some elementary facts about matrix norms that we may use later without explicit mention. Given $u = (u_1, \ldots, u_n) \in \mathbb{R}^n$, we define $\|u\|_2 = \langle u, u \rangle^{\frac{1}{2}}$, where $\langle \cdot, \cdot \rangle$ denotes the standard dot product on $\mathbb{R}^n$. We also set $\|u\|_1 := \sum_{i=1}^{n} |u_i|$.

Any two vector norms on $\mathbb{R}^n$ are equivalent. It can easily be shown, for example, (by using that $\phi(x) = x^2$ is a convex function) that for all $u \in \mathbb{R}^n$,

$$\|u\|_2 \leq \|u\|_1 \leq \sqrt{n}\, \|u\|_2.$$

For an arbitrary norm $\| \cdot \|$ on $\mathbb{R}^n$, one defines a norm on linear transformations T of $\mathbb{R}^n$ by $\|T\| := \max\{\|Tu\| \mid \|u\| \leq 1\}$. Writing $T = (T_{ij})$, we have [13]

$$\|T\|_1 = \max\left\{ \sum_{i=1}^{n} |T_{ij}| \;\middle|\; j = 1, \ldots, n \right\},$$

$$\|T\|_2 = \max\{\sqrt{\lambda} \mid \lambda \text{ is an eigenvalue of } T^*T\},$$

where the dual T^* is defined by the equation $\langle T^*u, v \rangle = \langle u, Tv \rangle$, for all $u, v \in \mathbb{R}^n$.

Exercise 9.2.1 Let $\| \cdot \|$ denote either $\| \cdot \|_1$ or $\| \cdot \|_2$. For any $T \in GL(n, \mathbb{R})$, show that

$$-n \ln\|T^{-1}\| \leq \ln|\det T| \leq n \ln\|T\|.$$

(Hint: It suffices to show $|\det T| \leq \|T\|_2^n$ for all $T \in GL(n, \mathbb{R})$. Prove it first for a diagonal matrix and use the KAK decomposition to obtain the general case. If D is a diagonal matrix with positive eigenvalues, $\|D\|_2$ is the value of the maximal eigenvalue of D.)

We now wish to study the iteration of sequences of linear maps such as

$$\cdots \to \mathbb{R}^n \xrightarrow{L_{i-1}} \mathbb{R}^n \xrightarrow{L_i} \mathbb{R}^n \xrightarrow{L_{i+1}} \mathbb{R}^n \to \cdots$$

where each L_i belongs to $GL(n, \mathbb{R})$. More precisely, we wish to study the sequence of m-fold compositions, $A(m)$, defined by

$$A(m) : \ \mathbb{R}^n \xrightarrow{L_0} \mathbb{R}^n \xrightarrow{L_1} \cdots \xrightarrow{L_{m-1}} \mathbb{R}^n$$

$$A(-m) : \ \mathbb{R}^n \xrightarrow{(L_{-1})^{-1}} \mathbb{R}^n \xrightarrow{(L_{-2})^{-1}} \cdots \xrightarrow{(L_{-m})^{-1}} \mathbb{R}^n$$

$m \in \mathbb{N}$. We also set $A(0) = \mathrm{Id}$. If $L_i = L$ is constant, $A(m) = L^m$ for all $m \in \mathbb{Z}$.

We say that the sequence $\{L_i \mid i \in \mathbb{Z}\}$ is *tempered* if for $\delta = \pm 1$

$$\lim_{i \to \pm\infty} \frac{1}{i} \ln \left\| L_i^\delta \right\| = 0.$$

Notice that the notion does not depend on the norm used.

Two sequences $\{L_i^{(1)}\}$ and $\{L_i^{(2)}\}$ will be said to be *cohomologous* if there exists a tempered sequence $\{\Phi_i \in GL(n, \mathbb{R}) \mid i \in \mathbb{Z}\}$ making the following diagram commute:

$$\cdots \xrightarrow{L_{i-2}^{(1)}} \mathbb{R}^n \xrightarrow{L_{i-1}^{(1)}} \mathbb{R}^n \xrightarrow{L_i^{(1)}} \mathbb{R}^n \xrightarrow{L_{i+1}^{(1)}} \cdots$$

$$\downarrow{\Phi_{i-1}} \qquad \downarrow{\Phi_i} \qquad \downarrow{\Phi_{i+1}}$$

$$\cdots \xrightarrow{L_{i-2}^{(2)}} \mathbb{R}^n \xrightarrow{L_{i-1}^{(2)}} \mathbb{R}^n \xrightarrow{L_i^{(2)}} \mathbb{R}^n \xrightarrow{L_{i+1}^{(2)}} \cdots$$

It is easy to check that if two sequences are cohomologous and one of them is tempered, then the other one also is. We will refer to cohomologous sequences simply as *equivalent*. (Later on, our sequences of matrices will be interpreted as the values of a "cocycle over a group action," and in that context the term "cohomologous cocycles" is standard usage.)

Lemma 9.2.2 Let L_i be a tempered sequence of lower-triangular matrices in $GL(n, \mathbb{R})$, and let $A(m)$ be as defined before. We denote by $a_{ij}(m)$ the entries of $A(m)$. Suppose that, for some $\chi \in \mathbb{R}^n$ and each j, $1 \le j \le n$, we have $\lim_{m \to \pm\infty} \frac{1}{m} \ln|a_{jj}(m)| = \chi$. Then, for each nonzero $v \in \mathbb{R}^n$,

$$\lim_{m \to \pm\infty} \frac{1}{m} \ln \| A(m)v \| = \chi.$$

This holds for an arbitrary norm $\| \cdot \|$ on $\mathbb{R}^n$. Moreover, if $\{L_i^{(1)}\}$ is another sequence equivalent to $\{L_i\}$ and $A^{(1)}(m)$ is the sequence of m-fold products of the $L_i^{(1)}$, then the corresponding limit exists and takes the same value χ.

Proof. We denote by $l_{ij}(m)$, $\bar{l}_{ij}(m)$, and $a_{ij}(m)$ the entries of L_m, L_m^{-1}, and $A(m)$, respectively. Notice that, for $m \ge 1$, $A(m)$ is given recursively by $A(m) = L_{m-1} A(m-1)$ and $A(-m) = (L_{-m})^{-1} A(-m + 1)$. Since the matrices are lower triangular, it follows that

$$a_{ii}(m) = l_{ii}(m - 1)a_{ii}(m - 1), \qquad a_{ii}(-m) = \bar{l}_{ii}(-m)a_{ii}(-m + 1),$$

whereas, for $1 \le j < i \le n$, we obtain $a_{ij}(m) = \sum_{k=j}^{i} l_{ik}(m-1)a_{kj}(m-1)$ and $a_{ij}(-m) = \sum_{k=j}^{i} \bar{l}_{ik}(-m+1)a_{kj}(-m+1)$. Dividing these two equations by the previous two, we obtain

$$\frac{a_{ij}(m)}{a_{ii}(m)} = \frac{a_{ij}(m-1)}{a_{ii}(m-1)} + \sum_{k=j}^{i-1} \frac{l_{ik}(m-1)}{l_{ii}(m-1)} \frac{a_{kj}(m-1)}{a_{ii}(m-1)}, \tag{9.3}$$

$$\frac{a_{ij}(-m)}{a_{ii}(-m)} = \frac{a_{ij}(-m+1)}{a_{ii}(-m+1)} + \sum_{k=j}^{i-1} \frac{\bar{l}_{ik}(-m)}{\bar{l}_{ii}(-m)} \frac{a_{kj}(-m+1)}{a_{ii}(-m+1)}. \tag{9.4}$$

We claim that the absolute value of $a_{ij}(m)/a_{ii}(m)$ increases, as $|m|$ tends to ∞, slower than $e^{c|m|}$, for an arbitrary positive constant c. To show the claim, notice first that since $\lim_{i \to \pm\infty} \frac{1}{i} \ln \| L_i^\delta \| = 0$, there must be for each positive ϵ a constant $C \ge 1$ such that

$$\frac{1}{C} e^{-\epsilon|m|} \le \left\| L_m^\delta \right\|^{\pm 1} \le C e^{\epsilon|m|}$$

for all $m \in \mathbb{Z}$. Moreover, since $l_{ii}(m)^{-1}$ is the (i,i) entry of L_m^{-1}, we have $|l_{ii}(m)^{-1}| \le \| L_m^{-1} \| \le C e^{\epsilon|m|}$ and $|l_{ii}(m)| \le \| L_m \| \le C e^{\epsilon|m|}$. Therefore,

$$\frac{\| L_m \|}{|l_{ii}(m)|} \le C^2 e^{2\epsilon|m|}, \qquad \frac{\| L_m^{-1} \|}{|\bar{l}_{ii}(m)|} \le C^2 e^{2\epsilon|m|}$$

for all $m \in \mathbb{Z}$. Consequently, it follows from (9.3) and (9.4) that

$$\left| \frac{a_{ij}(m)}{a_{ii}(m)} - \frac{a_{ij}(m-1)}{a_{ii}(m-1)} \right| \le C^2 e^{2\epsilon|m|} \sum_{k=j}^{i-1} \left| \frac{a_{kj}(m-1)}{a_{ii}(m-1)} \right|, \tag{9.5}$$

$$\left| \frac{a_{ij}(-m)}{a_{ii}(-m)} - \frac{a_{ij}(-m+1)}{a_{ii}(-m+1)} \right| \le C^2 e^{2\epsilon|m|} \sum_{k=j}^{i-1} \left| \frac{a_{kj}(-m+1)}{a_{ii}(-m+1)} \right| \tag{9.6}$$

The claim can be proved by induction. First notice that $|\frac{a_{jj}(m)}{a_{ii}(m)}|$ grows slower than any exponential $e^{c|m|}$, $c > 0$, as $m \to \pm\infty$, since by assumption

$$\lim_{m \to \pm\infty} \frac{1}{m} \ln \left| \frac{a_{jj}(m)}{a_{ii}(m)} \right| = \lim_{m \to \pm\infty} \frac{1}{m} \ln|a_{jj}(m)| - \lim_{m \to \pm\infty} \frac{1}{m} \ln|a_{ii}(m)|$$
$$= \chi - \chi = 0.$$

Using inequalities (9.5) and (9.6) and $\sum_{i=0}^{|m|} e^{\epsilon i} = C' e^{\epsilon|m|}$, we obtain the claim by an induction argument, having in mind that $\epsilon > 0$ is arbitrary.

Fix an index i and define the matrix $\alpha(m) = (a_{kj}(m)/a_{ii}(m))$. The next claim is that $\lim_{m \to \pm\infty} \frac{1}{m} \ln \|\alpha(m)\| = 0$. Recall that, for some constant $C \geq 1$, we have by exercise 9.2.1

$$\frac{1}{C} |\det \alpha(m)|^{\frac{1}{n}} \leq \|\alpha(m)\| \leq C \sum_{k=1}^{n} \max\{|\alpha_{kj}(m)| \mid 1 \leq j \leq n\}.$$

Since $\alpha(m)$ is lower triangular, for any positive ϵ there is a constant $C_\epsilon \geq 1$ such that

$$\frac{1}{C} \left(\prod_{j=1}^{n} \frac{|a_{jj}(m)|}{|a_{ii}(m)|} \right)^{\frac{1}{n}} \leq \|\alpha(m)\| \leq C_\epsilon e^{\epsilon |m|}.$$

Therefore,

$$0 \leq \liminf_{m \to \pm\infty} \frac{1}{|m|} \ln \|\alpha(m)\| \leq \limsup_{m \to \pm\infty} \frac{1}{|m|} \ln \|\alpha(m)\| \leq \epsilon.$$

Since ϵ is arbitrary, the second claim follows. In particular,

$$\lim_{m \to \pm\infty} \frac{1}{m} \ln \|A(m)\| = \chi.$$

We repeat the preceding argument for $A(m)^{-1}$. Let $\bar{a}_{ij}(m)$ denote the entries of $A(m)^{-1}$. Notice that

$$\frac{\bar{a}_{ij}(m)}{\bar{a}_{jj}(m)} = \frac{\bar{a}_{ij}(m-1)}{\bar{a}_{jj}(m-1)} + \sum_{k=j+1}^{i} \frac{\bar{a}_{ik}(m-1)}{\bar{a}_{jj}(m-1)} \frac{\bar{l}_{kj}(m-1)}{\bar{l}_{jj}(m-1)},$$

$$\frac{\bar{a}_{ij}(-m)}{\bar{a}_{jj}(-m)} = \frac{\bar{a}_{ij}(-m+1)}{\bar{a}_{jj}(-m+1)} + \sum_{k=j+1}^{i} \frac{\bar{a}_{ik}(-m+1)}{\bar{a}_{jj}(-m+1)} \frac{l_{kj}(-m)}{l_{jj}(-m)}$$

replace (9.3) and (9.4) is this case. We proceed along the lines of the foregoing argument, with the obvious modifications. For example, in the induction process the index j starts at $i - 1$ and decreases by 1 at each step. Since the diagonal entries of $A(m)^{-1}$ are $a_{ii}(m)^{-1}$, we obtain

$$\lim_{m \to \pm\infty} \ln \|A(m)^{-1}\| = -\chi.$$

Finally, the inequality

$$\|A(m)^{-1}\|^{-1} \leq \frac{\|A(m)v\|}{\|v\|} \leq \|A(m)\|,$$

$v \neq 0$, implies $\lim_{m \to \pm\infty} \frac{1}{m} \ln \frac{\|A(m)v\|}{\|v\|} = \chi$, concluding the proof. $\qquad \square$

We now consider a tempered sequence of lower-triangular $L_i \in GL(n, \mathbb{R})$ and $A(m) = (a_{ij}(m))$ as before, such that the limit

$$\lim_{m \to \pm\infty} \frac{1}{m} \ln|a_{jj}(m)| = \ln|\lambda_j|$$

exists for each $1 \le j \le n$, but may take different values for different j. If χ_i, $1 \le i \le k$, are the distinct values of $\ln|\lambda_j|$, we would like to know whether there exists an equivalent lower-triangular sequence that has block-diagonal form, with k block matrices on the diagonal, such that for all the diagonal entries of the ith block on the diagonal the preceding limit is χ_i. As an example, notice that the constant sequence $L_i^{(1)} = \left(\begin{smallmatrix} 2 & 0 \\ 7 & 3 \end{smallmatrix}\right)$ is equivalent to the sequence given by $L_i^{(2)} = \left(\begin{smallmatrix} 2 & 0 \\ 0 & 3 \end{smallmatrix}\right)$ and the equivalence is realized by $\Phi_i = \left(\begin{smallmatrix} 1 & 0 \\ 7 & 1 \end{smallmatrix}\right)$.

We first consider the two-dimensional case.

Lemma 9.2.3 Let $L_i = \left(\begin{smallmatrix} l_{11}(i) & 0 \\ l_{21}(i) & l_{22}(i) \end{smallmatrix}\right) \in GL(2, \mathbb{R})$ be a tempered sequence such that

$$\lim_{i \to \pm\infty} \frac{1}{i} \ln|l_{11}(i)| = \lim_{i \to \pm\infty} \frac{1}{i} \ln|l_{22}(i)| = 0,$$

and denote by L_i' the diagonal matrix with the same diagonal entries as L_i. Let $A(m) = (a_{ij}(m))$ be the m-fold product of the L_i, as defined before. Suppose that the limits

$$\lim_{m \to \pm\infty} \frac{1}{m} \ln|a_{11}(m)| = \chi_1, \qquad \lim_{m \to \pm\infty} \frac{1}{m} \ln|a_{22}(m)| = \chi_2$$

exist and that $\chi_1 < \chi_2$. (If $\chi_1 > \chi_2$ we consider instead the sequence L_i^{-1}.) Then L_i and L_i' are equivalent and there exists a unique tempered sequence $\Phi_i = \left(\begin{smallmatrix} 1 & 0 \\ c_i & 1 \end{smallmatrix}\right)$ that realizes the equivalence, that is, $L_i'\Phi_i = \Phi_{i+1}L_i$ for each $i \in \mathbb{Z}$. Furthermore, $c_0 = \lim_{i \to +\infty}(a_{21}(m)/a_{22}(m))$.

Proof. By induction, using $L_i'\Phi_i = \Phi_{i+1}L_i$, the sequence c_i, if it exists, is uniquely determined by the value of c_0. In fact, we must have for $i \in \mathbb{Z}$,

$$c_i = \frac{a_{22}(i)}{a_{11}(i)}\left(c_0 - \frac{a_{21}(i)}{a_{22}(i)}\right). \tag{9.7}$$

Conversely, if Φ_i is defined by means of a sequence c_i satisfying (9.7), then the commutative diagram that defines equivalent sequences is satisfied, so the proposition will be proven if we find c_0 for which Φ_i, defined by (9.7), is tempered. Notice that if another tempered $\Phi_i' = \left(\begin{smallmatrix} 1 & 0 \\ e_i & 1 \end{smallmatrix}\right)$ also realizes the equivalence between L_i and L_i', then

$$|c_0 - e_0| = \left| \frac{a_{11}(m)}{a_{22}(m)} (c_m - e_m) \right|.$$

Therefore, $c_0 = e_0$ and there can be at most one sequence Φ_i.

To prove the existence of a tempered Φ_i, we first show that the limit of $a_{21}(m)/a_{22}(m)$, as m goes to $+\infty$, exists. From $A(m+1) = L_m A(m)$, $m \in \mathbb{Z}$, it follows that

$$\frac{a_{21}(m+1)}{a_{22}(m+1)} = \frac{l_{21}(m)}{l_{22}(m)} \frac{a_{11}(m)}{a_{22}(m)} + \frac{a_{21}(m)}{a_{22}(m)}. \tag{9.8}$$

By assumption,

$$\limsup_{m \to \pm\infty} \frac{1}{|m|} \ln\left| \frac{l_{21}(m)}{l_{22}(m)} \right| \le \lim_{m \to \pm\infty} \frac{1}{|m|} \ln\|L_m\| - \lim_{m \to \pm\infty} \frac{1}{|m|} \ln|l_{22}(m)| = 0.$$

From this and (9.8), it follows that

$$\limsup_{m \to \pm\infty} \frac{1}{|m|} \ln\left| \frac{a_{21}(m+1)}{a_{22}(m+1)} - \frac{a_{21}(m)}{a_{22}(m)} \right| \le \pm(\chi_1 - \chi_2). \tag{9.9}$$

Therefore, $a_{21}(m)/a_{22}(m)$ is a Cauchy sequence for $m \to +\infty$. We denote its limit by c_0. Notice that

$$\limsup_{m \to +\infty} \frac{1}{m} \ln\left| \frac{a_{21}(m)}{a_{22}(m)} - c_0 \right| \le \chi_1 - \chi_2.$$

Therefore

$$\limsup_{m \to +\infty} \frac{1}{m} \ln|c_m| = \limsup_{m \to +\infty} \frac{1}{m} \ln\left| \frac{a_{22}(m)}{a_{11}(m)} \left(c_0 - \frac{a_{21}(m)}{a_{22}(m)} \right) \right| \le 0. \tag{9.10}$$

In order to prove that Φ_i is tempered, it remains to show that

$$\limsup_{m \to +\infty} \frac{1}{|m|} \ln|c_m| \le 0. \tag{9.11}$$

Using (9.9) and a telescoping series argument, we obtain

$$\limsup_{m \to -\infty} \frac{1}{|m|} \ln\left| \frac{a_{21}(m)}{a_{22}(m)} \right| \le \chi_2 - \chi_1. \tag{9.12}$$

From this and (9.7), we obtain

$$\limsup_{m \to -\infty} \frac{1}{|m|} \ln|c_m| \leq \limsup_{m \to -\infty} \frac{1}{|m|} \ln\left|c_0 - \frac{a_{21}(m)}{a_{22}(m)}\right|$$
$$+ \lim_{m \to -\infty} \frac{1}{|m|} \ln|a_{22}| - \lim_{m \to -\infty} \frac{1}{|m|} \ln|a_{11}|$$
$$\leq (\chi_2 - \chi_1) + \chi_1 - \chi_2 = 0.$$

Therefore, the sequence Φ_i is tempered, concluding the proof.   □

Next, we generalize the previous lemma for arbitrary n. The general case will, in fact, be reduced to $n = 2$, after we make the following definitions. Let $\mathcal{P} = (\mathcal{P}_1, \dots, \mathcal{P}_l)$ denote a partition of $\{1, 2, \dots, n\}$ into l disjoint subsets. We denote by $\mathcal{P}(i)$ the element of the partition that contains i and define

$G(\mathcal{P}, i) = \{(a_{rs}) \in GL(n, \mathbb{R})$ such that

$\quad a_{rs} = 0$ if $r < s$ or both $0 < r - s < i$ and $\mathcal{P}(r) \neq \mathcal{P}(s)\}.$

Exercise 9.2.4 Show that $G(\mathcal{P}, i)$ is a closed subgroup of the group of lower-triangular matrices in $GL(n, \mathbb{R})$ and that

$$G(\mathcal{P}, 1) \supset G(\mathcal{P}, 2) \supset \cdots \supset G(\mathcal{P}, n).$$

Notice that $G(\mathcal{P}, 1)$ is the full group of lower-triangular matrices and $G(\mathcal{P}, n)$ is the subgroup of lower-triangular matrices (a_{rs}) such that $a_{rs} = 0$ if $\mathcal{P}(r) \neq \mathcal{P}(s)$.

Let r, s, $1 \leq s < r \leq n$, be such that $r - s = i_0$ and $\mathcal{P}(r) \neq \mathcal{P}(s)$. We define the map

$$\rho_{rs} : A = (a_{ij}) \in G(\mathcal{P}, i_0) \mapsto \begin{pmatrix} a_{ss} & 0 \\ a_{rs} & a_{rr} \end{pmatrix} \in GL(2, \mathbb{R}).$$

For the same r and s, we also define the map

$$\iota_{rs} : \left\{ \begin{pmatrix} 1 & 0 \\ a & 1 \end{pmatrix} \middle| a \in \mathbb{R} \right\} \to G(\mathcal{P}, i_0)$$

that sends $\begin{pmatrix} 1 & 0 \\ a & 1 \end{pmatrix}$ to the matrix with 1s on the diagonal, a in the (r, s) entry, and 0 in all other places.

Exercise 9.2.5 Show that ρ_{rs} is a group homomorphism. Also show that left or right multiplication of any $(u_{ij}) \in G(\mathcal{P}, i_0)$ by $\iota(A)$ can only affect the entries u_{rj}, $j \leq s$, or u_{is}, $i \geq r$. In particular, right and left multiplication by $\iota(A)$ map $G(\mathcal{P}, i)$ onto itself for each $i < i_0$.

Lemma 9.2.6 Given a tempered sequence L_m of lower-triangular matrices in $GL(n, \mathbb{R})$, let $A(m) = (a_{ij}(m))$ be the m-fold product of the L_i as defined earlier. Suppose that $\lim_{m \to \pm\infty} (\ln|a_{ii}(m)|)/m$ exists for each i, $1 \le i \le n$, and let $\chi_1 < \chi_2 < \cdots < \chi_k$ be the different values of these limits. Also suppose that $\lim_{m \to \pm\infty} (\ln|l_{ii}(m)|)/m = 0$ for each i. Let $\mathcal{P} = (\mathcal{P}_1, \ldots, \mathcal{P}_k)$ denote the partition of $\{1, 2, \ldots, n\}$ such that

$$r \in \mathcal{P}_i \text{ iff } \lim_{m \to \pm\infty} (\ln|a_{rr}(m)|)/m = \chi_i.$$

Then L_m is cohomologous to a tempered sequence in $G(\mathcal{P}, n)$ and the tempered sequence Φ_m that realizes the equivalence can be taken so as to lie in the group of lower-triangular matrices with 1s on the diagonal.

Proof. The proof is by induction, making use of the previous lemma and exercises. Clearly, $L_i \in G(\mathcal{P}, 1)$, since this is the full group of lower-triangular matrices. Suppose that $L_i \in G(\mathcal{P}, k)$ for some k, $1 \le k \le n - 1$, and fix s, r, $1 \le s < r \le n$, such that $r - s = i$ and $\mathcal{P}(r) \ne \mathcal{P}(s)$. Applying the previous lemma to the sequence $\rho_{rs}(L_i)$, we obtain a tempered sequence $\bar{\Psi}_i = \left(\begin{smallmatrix} 1 & 0 \\ c_i & 1 \end{smallmatrix}\right)$ that realizes the equivalence between $\rho_{rs}(L_i)$ and the sequence of diagonal matrices with the same diagonal elements as $\rho_{rs}(L_i)$. Set $\Psi_i = \iota_{rs}(\bar{\Psi}_i)$. Then Ψ_i realizes the equivalence between L_i and a sequence $L_i^{(1)}$ whose (r, s) entry is 0 and entries other than (r, j), for $j \le s$, and (i, s), for $i \ge r$, are the same as the corresponding entries of L_i. By repeating this operation a finite number of times we obtain an equivalence between L_i and a tempered sequence in $G(\mathcal{P}, i + 1)$, the equivalence being realized by a sequence with 1s on the diagonal. The proof ends when $i + 1 = n$.   □

Let L_i and $\mathcal{P}$ be as in the previous lemma and denote by $\bar{L}_i$ the equivalent sequence that the lemma yields. Let $\bar{A}(m)$ be the corresponding m-fold product. By permuting the elements in the standard basis $\{e_1, \ldots, e_n\}$ of $\mathbb{R}^n$, we may assume that

$$\mathcal{P}_1 = \{1, \ldots, l_1\}, \qquad \mathcal{P}_2 = \{l_1 + 1, \ldots, l_2\} \ldots, \qquad \mathcal{P}_k = \{l_{k-1} + 1, \ldots, n\}.$$

Therefore, we may assume that $\bar{L}_i$, hence also $\bar{A}(m)$, has block-diagonal form and that within each diagonal block of $\bar{A}(m)$, $\lim_{m \to \pm\infty} (\ln|a_{ii}(m)|)/m$ has the same value for all i.

The results obtained so far imply the following proposition.

Proposition 9.2.7 Let L_m be a tempered sequence of lower-triangular matrices in $GL(n, \mathbb{R})$, and let $A(m) = (a_{ij}(m))$ be, as before, the m-fold product of the L_i. Suppose that $\lim_{m \to \pm \infty} (\ln|a_{ii}(m)|)/m$ exists for each i and let $\chi_1 <$ $\chi_2 < \cdots < \chi_k$ be the different values of these limits. Also suppose that $\lim_{m \to \pm \infty} (\ln|l_{ii}(m)|)/m = 0$ for each i. Then we can find for each $1 \le l \le k$ a sequence of subspaces $V_l(i) \subset \mathbb{R}^n$ such that the following hold

1. $L_i V_l(i) = V_l(i+1)$, for each l and each $i \in \mathbb{Z}$.
2. $\mathbb{R}^n = V_1(i) \oplus \cdots \oplus V_k(i)$, for each $i \in \mathbb{Z}$.
3. For each $1 \le l \le k$ and each nonzero $v \in V_l(0)$, we have

$$\lim_{m \to \pm \infty} \frac{1}{m} \ln \| A(m)v \| = \chi_l.$$

4. Given a nontrivial partition of $\{1, 2, \ldots, k\}$ into sets $\mathcal{I}$ and $\mathcal{I}^c$, let θ_i be the angle between $\bigoplus_{l \in \mathcal{I}} V_l(i)$ and $\bigoplus_{l \in \mathcal{I}^c} V_l(i)$. Then

$$\lim_{m \to \pm \infty} \frac{1}{m} \ln \sin \theta_m = 0.$$

Exercise 9.2.8 Show that the previous proposition indeed follows from the lemmas proved earlier.

9.3 Proof of Oseledec's Theorem

Let $\bar{p} : \bar{M} \to M$ be a fiber bundle over M with compact fibers, and suppose that $\bar{f} : \bar{M} \to \bar{M}$ is an automorphism of $\bar{M}$ covering a map $f : M \to M$, that is, such that $f \circ \bar{p} = \pi \circ \bar{f}$. We call $\bar{f}$ a *compact extension* of f. By exercise 2.1.5, if μ is an f-invariant probability measure on M, there exists an $\bar{f}$-invariant probability measure $\bar{\mu}$ on $\bar{M}$ such that $\bar{p}_* \bar{\mu} = \mu$. If μ is ergodic, we can choose $\bar{\mu}$ to be ergodic.

If $\pi : E \to M$ is a vector bundle, we define the *pull-back* of E to $\bar{M}$ as follows (see section 6.1):

$$\bar{\pi} : \bar{E} := \bar{p}^* E = \{ (\bar{x}, v) \in \bar{M} \times E \mid \bar{p}(\bar{x}) = \pi(v) \} \to \bar{M},$$

where $\bar{\pi}(\bar{x}, v) = \bar{x}$. A norm or inner product on E induces a norm or inner product on $\bar{E}$ in a natural way, since the fibers of $\bar{E}$ are actually fibers of E. Moreover, if $F : E \to E$ is an automorphism of E covering $f : M \to M$ (so that $\pi \circ F = f \circ \pi$), we obtain an automorphism $\bar{F}$ of $\bar{E}$ defined by $\bar{F}_{\bar{x}}(\bar{x}, v) := (\bar{f}(\bar{x}), F_x v)$, where $x = \bar{p}(\bar{x})$.

Lemma 9.3.1 We assume that $E, \bar{E}, F, \bar{F}, M, \bar{M}, f, \bar{f}, \mu$, and $\bar{\mu}$ are as before. Suppose that the conclusion of Oseledec's theorem holds for the action of $\bar{F}$ on $\bar{E}$. Then the same conclusion holds for the action of F on E.

Proof. The assumption is that there exists a subset $\bar{X} \subset \bar{M}$ of full $\bar{\mu}$-measure such that for each $\bar{x} \in \bar{X}$ the following holds. There exist real numbers $\chi_1(x) < \cdots < \chi_{k(\bar{x})}(\bar{x})$, $1 \leq k(\bar{x}) \leq n$, and subspaces $\bar{S}_i(\bar{x}) \subset \bar{E}(\bar{x})$, $1 \leq i \leq k(\bar{x})$, such that conditions 1, 2, 3, and 4 in the definition of regular points hold.

Notice that for each $\bar{v} = (\bar{x}, v) \in \bar{E}(\bar{x})$, $\bar{p}(\bar{x}) = x$, we have

$$\| \bar{F}_{\bar{x}} \bar{v} \| = \| (\bar{f}(\bar{x}), F_x v) \| = \| F_x v \|.$$

Consequently, the set $\bar{X}$ of regular points for $\bar{F}$ is saturated by the fibers of $\bar{p}$, and the subspaces $\bar{S}_i(\bar{x})$ have the form $\{\bar{x}\} \times S_i(x)$ for a subspace $S_i(x)$ independently of the element $\bar{x}$ in the fiber of x. Therefore, the set $X = \bar{p}(\bar{X})$ has full μ-measure and consists of regular points for F. Notice that the exponents $\chi_i(x)$ for F are the same as those for $\bar{F}$. □

The assumptions and notation of theorem 9.1.5, as well as the notation introduced in the remarks after the statement of the theorem, are now in force. In particular, let σ be a measurable section of the orthogonal frame bundle of E, and denote by $L : M \to GL(n, \mathbb{R})$ the measurable function such that $L(x)$ is the matrix representing F_x with respect to $\sigma(x)$ and $\sigma(f(x))$. More precisely,

$$F_x \sigma(x) = \sigma(f(x)) L(x),$$

where $\sigma(x) : \mathbb{R}^n \to E(x)$ is a linear isometry for each $x \in M$ and $\sigma(f(x)) L(x)$ denotes ordinary composition of linear maps.

Notice that $\| F_x \| = \| L(x) \|$, where the norm on the right-hand side comes from the Euclidean metric on $\mathbb{R}^n$ and the one on the left-hand side comes from the Riemannian metric on E. Therefore, the integrability condition of Oseledec's theorem can be restated thus:

$$\int_M \max\{ \ln\| L(x) \|, \| L(x)^{-1} \|, 0 \}\, d\mu(x) < \infty. \qquad (9.13)$$

Exercise 9.3.2 Show that condition (9.13) is equivalent to $\ln\| L \|$ and $\ln\| L^{-1} \|$ being in $L^1(M, \mu)$. (Hint: Notice that $\ln\| L(x) \| + \ln\| L(x)^{-1} \| \geq 0$ and use the f-invariance of μ.) Also show that for μ-a.e. $x \in M$ the sequence $L_i := L(f^i(x))$, $i \in \mathbb{Z}$, is tempered. (Hint: If $l \in L^1(M, \mu)$, apply Birkhoff's ergodic theorem to $g = l - l \circ f$ to conclude that $(\ln|l(f^m(x))|)/m \to 0$ as $m \to \pm\infty$ for μ-a.e. $x \in M$.)

Our goal is to obtain a compact extension $\bar{F}$ of F such that the matrix-valued function $\bar{L}$ representing $\bar{F}$ can be assumed to take values in the group of lower-triangular matrices. That will allow us to use the results of the previous section.

Denote by $SL'(n, \mathbb{R})$ the subgroup of $GL(n, \mathbb{R})$ consisting of matrices with determinant equal to $+1$ or -1, let B be the group of lower-triangular matrices with positive diagonal elements, and let B_0 be the subgroup of B consisting of matrices with determinant 1. Let D be the coset space $SL'(n, \mathbb{R})/B_0 = GL(n, \mathbb{R})/B$.

Exercise 9.3.3 Show that D is a compact manifold (with two connected components) and that there exists a smooth map $\kappa : D \to O(n)$ such that $\kappa(x)B = x$ for each $x \in D$. Hint: Show that the product map $O(n) \times B_0 \to SL'(n, \mathbb{R})$ is a diffeomorphism. Notice that by the Iwasawa decomposition $SL(n, \mathbb{R})$ is diffeomorphic to $SO(n) \times B_0^{\mathfrak{t}}$, where $B_0^{\mathfrak{t}}$ is the group of upper-triangular matrices of determinant 1 with positive diagonal elements, and that B_0 and $B_0^{\mathfrak{t}}$ are diffeomorphic under conjugation by $C = (c_{ij}) \in SO(n)$, where $c_{ij} = (-1)^{i+1}$ if $i + j = n + 1$ and 0 otherwise.

We now let $\mathcal{F}(E)$ denote the bundle of frames of E (see section 6.1). Thus, for each $x \in M$, the fiber $\mathcal{F}(E)_x$ consists of all linear isomorphisms $\sigma : \mathbb{R}^n \to E(x)$. Set $\bar{M} = \mathcal{F}(E)/B$, the space of orbits of the right action of B on the frame bundle. The natural projection $\bar{p} : \bar{M} \to M$, which maps a B-orbit to the base point in M, makes $\bar{M}$ into a fiber bundle over M with fibers diffeomorphic to D.

The pull-back of $\mathcal{F}(E)$ under $\bar{p}$ is the bundle of frames of $\bar{E}$, that is,

$$\bar{p}^*(\mathcal{F}(E)) = \mathcal{F}(\bar{E}) = \{(\bar{x}, \sigma) \in \bar{M} \times \mathcal{F}(E)_x \mid \bar{p}(\bar{x}) = x\}.$$

F induces in a natural way automorphisms of the bundles $\mathcal{F}(E)$ and $\mathcal{F}(\bar{E})$, which we also denote by F and $\bar{F}$, respectively.

We define a map $\mathcal{G} : \mathcal{F}(\bar{E}) \to D$ as follows: Given $(\bar{x}, \sigma)$, there exists a unique $\mathcal{G}(\bar{x}, \sigma) := hB \in D$ such that $\bar{x} = \sigma hB$. Then $\mathcal{G}$ is $\bar{F}$-invariant and $GL(n, \mathbb{R})$-equivariant, that is, $\mathcal{G}(\bar{\sigma}h) = h^{-1}\mathcal{G}(\bar{\sigma})$.

Let $\sigma : M \to \mathcal{F}(E)$ be a section of the frame bundle of E for which the function L representing F is tempered (for example, choose a measurable section consisting of orthonormal frames with respect to the Riemannian metric for which the integrability condition of theorem 9.1.5 holds). Define a section $\bar{\sigma}$ of $\mathcal{F}(\bar{E})$ by setting $\bar{\sigma}(\bar{x}) = (\bar{x}, \sigma(x))$. The matrix-valued function that describes $\bar{F}$ in terms of $\bar{\sigma}$ is then $L \circ \bar{p}$, hence also tempered.

Define a new section σ_0 of $\mathcal{F}(\bar{E})$ by $\sigma_0(\bar{x}) = \bar{\sigma}(\bar{x})\kappa(\mathcal{G}(\bar{\sigma}(\bar{x})))$. Then, by definition, $\mathcal{G}(\sigma_0(\bar{x})) = B \in GL(n, \mathbb{R})/B$. Since $\mathcal{G}$ is $\bar{F}$-invariant, the matrix-valued function representing $\bar{F}$ relative to the frame σ_0 takes values in B.

Exercise 9.3.4 For each nonsingular linear map $L : \mathbb{R}^n \to \mathbb{R}^n$, and for each eigenvalue λ of L, let $V_\lambda^{\mathbb{C}}$ denote the generalized eigenspace of λ, defined as the subspace of $\mathbb{C}^n$ consisting of all $v \in \mathbb{C}^n$ such that $(L - \lambda I)^k v = 0$ for some positive integer k. If λ is real, set $V_\lambda := V_\lambda^{\mathbb{C}} \cap \mathbb{R}^n$, and if λ and $\bar{\lambda}$ are a pair of complex conjugate eigenvalues of L, set $V_{\lambda, \bar{\lambda}} := (V_\lambda^{\mathbb{C}} \oplus V_{\bar{\lambda}}^{\mathbb{C}}) \cap \mathbb{R}^n$. If $\chi_1 < \cdots < \chi_r$ are the distinct values of $\ln|\lambda|$, we let E_{χ_i} denote the direct sum of subspaces V_λ or $V_{\lambda, \bar{\lambda}}$ such that $\ln|\lambda| = \chi_i$. Show that the subspaces E_{χ_i}, $1 \le i \le r$, are L-invariant and

$$\mathbb{R}^n = E_{\chi_1} \oplus \cdots \oplus E_{\chi_r}.$$

Moreover, for any norm $\|\cdot\|$ on $\mathbb{R}^n$ and any nonzero $v \in E_{\chi_i}$, show that

$$\lim_{m \to \pm\infty} \frac{1}{m} \ln\|L^m v\| = \chi_i.$$

Exercise 9.3.5 Given $A \in SL(n, \mathbb{Z})$, define a diffeomorphism f of the n-torus $\mathbb{T}^n = \mathbb{R}^n / \mathbb{Z}^n$ such that $f([x]) = [Ax]$ for all $[x] = x + \mathbb{Z}^n \in \mathbb{T}^n$. Show that the tangent bundle $T\mathbb{T}^n$ decomposes as a direct sum of smooth subbundles L_{χ_i}, where χ_i are the distinct values of $\ln|\lambda|$ and λ ranges over the eigenvalues of A, such that for each x and each nonzero $v \in L_i(x)$, we have $\lim_{m \to \pm\infty} \frac{1}{m} \ln\|Df_x^m v\| = \chi_i$. (The norm is induced by an arbitrary Riemannian metric on the torus.)

Exercise 9.3.6 Let $M = G / \Gamma$, where G is a connected semisimple Lie group and Γ is a uniform lattice in G. To each X in the Lie algebra $\mathfrak{g}$ of G, associate the vector field $\tilde{X}$ on M obtained by projecting to the quotient the right-invariant vector field on G determined by X. Let $\mathfrak{a}$ be a maximal abelian $\mathbb{R}$-diagonalizable subalgebra of $\mathfrak{g}$. Denote by $\mathfrak{g}_\alpha$ the root space associated to a root α of $(\mathfrak{a}, \mathfrak{g})$. For each $X \in \mathfrak{a}$, let Ψ_t denote the flow of $\tilde{X}$ and let E_α be the vector subbundle of TM induced by $\mathfrak{g}_\alpha$. Show that

$$\lim_{t \to \pm\infty} \frac{1}{t} \ln\|(D\Psi_t)_x v\| = \alpha(X)$$

for all nonzero $v \in E_\alpha(x)$, $x \in M$. Therefore, the Oseledec decomposition for Ψ_t is induced by the root space decomposition of $\mathfrak{g}$.

Let $\pi : E \to M$ be a measurable vector bundle with fiber dimension n over a Borel probability space $(M, \mathcal{B}, \mu)$. Given an automorphism F of E and an F-invariant measurable subbundle E_0 of E, we may consider the restriction $F_0 = F|_{E_0}$, as well as the induced automorphism $\bar{F}$ on the quotient $\bar{E} = E/E_0$. The latter is defined as follows: For each $x \in M$, the fiber $\bar{E}(x)$ is the quotient of vector spaces, $E(x)/E_0(x)$, and

$$\bar{F}(v + E_0(x)) = Fv + E_0(f(x)) \in \bar{E}(f(x))$$

for each $v + E_0(x) \in \bar{E}(x)$. We denote by $p : E \to \bar{E}$ the natural bundle map. We also assume that the map $f : M \to M$ such that $\pi \circ F = f \circ \pi$ is a measure-preserving transformation of M and that the integrability condition on

F needed for theorem 9.1.5 is satisfied. It is clear that the same condition will also hold for F_0. It is also immediate that the integrability condition holds for $\bar{F}$, if we define on $\bar{E}$ a norm $\|\bar{v}\|_x$ as the infimum of $\|v\|_x$ over all $v \in E(x)$ such that $p(v) = \bar{v}$. (Notice that $\|p(v)\| \leq \|v\|$.) Therefore, Oseledec's theorem applies to F_0 and $\bar{F}$, and it is natural to ask how the Lyapunov spectra of $F, \bar{F}$, and F_0 are related. The answer is given by the next proposition [19, 2.3].

Proposition 9.3.7 Let $(M, \mathcal{B}, \mu)$, $F, \bar{F}$, F_0, $E, \bar{E}$, E_0, and p be as before, and suppose that F satisfies the integrability condition of theorem 9.1.5. Then the same is true for $\bar{F}$ and F_0. Let $\chi_1(x) < \cdots < \chi_{k(x)}(x)$, $1 \leq k(x) \leq n$, be the Lyapunov exponents of F, and $S_i(x)$ the subspace corresponding to $\chi_i(x)$, $1 \leq k(x) \leq n$. Then, for each $x \in M$ regular for $F, \bar{F}$, and F_0, the following hold.

1. $E_0(x) = \bigoplus_{1 \leq i \leq k(x)} E_0(x) \cap S_i(x)$ is the Oseledec decomposition of F_0; the Lyapunov exponents for F_0 are the exponents $\chi_i(x)$ of F such that $E_0(x) \cap S_i(x) \neq 0$.
2. Denote by $\chi'_j(x)$ and $S'_j(x)$ the Lyapunov exponents and corresponding invariant subspaces for $\bar{F}$. Then for each j, $\chi'_j(x) = \chi_i(x)$ and $p(S_i(x)) = S'_j(x)$ for some i and a.e. $x \in M$.
3. If, for some $1 \leq i \leq k(x)$, $\chi_i(x) \neq \chi'_j(x)$ for all j, then $S_i(x)$ is a subspace of $E_0(x)$, for a.e. $x \in M$.

Proof. For each $a \in \mathbb{R}$ and $x \in M$, we define

$$E_a(x) = \bigoplus_{\chi_i(x) \leq a} S_i(x),$$

$$E^a(x) = \bigoplus_{\chi_i(x) \geq a} S_i(x),$$

$$\bar{E}_a(x) = \bigoplus_{\chi'_i(x) \leq a} S'_i(x),$$

$$\bar{E}^a(x) = \bigoplus_{\chi'_i(x) \geq a} S'_i(x).$$

Using property 4 in the definition of regular points (and the remark made immediately after it), we obtain that for a.e. $x \in M$

$$p(E_a(x)) \subset \bar{E}_a(x), \qquad p(E_a(x)) \subset \bar{E}_a(x).$$

On the other hand,

$$S_i(x) = E_{\chi_i(x)}(x) \cap E^{\chi_i(x)}(x), \qquad S_i'(x) = \bar{E}_{\chi_i(x)}(x) \cap \bar{E}^{\chi_i(x)}(x),$$

and $\bar{E}_a(x) \cap \bar{E}^a(x) = \{0\}$ whenever a is not one of the exponents $\chi_j'(x)$. Therefore, $p(S_i(x)) \subset S_j'(x)$ if $\chi_i(x) = \chi_j'(x)$, and $p(S_i(x)) = 0$ if $\chi_i(x)$ is not among the $\chi_j'(x)$. This last remark implies item 3 in the proposition. Since $E(x)$ is the direct sum of the $S_i(x)$ and $\bar{E}$ is the direct sum of the $S_j'(x)$, item 2 also follows. Item 1 is immediate. $\qquad\square$

9.4 Nonuniform Hyperbolicity and Entropy

In section 1.5, an Anosov diffeomorphism f of a compact manifold M was defined by a "uniform hyperbolicity" condition; namely, at each $x \in M$, $T_x M$ decomposes into a direct sum of subspaces $E^+(x)$ and $E^-(x)$ such that for any Riemannian metric on TM with norm $\|\cdot\|$, there are constants $C > 0$ and $0 < \lambda < 1$ (where λ does not depend on the metric chosen) such that

$$\left\| Df_x^{\mp m} v \right\| \le C\lambda^m \|v\|$$

for each $x \in M$, $v \in E^{\pm}(x)$, and $m \in \mathbb{N}$.

Although this is a condition on f that can be checked locally, uniform hyperbolicity implies a number of global properties of the $\mathbb{Z}$-action, such as topological transitivity and the existence of a dense set of periodic points (when $NW(f) = M$; see theorem 1.5.4). The key fact that was used in the proof of theorem 1.5.4 is the existence of stable and unstable foliations, which integrate the (continuous) subbundles E^+ and E^-.

A much weaker condition of hyperbolicity may be defined as follows. Given positive constants λ, μ, ϵ (where ϵ should be thought of as being much smaller than λ and μ), and for each $k \in \mathbb{N}$, define $\Lambda_k := \Lambda_k(\lambda, \mu, \epsilon)$ to be the set of all $x \in M$ for which $T_x M$ has a direct sum decomposition $E^-(x) \oplus E^+(x)$ such that

1. $\|Df^n|_{E^-(f^m(x))}\| \le e^{\epsilon k} e^{-(\lambda - \epsilon)n} e^{\epsilon |m|}$ for each $m \in \mathbb{Z}$ and $n \in \mathbb{N}$;
2. $\|Df^{-n}|_{E^+(f^m(x))}\| \le e^{\epsilon k} e^{-(\mu - \epsilon)n} e^{\epsilon |m|}$ for each $m \in \mathbb{Z}$ and $n \in \mathbb{N}$;
3. $\tan \alpha(f^m(x)) \ge e^{-\epsilon k} e^{-\epsilon |m|}$ for each $m \in \mathbb{Z}$;
4. E^+ and E^- are f-invariant in the following sense: For each $x \in \Lambda_k$ and $m \in \mathbb{Z}$ such that $f^m(x) \in \Lambda_k$, we have

$$Df_x^m E^{\pm}(x) = E^{\pm}(f^m(x)).$$

We write $\Lambda = \Lambda(\lambda, \mu, \epsilon) := \bigcup_{k=1}^{\infty} \Lambda_k$. It is immediate that

$$f^{\pm 1}(\Lambda_k) \subset \Lambda_{k+1} \quad \text{and} \quad \Lambda_1 \subset \Lambda_2 \subset \Lambda_3 \subset \cdots.$$

Therefore, Λ is f-invariant: $f(\Lambda) = \Lambda$.

Properties 1–4 characterize a *nonuniform (completely) hyperbolic* system on Λ. If we allow μ to be nonnegative ($\mu \geq 0$), we have the notion of a *nonuniform partially hyperbolic* diffeomorphism. The set Λ will be called (in the nonuniform completely or partially hyperbolic cases) a *Pesin set*.

In the next exercise, we allow $\mu \geq 0$.

Exercise 9.4.1 Show that for each k, E^+ and E^- are continuous over Λ_k and Λ_k is a compact set.

In spite of the previous exercise, the Pesin set need not be compact and $E^\pm$ may fail to be continuous on it.

Of course, Λ could be empty for a particular f. Suppose, however, that ν is an ergodic Borel probability measure such that some of the Lyapunov exponents for ν are negative. In this case, Oseledec's theorem implies the following proposition. (The proposition is not an immediate corollary of Oseledec's theorem. The details can be found in [22, theorem 1.1]. See also [23].)

Proposition 9.4.2 (Pesin) Let $f : M \to M$ be a smooth diffeomorphism of a compact manifold. Let ν be an ergodic Borel probability measure, invariant under f. Let

$$\chi_1 \geq \cdots \geq \chi_r \geq 0 > \chi_{r+1} \geq \cdots \geq \chi_k$$

be the Lyapunov exponents, and write $\mu = \chi_r$ and $\lambda = |\chi_{r+1}|$. Then, for each sufficiently small positive ϵ, we have $\nu(\Lambda(\lambda, \mu, \epsilon)) = 1$.

Notice that all that has been defined and stated also applies to a smooth $\mathbb{Z}$-action on a vector bundle E over a compact manifold M, instead of the action of Df on TM. E could be, for example, an f-invariant subbundle of TM and the $\mathbb{Z}$-action is that which is generated by the restriction of Df to E.

Ya. Pesin also shows in [22] that nonuniform hyperbolic diffeomorphisms also have stable and unstable manifolds. More precisely, the following holds. Suppose that $\Lambda(\lambda, \mu, \epsilon)$ is a nonempty Pesin set, with $\lambda > 0$, $\mu > 0$, and

$0 < \epsilon < \min\{\lambda, \mu\}$. Define, for each $x \in \Lambda$ and some $\delta > 0$, the sets

$$W_\delta^u(x) = \{y \in M \mid d(f^{-n}(x), f^{-n}(y)) \leq \delta e^{-(\mu-\epsilon)n}, n \in \mathbb{N}\},$$

$$W_\delta^s(x) = \{y \in M \mid d(f^n(x), f^n(y)) \leq \delta e^{-(\lambda-\epsilon)n}, n \in \mathbb{N}\}.$$

Theorem 9.4.3 (Local stable manifolds) Let $f : M \to M$ be a smooth diffeo-morphism, and let $\Lambda = \Lambda(\lambda, \mu, \epsilon)$ be a Pesin set. Then there exists $\epsilon_0 > 0$ such that for each $x \in \Lambda_k$, $k \in \mathbb{N}$, and $\delta = \epsilon_0 e^{-\epsilon k}$, the following hold:

1. $W_\delta^s(x)$ and $W_\delta^u(x)$ are smooth submanifolds of M diffeomorphic to disks of dimensions $\dim E^-(x)$ and $\dim E^+(x)$, respectively.
2. $T_x W_\delta^s(x) = E^-(x)$, $T_x W_\delta^s(x) = E^+(x)$.

As in the remark after theorem 1.5.1, the local stable and unstable manifolds $W_\delta^s(x)$ and $W_\delta^u(x)$ give rise to (global) stable and unstable manifolds:

$$W^s := \bigcup_{n=0}^{\infty} f^{-n} W_\delta^s(f^n(x)) \quad \text{and} \quad W^u := \bigcup_{n=0}^{\infty} f^n W_\delta^u(f^{-n}(x)).$$

It is possible to use the stable and unstable manifolds in the way they were used in theorem 1.5.4, to prove the existence of fixed points for f. The precise statement is as follows (see [16]).

Theorem 9.4.4 Let $f : M \to M$ be a smooth diffeomorphism of a compact manifold M, and let ν be an f-invariant Borel probability measure. Suppose that ν is a hyperbolic measure, that is, at ν-a.e. $x \in M$, none of the Lyapunov exponents is zero. Then the support of ν is contained in the closure of the set of periodic points for f.

We now return to the more general situation, of measures possibly having zero exponent. The measure-theoretic entropy can also be estimated using the Lyapunov spectrum, as follows. Let f be a smooth diffeomorphism of a compact manifold M. It can be shown, using Oseledec's theorem, that there exists an f-invariant Borel set $\Lambda \subset M$ consisting of regular points, such that for each f-invariant Borel probability measure μ on M, we have $\mu(\Lambda) = 1$. Define a function $\chi : \Lambda \to \mathbb{R}$ by

$$\chi(x) := \sum_{\chi_i(x) > 0} l_i(x) \chi_i(x),$$

where the sum ranges over the set of positive Lyapunov exponents of f and $l_i(x)$ is the multiplicity of the exponent $\chi_i(x)$ (see section 9.1). Then χ is a measurable f-invariant function.

Theorem 9.4.5 Let f be a smooth diffeomorphism of a compact manifold M, and let Λ and χ be as defined in the previous paragraph.

1. (Ruelle) If ν is an f-invariant Borel probability measure on M,

$$h_\nu(f) \leq \int_M \chi(x)\, d\nu(x).$$

2. (Pesin) If ν is absolutely continuous with respect to the natural smooth measure class on M, then

$$h_\nu(f) = \int_M \chi(x)\, d\nu(x).$$

A proof of the theorem can be found in [16] and [23].

10

Rigidity Theorems

Although the existence of the Oseledec decomposition for a $\mathbb{Z}$-action on a vector bundle E is very useful by itself, it is certainly desirable to know the precise values of the Lyapunov exponents.

When the $\mathbb{Z}$-action is part of a G-action with an invariant probability measure, where G is a connected simple Lie group of real rank at least 2, it is possible to give a very detailed description of the Lyapunov spectrum in terms of the n-dimensional linear representations of G, where n is the fiber dimension of E.

It will be shown, in fact, that E admits a measurable trivialization with respect to which the action, in a sense, "reduces" to a homomorphism from the universal covering group of G into a subgroup $H \subset GL(n, \mathbb{R})$. (This is actually true only modulo a compact normal subgroup of H, as explained later.) This is the content of theorem 10.7.3. The theorem does not tell which representation arises; that has to be studied separately, in any particular situation to which the theorem is applied; this is illustrated by theorem 10.5.3. Moreover, as theorem 10.5.3 also shows, one can sometimes prove that the trivialization is continuous or even differentiable.

The main technical result of the chapter is the C^r rigidity theorem, given in section 10.4.

10.1 Straightening Sections

The main results of the chapter can be cast in terms of finding sections of a principal bundle P that transform in a special way under the action of a group of automorphisms of P.

These special sections are defined as follows. Let $p : P \to M$ be a smooth principal H-bundle over a manifold M, and G a Lie group acting smoothly on P by bundle automorphisms. Let $\sigma : M \to P$ be a C^r section of P, where, as

196

usual, we allow $r =$ meas. Then σ will be called a *straightening section*, or a ρ-*section*, if there exists a smooth homomorphism $\rho : G \to H$ such that for each $g \in G$ and $x \in M$,

$$g\sigma(x) = \sigma(gx)\rho(g).$$

Notice that, for a general C^r section σ that is not necessarily straightening, $g\sigma(x)$ is an element of P_{gx}, and hence it can be written as a right translation of $\sigma(gx)$ by some element $\alpha(g, x) \in H$. The map $\alpha : G \times M \to H$ is clearly C^r. It also satisfies the *cocycle identity*:

$$\alpha(g_1 g_2, x) = \alpha(g_1, g_2 x)\alpha(g_2, x)$$

for all $g_1, g_2 \in G$ and all $x \in M$. In fact,

$$\sigma(g_1 g_2 x)\alpha(g_1 g_2, x) = (g_1 g_2)\sigma(x)$$

$$= g_1\sigma(g_2 x)\alpha(g_2, x)$$

$$= \sigma(g_1 g_2 x)\alpha(g_1, g_2 x)\alpha(g_2, x)$$

and the identity follows since H acts freely on P. We say that α is a C^r cocycle over the G-action on M.

Therefore, a section σ of P is straightening exactly when the cocycle α does not depend on x, so $\alpha(g, x) = \rho(g)$ for some homomorphism $\rho : G \to H$.

Exercise 10.1.1 Let G be a connected Lie group with Lie algebra $\mathfrak{g}$, L a subgroup of G, and $M = G/\Gamma$, where Γ is a lattice in G. Consider the action of L on M by left translations. Denote by P the frame bundle of M, and consider the natural action of L on P. Show that P admits a smooth straightening section and that the homomorphism ρ is (equivalent to) the adjoint representation $\mathrm{Ad} : G \to GL(\mathfrak{g})$.

Straightening sections, when they exist at all, are very special, even when we only require that they be measurable, or that they exist only over some open G-invariant subset of M. The following proposition makes this claim precise. We state the proposition for the measurable case only.

Proposition 10.1.2 Let G be a connected simple noncompact Lie group, acting on a principal H-bundle $p : P \to M$ by bundle automorphisms. Suppose that H is a real linear algebraic group and that the G-action on M leaves invariant an ergodic probability measure μ. Let σ_i, $i = 1, 2$, be measurable straightening sections of P. Then there exists $h_0 \in H$ such that $\sigma_2(x) = \sigma_1(x)h_0$ for μ-a.e. $x \in M$.

Proof. Suppose that σ_i is a measurable ρ_i-section of P, where $\rho_i : G \to H$ is a smooth homomorphism, $i = 1, 2$. Define a measurable map $A : M \to H$ by the property

$$\sigma_2(x) = \sigma_1(x)A(x).$$

Our goal is to prove that A is essentially constant.

The map A satisfies the following identity:

$$A(gx) = \rho_1(g)A(x)\rho_2(g)^{-1}$$

for all $g \in G$ and $x \in M$. In fact,

$$\begin{aligned}
\sigma_1(gx)A(gx)\rho_2(g) &= \sigma_2(gx)\rho(g) \\
&= g\sigma_2(x) \\
&= g\sigma_1(x)A(x) \\
&= \sigma_1(gx)\rho_1(g)A(x).
\end{aligned}$$

We know that G contains a group locally isomorphic to $SL(2, \mathbb{R})$, and in this subgroup we select a diagonalizable one-parameter subgroup, which we denote by C. The group C acts on H by

$$(c, h) \mapsto T(c, h) := \rho_1(c)h\rho_2(c)^{-1}.$$

This action is algebraic, which can be seen by the explicit description of the linear representations of $SL(2, \mathbb{R})$ given in section 7.7.

By Moore's ergodicity theorem, the action of C on M is also ergodic. Moreover, the action of C on H leaves invariant the probability measure $\nu := A_*\mu$. In fact, the identity $A \circ c = T_c \circ A$, where $T_c = T(c, \cdot)$, gives $A_*c_*\mu = (T_c)_*A_*\mu$, and, since $g_*\mu = \mu$ for each $g \in G$, we conclude that $\nu = (T_c)_*\nu$.

We can now apply corollary 4.9.5 to conclude that ν is supported on the set of fixed points for the action of C on H. This means that for each c and μ-a.e. x, $T_cA(x) = A(x)$. Therefore, $A(cx) = A(x)$ for each c and almost every x. Applying proposition 2.3.1 (and 2.3.5), we conclude that A is constant μ-almost everywhere.   □

Exercise 10.1.3 Let $\pi : E \to M$ be a vector bundle over a manifold M with fiber dimension n, and let G be a group that acts on E by bundle automorphisms such that the action on M leaves invariant a finite measure μ. Denote by $\mathcal{F}(E)$ the frame bundle of E and suppose that there exist a measurable section σ of $\mathcal{F}(E)$, a homomorphism

$\rho : G \to GL(n, \mathbb{R})$, and a measurable cocycle $k : G \times M \to K$, where K is a compact subgroup of $GL(n, \mathbb{R})$ centralizing $\rho(G')$, such that the following holds:

$$g\sigma(x) = \sigma(gx)k(g, x)\rho(g)$$

for each $g \in G$ and μ-a.e. $x \in M$. Show that, for each $g \in G$, the Lyapunov exponents for the $\mathbb{Z}$-action generated by g are the numbers $\ln |\lambda|$, where the λ are the eigenvalues of $\rho(g)$. Moreover, by identifying the fiber $E(x)$ with $\mathbb{R}^n$ at each x via the isomorphism $\sigma(x)$, conclude that the Oseledec decomposition corresponds to the decomposition of $\mathbb{R}^n$ given in exercise 9.3.4. (Notice that there would be no difficulty if, with respect to some norm $\| \cdot \|$ on E, the norm of the isomorphism $\sigma(x) : \mathbb{R}^n \to E(x)$ was bounded uniformly over x. Although that may not be true, by Poincaré recurrence almost every orbit will visit a set where $\|\sigma(x)\|$ is bounded infinitely many times.)

10.2 H-Pairs

The reader is advised at this point to review the general definitions in the first two sections of chapter 6. The results of this chapter generally will have two versions, one topological and the other measurable. Whenever possible, a common language will be used for both cases. For example, by a C^r map, $r = $ meas., we mean a Borel measurable map. When we say that a C^r map is defined over a set U, we implicitly assume that U is open, if $r \geq 0$, or Borel measurable, if $r = $ meas. In the measurable case we will implicitly assume, and often will make explicit mention to, a measure class represented by some measure μ. U will usually denote a subset of the base M of some fiber bundle and μ will usually refer to a measure on M. (Even in the measurable case, M will always be at least a topological manifold, even though we could allow for much greater generality. The reader will have no difficulty formulating the main (measurable) results of the chapter for standard Borel spaces rather than topological manifolds.)

If a group B acts by bundle automorphisms on a fiber bundle with base M, such that the action is C^r, then B naturally acts on M and the action is also C^r. When $r = $ meas., we assume that the measure class of μ is preserved by this action. We say that $U \subset M$ is a *full subset* if U is open and dense, when $r \geq 0$, or $\mu(M - U) = 0$, when $r = $ meas.

Let H be a real algebraic group and P a principal H-bundle over a manifold M. The projection map will be denoted $p : P \to M$. We suppose that P is a C^r bundle for $r \geq 0$ or $r = $ meas. (We refer to the first sections of chapter 6 for the basic definitions. In the measurable case, P is isomorphic to $M \times H$.)

Let V be a real algebraic H-space, that is, a real variety equipped with a real algebraic action of H. Recall from section 6.2 that a C^r geometric structure of type V may be defined as a C^r H-equivariant map $\varphi : P \to V$. We will keep in mind the example of a vector field on M, as we go through some of the more abstract definitions given later. Thus suppose that M is a smooth manifold of dimension n, and let X be a C^r vector field on M. Let $P = \mathcal{F}(TM)$ be the frame bundle of M, which is a principal $GL(n, \mathbb{R})$-bundle. We suppose that $V = \mathbb{R}^n$ is a $GL(n, \mathbb{R})$-space via the standard $GL(n, \mathbb{R})$-action given by matrix multiplication of column vectors. Each $\xi \in \mathcal{F}(M)$ is, by definition, a linear isomorphism from $\mathbb{R}^n$ to $T_x M$, with $x = p(\xi)$. If X is a C^r vector field on M, that is, a C^r section of the tangent bundle TM, we define $\varphi : P \to V$ by

$$\varphi(\xi) := \xi^{-1} X(p(\xi)).$$

The H-equivariance property of $\mathcal{G}$ corresponds, in this case, to the identity $\varphi(\xi h) = h^{-1}\xi$.

Back to the general case, the space of all H-equivariant C^r maps from P into V will be denoted $C^r(P, V)^H$. If U is an open (resp., measurable) subset of M and $r \geq 0$ (resp., $r =$ meas.), $C^r(P|_U, V)^H$ consists of those C^r H-equivariant maps that are defined over U. If B is a group of C^r bundle automorphisms of P, then B acts on $C^r(P, V)^H$ by $b \cdot \varphi := \varphi \circ b^{-1}$.

For each $\xi \in P$, let $e_\xi : C^r(P, V)^H \to V$ denote the evaluation map. Namely, if $\varphi \in C^r(P, V)^H$, then $e_\xi(\varphi) = \varphi(\xi)$. Notice that $e_{\xi h} = h^{-1}e_\xi$ for each $h \in H$ and $\xi \in P$. Also notice that $e_{b\xi} = e_\xi \circ b^{-1}$ for each $\xi \in P$ and $b \in B$.

The proof of the main result of this chapter revolves around one technical notion, which we call an H-pair, associated to the action of a group B by C^r automorphisms of a principal H-bundle (such that the action is C^r). An H-pair for B over $U \subset M$ is a pair (W, V) such that U is a full B-invariant subset of M, W is a B-invariant subset of $C^r(P|_U, V)^H$, and the following conditions hold.

1. For each $p \in P|_U$, $e_\xi : W \to V$ is injective and $W_\xi := e_\xi(W)$ is a real subvariety of V.
2. For all $\xi, \xi' \in P|_U$, $\tau_{\xi,\xi'} := e_\xi \circ e_{\xi'}^{-1} : W_{\xi'} \to W_\xi$ is an H-translation, that is, for some $h \in H$ and all $v \in W_{\xi'}$, we have $\tau_{\xi,\xi'}(v) = hv$.
3. H acts transitively and effectively on V.

The following example may help to clarify the content of the preceding definition. Let $X_1, \ldots, X_m$ be C^r vector fields defined on an open dense subset U of M, where $m \leq n = \dim M$. Suppose that these fields are linearly independent everywhere on U. Also suppose that a group B acts on M by diffeomorphisms, leaving U invariant, so that for each $b \in B$ we have

$$b_* X_i = \sum_{j=1}^m a_{ij}(b) X_j,$$

where the $a_{ij}(b)$, $1 \leq i, j \leq m$, are real constants. Let

$$\varphi_i : \mathcal{F}(TM)|_U \to \mathbb{R}^n$$

be the $GL(n, \mathbb{R})$-equivariant maps associated to the X_i. Let W be the complement of 0 in the real vector space of dimension m spanned by the φ_i, and write $V = \mathbb{R}^n - \{0\}$. Notice that $GL(n, \mathbb{R})$ acts transitively and effectively on V. Moreover, W is B-invariant and the evaluation maps e_ξ, $\xi \in \mathcal{F}(M)|_U$, are injective since the vector fields X_i are linearly independent. Condition 2 is also immediate since, for each ξ and ξ' in $\mathcal{F}(M)|_U$, we can find a linear isomorphism of $\mathbb{R}^n$ that maps $\varphi_i(\xi')$ to $\varphi_i(\xi)$ for each i, $1 \le i \le m$.

The m-by-m matrix $\rho(b) = (a_{ij}(b))$ obtained in the previous example gives a representation of B in dimension m, that is,

$$\rho : B \to GL(m, \mathbb{R})$$

is a group homomorphism. It will be seen shortly that there is also a similar homomorphism of B built into the definition of a general H-pair. First notice that since, by condition 3, H acts transitively on V, we can write $V = Hv_0$ for some $v_0 \in V$. Therefore, given v_0, we can naturally identify V with the quotient H/H_0, where H_0 is the isotropy subgroup of v_0 in H. Also notice that H_0 does not contain a nontrivial subgroup that is normal in H. This, in fact, is equivalent to the condition that H acts effectively on V.

Exercise 10.2.1 Show that the action of H by left translations of the homogeneous space H/H_0, where H_0 is a subgroup of H, is effective if and only if no nontrivial subgroup of H_0 is normal in H.

We now assume that an H-pair for B is given and construct, for each $\xi \in P|_U$, a homomorphism of B. Define

$$F_\xi := \{h \in H \mid hv = v \text{ for each } v \in W_\xi\}.$$

F_ξ is a subgroup of H and it is a normal subgroup of the group

$$N_\xi := \{h \in H \mid hv \in W_\xi \text{ for each } v \in W_\xi\}.$$

Moreover, F_ξ and N_ξ are real algebraic groups. (This is clear for F_ξ, since this group is the intersection of the isotropy subgroups of v for all $v \in W_\xi$. As for N_ξ, notice that this is the subgroup consisting of the $h \in H$ that are the zeros of all $p_w \in \mathbb{R}[V]$ of the form $p_w(h) = p(hw)$, where $w \in W_\xi$ and $p \in \mathbb{R}[V]$ lies in the ideal that vanishes on W_ξ.)

The collection of all H-translations from W_ξ into V can naturally be identified with the quotient H/F_ξ, since $hv = h'v$ for all $v \in W_\xi$ if and only if $h \in F_\xi$. Moreover, for each $\xi \in P|_U$ and each $b \in B$, $\tau_{b^{-1}\xi,\xi} : W_\xi \to V$ is an H-translation (property 2) so there is $h \in H$ such that $\tau_{b^{-1}\xi,\xi}(v) = hv$ for all

$v \in W_\xi$. Such h is characterized by the identity

$$he_\xi(\varphi) = e_{b^{-1}\xi}(\varphi) = \varphi(b^{-1}\xi) = e_\xi(b \cdot \varphi)$$

for all $\varphi \in W$. Therefore, since W is B-invariant, we conclude that $h \in N_\xi$. In other words, to each $b \in B$ and $\xi \in P|_U$, we obtain in a unique way an element of N_ξ/F_ξ, which we denote by $\rho_\xi(b)$.

Lemma 10.2.2 Let $H_1, \ldots, H_k$ be closed subgroups of a Lie group H and write $F = \bigcap_{i=1}^{k} H_i$. Denote by $\pi_i : H/F \to H/H_i$ the natural projection. Then a map Ψ into H/F is C^r if and only if $\pi_i \circ \Psi$ is C^r for each i.

Proof. Define a map

$$\mathcal{H} : H/F \to H/H_1 \times \cdots \times H/H_k$$

by $\mathcal{H}(hF) = (hH_1, \ldots, hH_k)$. Notice that $\mathcal{H}$ is smooth and injective. Moreover, it is immediate that if $X \in T_e H$ is mapped to 0 under the differential at e of the natural projection $H \to H/H_i$, then $X \in T_e H_i$. If this is true for each i, then $X \in T_e F$. It follows from this remark that $(D\mathcal{H})_{eF}$ is injective. From the fact that $\mathcal{H}$ commutes with left translation by $h \in H$ on H/F and on the product, it follows that $(D\mathcal{H})_{hF}$ is injective for each hF. Therefore, $\mathcal{H}$ is an injective smooth immersion of H/F into the product, so the conclusion of the lemma follows. $\square$

Proposition 10.2.3 The map $\rho_\xi : B \to N_\xi/F_\xi$ associated to an H-pair (W, V) is a smooth homomorphism.

Proof. That ρ_ξ is a homomorphism is immediate from the fact that $\rho_\xi(b)$ is uniquely characterized by the identity $\rho_\xi(b)e_\xi(\varphi) = e_\xi(b \cdot \varphi)$ for all $\varphi \in W$, since for all $b_1, b_2 \in B$, we have

$$\rho_\xi(b_1 b_2)e_\xi(\varphi) = e_\xi((b_1 b_2) \cdot \varphi) = \rho_\xi(b_1)e_\xi(b_2 \cdot \varphi) = \rho_\xi(b_1)\rho_\xi(b_2)e_\xi(\varphi).$$

We show next that ρ_ξ is C^r. Once this is proved, it will follow from theorems 3.7.1 and 8.2.3 that ρ_ξ is smooth.

Denote by H_w the real algebraic subgroup of H consisting of $h \in H$ such that $hw = w$. Then $F_\xi = \bigcap_{w \in W_\xi} H_w$. By the descending chain condition (section 4.3) we can find a finite subset $\{w_i \mid i = 1, \ldots, k\} \subset W_\xi$ such that

$$F_\xi = H_{w_1} \cap \cdots \cap H_{w_k}.$$

We identify H/H_{w_i} with the orbit Hw_i and denote by φ_i the unique element of W such that $e_\xi(\varphi) = w_i$. By the previous lemma, to show that ρ_ξ is C^r it suffices to show that the map

$$\rho_i : b \in B \mapsto \rho_\xi(b)w_i = \varphi_i(b^{-1}\xi),$$

obtained by composing ρ_ξ on the left with the projection $H/F \to H/H_{w_i}$, is C^r. But φ_i is C^r so the same is true for ρ_i.   □

Exercise 10.2.4 Show that the H-translations $\tau_{\xi,\xi'} : W_{\xi'} \to W_\xi$ associated to an H-pair (W, V) for a group B, defined over $U \subset M$, satisfy the following identities:

1. $\tau_{b\xi,\xi'} = \tau_{\xi,b^{-1}\xi'}$,
2. $\rho_\xi(b)\tau_{\xi,\xi'} = \tau_{\xi,\xi'}\rho_{\xi'}(b)$,
3. $\rho_{\xi h}(b) = h^{-1}\rho_\xi(b)h$,
4. $\tau_{\xi h,\xi'h'} = h^{-1}\tau_{\xi,\xi'}h'$,
5. $\tau_{b\xi,b'\xi'} = \rho_\xi(b)^{-1}\tau_{\xi,\xi'}\rho_{\xi'}(b')$,

where $\xi, \xi' \in P|_U$, $b, b' \in B$, and $h, h' \in H$.

We check now a couple of the preceding identities to make sure that their meaning is clear. First take identity 2. For each $\varphi \in W$, we have

$$\begin{aligned}
\tau_{\xi,\xi'}(\rho_{\xi'}(b)e_{\xi'}(\varphi)) &= \left(e_\xi \circ e_{\xi'}^{-1}\right)(e_{\xi'}(b \cdot \varphi)) \\
&= e_\xi(b \cdot \varphi) \\
&= \rho_\xi(b)e_\xi(\varphi) \\
&= \rho_\xi(b)\left(e_\xi \circ e_{\xi'}^{-1}\right)(e_{\xi'}(\varphi)) \\
&= \rho_\xi(b)\tau_{\xi,\xi'}(e_{\xi'}(\varphi)).
\end{aligned}$$

Since each element in the domain of $\tau_{\xi,\xi'}$ is of the form $e_{\xi'}(\varphi)$ for some $\varphi \in W$, property 2 follows.

For property 1, we write

$$\begin{aligned}
\rho_\xi(b)^{-1}\tau_{\xi,b^{-1}\xi'}\rho_{\xi'}(b)e_{\xi'}(\varphi) &= \rho_\xi(b)^{-1}\left(e_\xi \circ e_{b^{-1}\xi'}^{-1}\right)(e_{\xi'}(b \cdot \varphi)) \\
&= \rho_\xi(b)^{-1}\left(e_\xi \circ e_{b^{-1}\xi'}^{-1}\right)(e_{b^{-1}\xi'}(\varphi)) \\
&= \rho_\xi(b)^{-1}e_\xi(\varphi) \\
&= (b^{-1} \cdot \varphi)(\xi) \\
&= \varphi(b\xi) \\
&= e_{b\xi}(\varphi) \\
&= \left(e_{b\xi} \circ e_{\xi'}^{-1}\right)(e_{\xi'}(\varphi)) \\
&= \tau_{b\xi,\xi'}e_{\xi'}(\varphi).
\end{aligned}$$

The remaining properties are proved in a similar way.

We say that an H-pair (W_1, V), defined over $U_1 \subset M$, is *contained* in another H-pair (W_2, V), defined over $U_2 \subset M$, if $W_i \subset C^r(P|_{U_i}, V)^H$, $i = 1, 2$, and for each $\varphi_1 \in W_1$ there is $\varphi_2 \in W_2$ such that φ_1 and φ_2 agree on $P|_{U_1 \cap U_2}$.

It will be convenient to identify pairs (W_1, V) and (W_2, V) if each one is contained in the other. This indeed defines an equivalence relation and we write $[W, V]$ for the equivalence class represented by (W, V). We say that $[W_1, V]$ is contained in $[W_2, V]$, and write $[W_1, V] < [W_2, V]$, if a representative of the former is contained in a representative of the latter. We also say that $[W, V]$ is a *maximal H-pair* for a group B if it is equal to any other H-pair in which it is contained.

A C^r H-pair (W, V) for B, defined over a full B-invariant set $U \subset M$, will be called an *invariant H-pair* if $b \cdot \varphi = \varphi$ for all $\varphi \in W$ and $b \in B$. If A is a subgroup of B, we say that the H-pair for B is $\mathbb{R}$-*split for A* if for each $\xi \in P|_U$ and any given linear representation $R : N_\xi / F_\xi \to GL(m, \mathbb{R})$, we have that $R \circ \rho_\xi|_A$ is a rational homomorphism and $(R \circ \rho_\xi)(a)$ is diagonalizable with real eigenvalues, for each $a \in A$.

H-pairs that are $\mathbb{R}$-split for a subgroup of B will arise in the following way. Suppose that B contains a group B' locally isomorphic to $SL(2, \mathbb{R})$, and let A be an $\mathbb{R}$-split Cartan subgroup of B'. Then the H-pair is $\mathbb{R}$-split for A due to corollary 7.7.6.

Proposition 10.2.5 Every C^r H-pair for B is contained in a maximal C^r H-pair for B. Furthermore, every invariant (resp., $\mathbb{R}$-split for a subgroup A of B) C^r H-pair for B is contained in a maximal invariant (resp., maximal $\mathbb{R}$-split for A) C^r H-pair for B.

Proof. We will construct for any given increasing sequence of C^r H-pairs for B,

$$[W^1, V] < [W^2, V] < \cdots < [W^i, V] < \cdots,$$

another such pair $[W^\infty, V]$, which contains each one in the sequence. The claim will then follow from Zorn's lemma.

Let (W^i, V) be a representative of $[W^i, V]$, where W^i is a subset of $C^r(P|_{U_i}, V)^H$, and U_i is a full B-invariant subset of M. The intersection of all the U_i is nonempty. This is due, in the measurable case, to the fact that the intersection of a countable family of sets of full measure is also of full measure. In the topological case, this is a consequence of Baire's property, namely, a countable intersection of open dense sets is also dense. Therefore, we can fix a $\xi_0 \in P$ that projects to a point in that intersection. For each $\xi \in P|_{U_i}$, we set

$W^i_\xi := e_\xi(W^i)$ and write

$$\psi^i_\xi := \tau_{\xi,\xi_0} : W^i_{\xi_0} \to W^i_\xi,$$

which is an H-translation. Therefore, as already noted, we may regard ψ^i_ξ as an element of $H/F^i_{\xi_0}$, where $F^i_{\xi_0}$ is the (real algebraic) subgroup of H that fixes $W^i_{\xi_0}$ pointwise. Thus, we obtain a C^r H-equivariant map

$$\Psi_i : P|_{U_i} \to H/F^i_{\xi_0}$$

defined by $\Psi_i(\xi) := \psi^i_\xi$. ($H$-equivariance is immediate from property 4 in exercise 10.2.4. The fact that the map is C^r follows from the same argument used in the proof of proposition 10.2.3.) Notice that for any given $w \in W^i_{\xi_0}$ we recover $\varphi \in W^i$ by the equation $\varphi(\xi) = \Psi_i(\xi)w$.

The sequence of subvarieties $W^i_{\xi_0}$ of V is increasing, so the sequence of real algebraic subgroups $F^i_{\xi_0}$ is decreasing. By the descending chain condition for algebraic groups, there must be a finite index i_0 such that $F^i_{\xi_0} = F^{i_0}_{\xi_0} =: F^\infty_{\xi_0}$ for all $i \geq i_0$. Consequently, for all i, $j \geq i_0$, the maps Ψ_i and Ψ_j agree on $P|_{U_i \cap U_j}$. (Observe that if, say, $j \geq i$, then the H-translation ψ^i_ξ is the restriction of ψ^j_ξ to W^i_ξ and is of the form $w \mapsto hw$ for some $h \in H$ and all $w \in W^j_{\xi_0}$, so $\psi^i_\xi = \psi^j_\xi$ if $F^i_{\xi_0} = F^j_{\xi_0}$.)

Therefore, a map Ψ_∞ can be defined on $P|_{\cup_{i \geq i_0} U_i}$ that extends all Ψ_i, for $i \geq i_0$, and is, in particular, a C^r map. Moreover, if $W^\infty_{\xi_0}$ denotes the Zariski closure of the union of all $W^i_{\xi_0}$, $i \geq i_0$, we have that $F^\infty_{\xi_0}$ must fix $W^\infty_{\xi_0}$ pointwise. In fact, $F^\infty_{\xi_0}$ is precisely the subgroup consisting of all $h \in H$ such that $hw = w$ for all $w \in W^\infty_{\xi_0}$, since if some h fixing $W^\infty_{\xi_0}$ pointwise were not already in $F^\infty_{\xi_0}$, it would give an element of H that fixes $W^{i_0}_{\xi_0}$ pointwise and is not in $F^{i_0}_{\xi_0}$, which is a contradiction.

We define on the B-invariant full set $U_\infty := \bigcup_{i \geq i_0} U_i$ an H-pair (W^∞, V) for B as follows. Let $W^\infty \subset C^r(P|_{U_\infty}, V)^H$ be the collection of C^r maps φ_w, $w \in W^\infty_{\xi_0}$, defined by

$$\varphi_w(\xi) := \Psi_\infty(\xi)w.$$

It is now a simple exercise to check that properties 1, 2, and 3 in the definition of an H-pair are satisfied. In fact, property 1 results from H-translations being injective; for property 2, notice that $\tau_{\xi,\xi'} : W^\infty_{\xi'} \to W^\infty_\xi$ is the H-translation that corresponds, with some abuse of notation, to $\Psi_\infty(\xi)\Psi_\infty(\xi')^{-1}$; and property 3 holds since V and the H-action on it have not changed.

The homomorphism ρ^∞_ξ associated to the just constructed H-pair can be described as follows. For each i and $\xi \in P|_{U_i}$, there is a smooth homomorphism

$\rho_\xi^i : B \to N_\xi^i / F_\xi^i$ such that $e_\xi(g \cdot \varphi) = \rho_\xi^i(g) e_\xi(\varphi)$ for all $\varphi \in W^i$ and $g \in B$. It follows that for the same i, ξ, g, we have

$$\Psi_i(g^{-1}\xi) = \rho_\xi^i(g) \Psi_i(\xi).$$

Since Ψ_i and Ψ_j agree on $P|_{U_i \cap U_j}$, for $i, j \geq i_0$, we must have, for ξ in this set, that $\rho_\xi^j(g) = \rho_\xi^i(g)$ for all $g \in B$. We thus obtain a family of homomorphisms ρ_ξ^∞ from B into $N_\xi^\infty / F_\xi^\infty$ extending $\xi \mapsto \rho_\xi^i$ over $P|_{U_\infty}$ such that $\Psi_\infty(g^{-1}\xi) = \rho_\xi^\infty(g) \Psi_\infty(\xi)$. Notice that the homomorphisms are C^r in ξ. Here, N_ξ^∞ denotes the algebraic subgroup of H that stabilizes W_ξ^∞ as a set. It contains F_ξ^∞ as a normal subgroup. It now follows from the definition of $\varphi_w \in W^\infty$ that $e_\xi(g \cdot \varphi_w) = \rho_\xi^\infty(g) e_\xi(\varphi_w)$. (The action of B on W^∞ is canonically defined by restricting the natural action on the space of H-equivariant maps over U.)

If each H-pair (W^i, V) is invariant, the same property is clearly inherited by (W^∞, V), since each ρ_ξ^i is trivial. Similarly, if each pair is $\mathbb{R}$-split for a subgroup A of B, the same property is inherited by (W^∞, V). $\qquad \square$

10.3 *H*-Pairs for Centralizers

The same setup as in section 2 is still in force. Let P be a principal H-bundle over a manifold M, where H is a real algebraic group, and let G be a Lie group that acts on P by bundle automorphisms via a C^r action. Let B be a subgroup of G, and suppose that (W, V) is a C^r H-pair for B. Given a subgroup Z of G that centralizes B, that is, such that each automorphism from Z commutes with each automorphism from B, we would like to know whether the H-pair for B is also an H-pair for Z. The next lemma gives sufficient conditions for this to happen. It should be noticed that all that we need to show in order to guarantee that (W, V) is also an H-pair for Z is that W is Z-invariant.

We say that an action on M is C^r-*transitive* if it is topologically transitive, for $r \geq 0$, or ergodic, for $r = $ meas. In the latter case, it is implicitly assumed that we have selected a measure class that is invariant under the action of G, with respect to which the B-action is ergodic.

Proposition 10.3.1 Suppose that (W, V) is a maximal invariant C^r H-pair for B, $W \subset C^r(P|_U, V)^H$, and that the B-action on M is C^r-transitive. Then (W, V) is a (not necessarily maxmimal) H-pair for Z. More precisely, the elements of W can be extended to $P|_{U'}$, where U' is a full Z-invariant subset of M, so that (W, V), now defined above U', is a C^r H-pair for the subgroup of G generated by Z and B. (If $r \geq 0$, U' can be taken to be $Z \cdot U$, i.e., the saturation of U by the Z-action.) The same conclusion holds if (W, V) is maximal $\mathbb{R}$-split for some $T \subset B$ such that the T-action on M is C^r-transitive.

Proof. Fix $z \in Z$ and define $z \cdot W$ to be the set of all maps $z \cdot \varphi := \varphi \circ z^{-1}$, $\varphi \in W$. It is immediate to check that $(z \cdot W, V)$ is a C^r H-pair for B defined over $z(U)$. In fact, notice that property 1 in the definition of an H-pair follows from $e_{z \cdot \xi}(z \cdot \varphi) = e_\xi(\varphi)$, and, since $(z \cdot W)_{z \cdot \xi} = W_\xi$, we have that $\tau_{z \cdot \xi, z \cdot \xi'} : (z \cdot W)_{z \cdot \xi'} \to (z \cdot W)_{z \cdot \xi}$ is the H-translation $\tau_{\xi, \xi'}$ for all $\xi, \xi' \in P|_U$, proving property 2. Property 3 is clear since H and V and the H-action on V are still the same.

Set $W^z := z \cdot W \cup W$ and $W^z_\xi := e_\xi(W^z) = W_{z^{-1}\xi} \cup W_\xi$. We claim that (W^z, V) is a C^r H-pair for B, defined over $z(U) \cap U$. To show this, we first check that the evaluation maps $e_\xi : W^z \to V$ are injective.

For any $\xi, \eta_0 \in P|_{z(U) \cap U}$, we recall that the H-translation τ_{ξ, η_0} on W_{η_0} may be regarded as an element of H/F_{η_0}, where F_{η_0} is the subgroup of H that fixes each element of W_{η_0}. Define

$$\Psi^z_{\eta_0} : P|_{z(U) \cap U} \to H/F_{\eta_0} \times H/F_{\eta_0}, \qquad \Psi^z_{\eta_0}(\xi) = (\tau_{\xi, \eta_0}, \tau_{z^{-1}\xi, \eta_0}).$$

Notice that $H \times B$ acts on $H/F_{\eta_0} \times H/F_{\eta_0}$ diagonally, H acting on the left and B on the right via the homomorphism $\rho_{\eta_0} : B \to N_{\eta_0}/F_{\eta_0}$. The map is $H \times B$-equivariant:

$$\Psi^z_{\eta_0}((h, b) \cdot \xi) := \Psi^z_{\eta_0}(b\xi h^{-1}) = h\Psi^z_{\eta_0}(\xi)\rho_{\eta_0}(b)^{-1} =: (h, b) \cdot \Psi^z_{\eta_0}(\xi),$$

where $\xi \in P|_{z(U) \cap U}$ and $(h, b) \in H \times B$. This is shown by means of the properties stated in exercise 10.2.4 and requires the commutativity of z and b.

We now apply proposition 6.6.5 (the third reduction lemma). It is at this point that we need (W, V) to be an invariant (or $\mathbb{R}$-split for T) H-pair. Notice that if T is $\mathbb{R}$-split, its image under ρ_{η_0} is an algebraic one-parameter subgroup of B.

It follows from proposition 6.6.5 that $\Psi^z_{\eta_0}$, restricted to $P|_{U_z}$ for some full B-invariant subset $U_z \subset z(U) \cap U$, takes values in a single H-orbit. We can conclude that the evaluation map $e_\eta : W^z \to W^z_\eta$ is injective, for all $\eta \in P|_{U_z}$, by arguing as follows. If $\eta \in P|_{U_z}$, $\varphi_1 \in W$, and $z \cdot \varphi_2 \in z \cdot W$, then

$$\begin{aligned}
(\varphi_1(\eta), z \cdot \varphi_2(\eta)) &= (\tau_{\eta, \eta_0}(\varphi_1(\eta_0)), \tau_{z^{-1}\eta, \eta_0}(\varphi_2(\eta_0))) \\
&= (h\tau_{\xi, \eta_0}(\varphi_1(\eta_0)), h\tau_{z^{-1}\xi, \eta_0}(\varphi_2(\eta_0))) \\
&= (h\varphi_1(\xi), h\varphi_2(z^{-1}\xi)) \\
&= h(\varphi_1(\xi), z \cdot \varphi_2(\xi)).
\end{aligned}$$

Therefore, if $\varphi_1(\eta) = z \cdot \varphi_2(\eta)$, we conclude that $\varphi_1(\xi) = z \cdot \varphi_2(\xi)$ for each $\xi \in P|_{U_z}$. This shows property 1 characterizing an H-pair.

It also follows from the equality $(\tau_{\eta,\eta_0}, \tau_{z^{-1}\eta,\eta_0}) = (h\tau_{\xi,\eta_0}, h\tau_{z^{-1}\xi,\eta_0})$ that $\tau_{\eta,\xi}$: $W_\xi^z \to W_\eta^z$ is an H-translation given by $\tau_{\eta,\xi}(w) = hw$. Such h is uniquely determined up to right translation by elements in the group

$$F_\xi^z := F_{z^{-1}\xi} \cap F_\xi = \{h \in H \mid hw = w \text{ for all } w \in W_\xi^z\}.$$

Therefore, property 2 also holds. The third property of an H-pair is, again, trivially satisfied since V and the H-action on it have not changed.

The foregoing discussion shows that (W^z, V) is an H-pair for B defined over a full subset of $z(U) \cap U$ and contains (W, V) and $(z \cdot W, V)$. By the maximality of (W, V), we conclude that

$$[W, V] = [W^z, V] = [z \cdot W, V].$$

We have shown, in fact, that for each $z \in Z$, there exists a full B-invariant set $U_z \subset z(U) \cap U$ such that for every $\varphi_1 \in W$ there exists $\varphi_2 \in W$ for which

$$(z \cdot \varphi_1)(\xi) = \varphi_2(\xi)$$

for all $\xi \in P|_{U_z}$.

We consider separately the measurable and continuous ($r \geq 0$) cases. In the continuous case, it follows from $[W, V] = [z \cdot W, V]$ that for each $\varphi \in W$ and each $z \in Z$, there is $\varphi' \in W$ such that $z \cdot \varphi = \varphi'$ above a full subset of $z(U) \cap U$. Therefore, by continuity, the equality holds above the entire $z(U) \cap U$, so φ has a continuous extension to $P|_{U \cup z^{-1}(U)}$. Since $z \in Z$ was arbitrary, we conclude that φ has a continuous extension above the saturation of U by Z. After extending all elements of W in this way, we have that W is Z-invariant. Therefore, (W, V) is an H-pair for Z as claimed. Notice that the saturation of U by Z is still a B-invariant set, and that W is still B-invariant, so (W, V) is an H-pair for the group generated by B and Z.

The measurable case requires a little more care. We need to show that there exists a fixed full (conull) subset U' of M, invariant under the group $B' \subset G$ generated by B and Z, such that each $\varphi \in W$ extends to a measurable function defined above U' and, so extended, W is invariant under B' and still satisfies the properties of an H-pair. Once this claim is verified, the proof of the proposition will be finished.

First, notice the following. For each $z \in B'$, each $x \in U_z$, and each $\xi \in P_x$, we have seen that $(z \cdot \varphi)(\xi) = \varphi'(\xi)$ for each $\varphi \in W$ and some $\varphi' \in W$. Therefore, $W_{z^{-1}\xi} \subset W_\xi$. But W_ξ is an H-translate of $W_{z^{-1}\xi}$, so $W_{z^{-1}\xi} = W_\xi$. Also notice that $(z^{-1} \cdot W)_{z^{-1}\xi} = W_\xi$. Consequently, for all $\xi_1, \xi_2 \in P_{U_z}$,

$$\tau_{z^{-1}\xi, z^{-1}\xi_2} = \tau_{\xi_1,\xi_2} : W_{z^{-1}\xi_2} = W_{\xi_2} \to W_{z^{-1}\xi_1} = W_{\xi_1}.$$

By the same argument, we have that for each $z, z_1, z_2 \in B'$, and for all $\xi, \eta \in P$ over a subset of full measure in U, the following hold:

1. $W_{z\xi} = W_\xi$, $W_{z\eta} = W_\eta$, and $\tau_{z\xi,z\eta} = \tau_{\xi,\eta}$;
2. $W_\eta = W_{z_2^{-1}\eta} = W_{z_1^{-1}z_2^{-1}\eta}$ and $\tau_{z_1^{-1}z_2^{-1}\eta,z_1^{-1}\eta} = \tau_{z_2^{-1}\eta,\eta}$, so

$$\tau_{z_1^{-1}z_2^{-1}\eta,\eta} = \tau_{z_2^{-1}\eta,\eta} \circ \tau_{z_1^{-1}\eta,\eta}.$$

Notice, in particular, that for a.e. $\eta \in P|_U$, the map

$$\rho_\eta(z) := \tau_{z^{-1}\eta,\eta} \in N_\eta/F_\eta$$

is an essential homomorphism, that is, for a.e. $z_1, z_2 \in B'$,

$$\rho_\eta(z_2z_1) = \rho_\eta(z_1)\rho_\eta(z_2).$$

By the exercise given right after this proof, such an essential homomorphism agrees a.e. with a smooth homomorphism.

Applying Fubini's theorem, we conclude that there exists an $\eta_0 \in P|_U$ such that for a.e. $z_1, z_2 \in Z$ and all ξ above a subset of full measure in U, properties 1 and 2 just written hold for all ξ above a subset of full measure of U and a.e. $z_1, z_2 \in B'$.

Define

$$\Psi : P_U \to N_{\eta_0}/F_{\eta_0}, \qquad \Psi(\xi) = \tau_{\xi,\eta_0}.$$

Properties 1 and 2 immediately imply that

$$\Psi(z\xi) = \Psi(\xi)\rho_{\eta_0}(z)^{-1}$$

for all $\xi \in P|_U$ above a subset of full measure, and for a.e. $z \in B'$. Therefore, Ψ is essentially B'-equivariant. By proposition 2.3.1, we can now modify Ψ on a set of measure 0 so as to make it strictly equivariant and defined over a set of the form $P|_{U'}$, where U' is B'-invariant.

After all this work, we can redefine each element of W over a set of measure zero so that W becomes a B'-invariant subset of $C^{\text{meas.}}(P|_{U'}, V)^H$. This is done as follows. For each $w \in W_{\eta_0}$, define for each $\xi \in P_{U'}$

$$\varphi_w(\xi) := \Psi(\xi)w.$$

It is immediate that (W, V) is still an H-pair, now for B', and that for each φ in the original W, there is w such that $\varphi = \varphi_w$ almost everywhere. This completes the proof. $\qquad\square$

Exercise 10.3.2 Let Z and N be Lie groups, and let $\rho : Z \to N$ be a Borel map such that $\rho(z_1 z_2) = \rho(z_1)\rho(z_2)$ for a.e. $(z_1, z_2) \in Z \times Z$, relative to a Haar measure on Z. Show that there exists a Borel (hence smooth) homomorphism $\rho_0 : Z \to N$ such that $\rho = \rho_0$ almost everywhere. The following sketch is taken from [36, B2]. Define

$$H := \{z_1 \in Z \mid z \mapsto \rho(z)^{-1}\rho(zz_1) \text{ is an essentially constant function}\}.$$

By Fubini's theorem, H is conull. If $z_1, z_2 \in H$, it follows from

$$\rho(z)^{-1}\rho(zz_1z_2) = \rho(z)^{-1}\rho(zz_1)\rho(zz_1)^{-1}\rho(zz_1z_2)$$

that $z_1 z_2 \in H$. But if a conull subset of Z is closed under multiplication, this subset must be Z, for the following reason. $H^{-1}u \cap H$ is conull for each $u \in Z$, so there is $z' \in H$ such that $z' = z^{-1}u$ for some $z \in H$. Therefore, $u = zz' \in H$. Now show that if $\rho_0(z_1)$ is the element of N such that $\rho(z)^{-1}\rho(zz_1)$ is constant a.e. on Z, then $\rho_0 : Z \to N$ is a homomorphism. Show that ρ_0 is Borel by Fubini's theorem and the fact that $\rho_0(z_1) = \int_Z \rho(z)^{-1}\rho(zz_1)\,dv(z)$, where v is a probability measure on Z in the Haar measure class.

The same setup as in the previous proposition is also assumed in the next one.

Proposition 10.3.3 Let Z be a group of H-bundle automorphisms commuting with the B-action. Assume that the action of B on M is C^r-transitive and that there exists a C^r H-pair $(W, H/H_0)$ for B. Then there exists a C^r H-pair $(W', H/F)$ for Z such that F is a subgroup of H_0. Moreover, if π is the natural projection from H/F onto H/H_0, there is $\varphi \in W'$ such that $\pi \circ \varphi \in W$.

Proof. The H-pair $(W, H/H_0)$ for B is defined over some full B-invariant subset $U \subset M$, and we recall that H_0 does not contain a nontrivial normal subgroup of H. Fix a point $\eta_0 \in P|_U$. By translating η_0 in its fiber in P by some appropriate element of H, we may assume without loss of generality that W_{η_0} (the image of W in H/H_0 under e_{η_0}) contains the coset H_0, that is, there is $\varphi_0 \in W$ such that $\varphi_0(\eta_0) = H_0$. Notice that if $\xi \in P|_U$, then

$$\varphi_0(\xi) = \tau_{\xi,\eta_0} H_0.$$

The group $F := F_{\eta_0}$ of elements of H that fix W_{η_0} pointwise is a subgroup of H_0 (since it fixes, in particular, the coset H_0), and hence it does not contain a nontrivial normal subgroup of H.

Let $N := N_{\eta_0}$ be, as before, the subgroup of H that stabilizes W_{η_0}, which contains F as a normal subgroup, and set $A := N/F$.

We regard $P_1 := P|_U \times A$ as a principal bundle with group $H_1 = H \times A$ and right action given by the product action. We define on H/F a right H_1-action given by

$$\tau \cdot (h, a) = h^{-1}\tau a,$$

for $\tau \in H/F$ and $(h, a) \in H_1$, and introduce the map

$$\psi : P_1 \to H/F, \qquad \psi(\xi, a) := \tau_{\xi, \eta_0} a.$$

Notice that $\pi(\psi(\xi, e)) = \varphi_0(\xi)$ for $\xi \in P|_U$. A simple calculation also shows that ψ is H_1-equivariant, that is, $\psi(\xi h, aa') = h^{-1}\psi(\xi, a)a'$ for $(\xi, a) \in P_1$ and $(h, a') \in H_1$. Moreover, with respect to the left B-action on P_1 given by

$$b \cdot (\xi, a) := (b\xi, \rho_{\eta_0}(b)a),$$

ψ is B-invariant. In fact,

$$\begin{aligned}
\psi(b \cdot (\xi, a)) &= \psi(b\xi, \rho_{\eta_0}(b)a) \\
&= \tau_{b\xi, \eta_0}\rho_{\eta_0}(b)a \\
&= \tau_{\xi, b^{-1}\eta_0}\rho_{\eta_0}(b)a \\
&= \tau_{\xi, \eta_0}a \\
&= \psi(\xi, a).
\end{aligned}$$

Therefore, $(\{\psi\}, H/F)$ defines an invariant C^r H_1-pair for B. By proposition 10.2.5, there exists a maximal invariant C^r H_1-pair for B containing $(\{\psi\}, H/F)$, which we denote by $(W_1, H/F)$.

Define a left Z-action on P_1 by

$$z \cdot (\xi, a) := (z\xi, a).$$

It is immediate to check that the Z-action commutes with both the B-action and the H_1-action on P_1. Therefore, by proposition 10.2.1, $(W_1, H/F)$ is also an H_1-pair for Z. Let now W' be the space of maps $\varphi : P \to H/F$ of the form $\varphi(\xi) := \varphi_1(\xi, e)$, for some $\psi_1 \in W_1$, where e is the identity element in A. It also follows that $(W', H/F)$ defines a C^r H-pair for Z. Finally, as we already noted, φ_0 is the image of some element of W' under the map from W' to W defined by composition with π. $\qquad \square$

10.4 The C^r Rigidity Theorem

The C^r rigidity theorem stated shortly is the main technical result of the chapter. It gives a precise answer to the following, somewhat vague, questions: Suppose that a noncompact simple Lie group G acts on a manifold M and that a subgroup G_0 of G leaves invariant some geometric structure on M. Does G also leave invariant the same structure? If not, what can we say about the possible structures invariant under the action of G? The answer to these questions given by the theorem will be expressed in terms of the algebraic hull of the actions. Vaguely stated, the "part" of the algebraic hull of the G-action that is "not already in" the algebraic hull of the G_0-action "corresponds" to a homomorphic image of G.

The key assumptions of the C^r rigidity theorem will be that certain subgroups of G act C^r-transitively on M and (if we want a nontrivial conclusion from the theorem) that the C^r algebraic hull of G_0 is a proper subgroup of the C^r algebraic hull of G. The latter condition can be interpreted in the language of chapter 6 by saying that there are geometric A-structures on M invariant under G_0 that are not invariant under G.

We denote by G a connected semisimple Lie group. Recall from section 7.5 that the real rank of G is the dimension of an $\mathbb{R}$-split Cartan subalgebra $\mathfrak{a}$ of the Lie algebra of G. We call the connected subgroup A of G with Lie algebra $\mathfrak{a}$ an $\mathbb{R}$-*split Cartan subgroup* of G. A one-parameter subgroup T of G will be called $\mathbb{R}$-*split* if for each linear representation ρ of G, $\rho(a)$ is diagonalizable with real eigenvalues for all $a \in T$. Any one-parameter subgroup of A is, of course, $\mathbb{R}$-split.

As always, when we say that a map is C^r we also include the case $r =$ meas. Also recall that an action is called (in this book) C^r-transitive if it is topologically transitive, when $r \geq 0$, or ergodic, when $r =$ meas.

Theorem 10.4.1 [11] Suppose that G is a connected semisimple Lie group with real rank at least 2 and that it acts by automorphisms on some C^r principal H-bundle P over a manifold M, such that the action is also C^r. If $r =$ meas., we assume that G preserves some measure class, represented by a Borel measure μ on M. (Any reference to ergodicity is to be understood relative to μ.) Assume, furthermore, that

1. H is the C^r algebraic hull of the G-action;
2. every $\mathbb{R}$-split one-parameter subgroup T of G acts C^r-transitively on M and each point in a full subset of M is a recurrent point for T.

Also assume that there is a subgroup $G_0 \subset G$ for which the following hold:

3. G_0 acts C^r-transitively on M;
4. G_0 commutes with some element of a maximal $\mathbb{R}$-split Cartan subgroup A of G;

5. the C^r algebraic hull of the G_0-action does not contain a nontrivial normal subgroup of H.

Then there exists a continuous surjective homomorphism $\rho : G \rightarrow H$, a full G-invariant subset $U \subset M$, and a C^r section σ of $P|_U$ such that for all $g \in G$ and $x \in U$,

$$g\sigma(x) = \sigma(gx)\rho(x).$$

Proof. We are going to use here the notation and the main result of section 7.9.

The C^r algebraic hull H_0 of G_0 determines a G_0-invariant C^r map $\varphi_0 : P|_{U_0} \rightarrow H/H_0$, where U_0 is some full G_0-invariant subset of M. It is immediate that $(\{\varphi_0\}, H/H_0)$ is a C^r H-pair for G_0 defined over U_0. In fact, properties 1 and 2 characterizing an H-pair are clearly satisfied since W is, in this case, a single point. The third property is one of the hypotheses of the theorem.

By proposition 10.3.3, there is a C^r H-pair for the subgroup $\langle h_0 \rangle$ generated by some element $h_0 \in A$ commuting with G_0. (It is an assumption of the theorem that h_0 exists.) The same proposition implies the existence of a C^r H-pair for A, which is clearly also an H-pair for each subgroup of A.

Let A_α be the connected subgroup of A with Lie algebra equal to $H_\alpha^\perp$, and let $\check{A}_\alpha$ be the one-dimensional connected subgroup of A with Lie algebra $\mathbb{R}H_\alpha$. Proposition 10.3.3, applied now to A_α, for $\alpha \in \Phi(\mathfrak{a}, \mathfrak{g})$, gives a C^r H-pair for the centralizer $H_\alpha^\perp$. We denote this centralizer by Z_α. This H-pair is also $\mathbb{R}$-split for $\check{A}_\alpha$. This is due to corollary 7.7.6 and the discussion prior to proposition 7.9.3, which says that H_α spans the $\mathbb{R}$-split Cartan subalgebra of a connected subgroup $S_\alpha \subset G$ that is locally isomorphic to $SL(2, \mathbb{R})$. By proposition 10.2.5, we obtain a C^r maximal $\mathbb{R}$-split H-pair (W, V).

We claim that there exists an H-pair for A that is also maximal $\mathbb{R}$-split for A. To obtain such a pair, fix $\alpha \in \Phi(\mathfrak{a}, \mathfrak{g})$ and a C^r maximal $\mathbb{R}$-split H-pair $(W, H/L)$ for $\check{A}_\alpha$, as obtained in the previous paragraph. We may assume that α and $(W, H/L)$ have been chosen so that L is minimal; that is, if $\beta \in \Phi(\mathfrak{a}, \mathfrak{g})$ and $(W', H/L')$ is an H-pair that is maximal $\mathbb{R}$-split for $\check{A}_\beta$ such that $L' \subset L$, then $L' = L$. By the descending chain condition for algebraic groups we conclude that α and $(W, H/L)$ indeed exist.

Proposition 10.3.1 implies that $(W, H/L)$ is also an H-pair for each element of A, since A centralizes $\check{A}_\alpha$. Let now Π be a hyperplane in $\mathfrak{a}^*$ (the dual space of $\mathfrak{a}$) spanned by roots and complementary to the direction spanned by α. Let $u \in \mathfrak{a}$ be a nonzero vector orthogonal to Π with respect to the Killing form. By applying proposition 10.3.3, we obtain a C^r H-pair $(W', H/F)$ for the centralizer of u in G, where $F \subset L$.

Notice, in particular, that $(W', H/F)$ is an H-pair for A. On the other hand, for any root $\beta \in \Pi$, S_β centralizes u, by the assumption that u is orthogonal to

Π. Since $\check{A}_\beta \subset S_\beta$, we conclude that $(W', H/F)$ is also $\mathbb{R}$-split for $\check{A}_\beta$.

Using the minimality of L, we conclude that $F = L$. Moreover, by the second part of proposition 10.3.3 and the fact that $F = L$, the intersection $W_0 := W \cap W'$ is not empty. Since both $(W, H/L)$ and $(W', H/L)$ are H-pairs for A, the intersection $(W_0, H/L)$ also defines an H-pair for A.

We claim that $(W_0, H/L)$ is the desired H-pair for A that is $\mathbb{R}$-split for A. First notice that since $(W, H/L)$ is $\mathbb{R}$-split for $\check{A}_\alpha$ and $(W', H/L)$ is $\mathbb{R}$-split for $\check{A}_\beta$, then $(W_0, H/L)$ is $\mathbb{R}$-split for both $\check{A}_\alpha$ and $\check{A}_\beta$ for all roots $\beta \in \Pi$. But $\check{A}_\alpha$ and all the $\check{A}_\beta$, for $\beta \in \Phi(\mathfrak{a}, \mathfrak{g}) \cap \Pi$, together span A. Therefore, as A is abelian, $(W_0, H/L)$ is $\mathbb{R}$-split for A.

Let (W, V) be a maximal $\mathbb{R}$-split C^r H-pair for A. We wish to prove that it is an H-pair for Z_α, for each root α. Let α be a root and let $\Pi' \subset \mathfrak{a}^*$ be a hyperplane containing α and spanned by roots. Let $u \in \mathfrak{a}$ be a nonzero element annihilated by all the elements of Π' (in particular, u lies in the Lie algebra of A_β, for all $\beta \in R \cap \Pi$) and consider an H-pair (W', V) for A that is maximal $\mathbb{R}$-split for the group whose Lie algebra is $\mathbb{R}u$. Notice that (W', V) contains (W, V). By proposition 10.3.1, (W', V) is an H-pair for the centralizer $Z_G(u)$ of u, whence it is $\mathbb{R}$-split for $\check{A}_\beta$, for each $\beta \in \Pi \cap \Phi(\mathfrak{a}, \mathfrak{g})$, since $\check{A}_\beta \subset S_\beta \subset Z_\beta \subset Z_G(u)$. But u, together with the groups $\check{A}_\beta$, generates A, so (W', V) is $\mathbb{R}$-split for A. By maximality, we obtain $W' = W$, so (W, V) is an H-pair for $Z_G(u)$. But u is in the Lie algebra of A_α, so $Z_\alpha \subset Z_G(u)$, whence (W, V) is an H-pair for Z_α, as claimed.

The next claim is that the H-pair (W, V) just obtained is, in fact, an H-pair for the whole group G, defined over some G-invariant full subset $U \subset M$. For that, all that is left to do is show that W is G-invariant, but this is a consequence of proposition 7.9.3.

The section of $P|_U$, whose existence is asserted by the theorem, is obtained as follows. Fix $\xi_0 \in P|_U$ and let $W_{\xi_0}, N_{\xi_0}, F_{\xi_0}, \rho_{\xi_0} : G \to N_{\xi_0}/F_{\xi_0}$ be as before. Denote by Ψ_{ξ_0} the map

$$\xi \in P|_U \mapsto \Psi_{\xi_0}(\xi) := \tau_{\xi,\xi_0} \in H/F_{\xi_0}$$

and by $\bar{\Psi}_{\xi_0}$ the composition of Ψ_{ξ_0} on the left with the projection from H/F_{ξ_0} onto H/N_{ξ_0}. Notice that Ψ_{ξ_0} satisfies

$$\Psi(g\xi) = \Psi(\xi)\rho_{\xi_0}(g)^{-1}$$

for all $g \in G$ and $\xi \in P|_U$, so $\bar{\Psi}_{\xi_0}$ is G-invariant. Therefore, we obtain a G-invariant N_{ξ_0}-reduction of P over some G-invariant open dense subset of M. But H is already the C^r algebraic hull of G, so $H = N_{\xi_0}$ and, since H is transitive on V, we conclude that $W_{\xi_0} = V$. Furthermore, H acts on V effectively, so F_{ξ_0} must be trivial. Therefore, Ψ is an H-equivariant map from $P|_U$ onto H, that is, $\Psi(\xi h) = h^{-1}\Psi(\xi)$, and is, in particular, a bijection on each fiber P_x, for $x \in U$.

We can now define a C^r section σ of $P|_U$ by setting $\Psi(\sigma(x)) = e$, where e is the identity element in H and x is any element in U. The equation $g\sigma(x) = \sigma(gx)\rho_{\xi_0}(g)$ can be checked as follows, using the injectivity of Ψ on fibers:

$$\Psi(g\sigma(x)) = \Psi(\sigma(x))\rho_{\xi_0}(g)^{-1}$$
$$= \rho_{\xi_0}(g)^{-1}$$
$$= \rho_{\xi_0}(g)^{-1}\Psi(\sigma(gx))$$
$$= \Psi(\sigma(gx)\rho_{\xi_0}(g)).$$

This completes the proof of the theorem. $\qquad\qquad\qquad\qquad\qquad\qquad$ □

10.5 Contracting Subbundles

One of the hypothesis of the C^r rigidity theorem is the existence of a subgroup G_0 of the semisimple group G such that the C^r algebraic hull of G_0 does not contain a nontrivial normal subgroup of the C^r algebraic hull of G. For this to happen, we first need to find some C^r geometric structure that is invariant under a subgroup of G but not under G, so that the algebraic hull of the subgroup will be strictly smaller than that of G.

For actions on vector bundles, it is very common to have elements in G that are partially hyperbolic, and in those cases we will see later that it is sometimes possible to use the contracting subbundle of a partially hyperbolic element as the geometric structure needed to satisfy the preceding hypothesis.

The general theory of hyperbolic and partially hyperbolic dynamical systems plays an important role in the study of smooth actions of semisimple groups and their lattices, and the results of this section should give a hint of that role. We will not, however, use these results later in the text, so the reader may skip to the next section without loss of continuity.

Let G be as in the C^r rigidity theorem. We also suppose that G has trivial center, which allows us to identify G with the connected component of the identity of a real algebraic group $\check{G}$ via the adjoint representation (see section 7.2). Suppose that A is an $\mathbb{R}$-split Cartan subgroup of G with Lie algebra $\mathfrak{a}$. Choose $X \in \mathfrak{a}$ such that $\alpha(X) \neq 0$ for every root α of $(\mathfrak{a}, \mathfrak{g})$, and define the nilpotent subalgebra $\mathfrak{n} := \bigoplus_{\alpha(X)<0} \mathfrak{g}_\alpha$. Let N be the connected subgroup of G with Lie algebra $\mathfrak{n}$, and let $Z_G(A)$ be the subgroup of G that centralizes A. Notice that $Z_G(A)$ normalizes N and that $Z_G(A) \cap N = \{e\}$, so we can form the semidirect product $B := Z_G(A)N$.

Lemma 10.5.1 B is the intersection with G of a real algebraic subgroup of $\check{G}$. Furthermore, G/B is compact.

Proof. Notice that N is closed in the ordinary topology by the Iwasawa decomposition, so B is a closed subgroup of G and, again by the Iwasawa decomposition, G/B is compact since B contains AN.

It is also clear that $Z_G(A)$ is the intersection of G with a real algebraic subgroup of $\check{G}$. N is a closed subgroup of G that corresponds under the identification of G with its image under Ad to a subgroup of the unipotent group of upper-triangular matrices with 1s on the diagonal.

Recall the remark made in section 4.10 that the exponential map of a unipotent group is an isomorphism of varieties. Therefore, N is also the intersection of G with a real algebraic subgroup of $\check{G}$.

B is the image of $Z_G(A) \times N$ under the multiplication morphism $m : G \times G \to G$, and m is a bijection from the product onto B. Also notice that B can be viewed as the orbit of $e \in G$ under the action of the product group $Z_G(A) \times N$ on G by $(z, n) \cdot g := zgn^{-1}$, so B is open (in the Zariski and ordinary topologies) in its Zariski closure, by corollary 4.9.3. But B is also closed in the ordinary topology, so it is the intersection with G of a Zariski closed subgroup. Therefore, B is the intersection of a real algebraic subgroup of $\check{G}$ with G.   □

We set $f := \exp(X) \in A$. Then for each $g \in B$ the set

$$\{f^n g f^{-n} \mid n \in \mathbb{N}\}$$

is relatively compact in G, since for each $u \in N$, $f^n u f^{-n} \to e$ as $n \to \infty$.

We now consider a smooth action of G on a compact manifold M, with an invariant Borel probability measure μ. Suppose that some of the Lyapunov exponents of the element f just introduced are negative. In the terminology of section 9.4, f is a nonuniform partially hyperbolic diffeomorphism. Denote by $E^-(x)$ the subspace of $T_x M$ consisting of vectors with negative Lyapunov exponent (together with the vector 0). By Oseledec's theorem, $x \mapsto E^-(x)$ is a Borel measurable subbundle of TM.

Lemma 10.5.2 Let f be a diffeomorphism of a compact smooth manifold M and μ an ergodic Borel probability measure on M invariant under f. Suppose that some of the Lyapunov exponents of f relative to μ are negative, and let E^- and Λ be as before. Then Λ can be assumed to be B-invariant and E^- is a

B-invariant subbundle, that is, $Db_x E^-(x) = E^-(bx)$ for each $x \in \Lambda$ and each $b \in B$. On the other hand, E^- is not G-invariant: There is $g \in G$ such that $Dg_x E^-(x) \neq E^-(gx)$ for μ-a.e. $x \in \Lambda \cap g^{-1}(\Lambda)$.

Proof. The invariance of E^- under $b \in B$ is clear since for each $x \in \Lambda$ and each $0 \neq v \in E^-(x)$, we have

$$\left\| Df_{bx}^m Db_x v \right\| = \left\| D(f^m b f^{-m})_{f^m x} Df_x^m v \right\| \leq C \left\| Df_x^m v \right\|,$$

where C is a constant that depends only on b.

To show that E^- is not G-invariant, we apply theorem 7.8.4. By that theorem, the Weyl group $W = W(\mathfrak{a}, \mathfrak{g})$ is a finite group of linear transformations of $\mathfrak{a}$ generated by $\mathrm{Ad}(k_\alpha)|_\mathfrak{a}$, where k_α is, for each root α, an element of the normalizer of $\mathfrak{a}$ in G. Since all the $\mathrm{Ad}(k_\alpha)(X)$ belong to $\mathfrak{a}$, the diffeomorphisms $k_\alpha f k_\alpha^{-1}$ commute.

Let w be an arbitrary element of W, and write $w = \mathrm{Ad}(k)|_\mathfrak{a} \in W$, for some $k = k_w$. We set $f_w := kfk^{-1}$. The diffeomorphisms f_w commute, so $(\prod_{w \in W} f_w)^m = \prod_{w \in W} f_w^m$. It follows that if each f_w contracts vectors of E^- exponentially, then the same holds for the product of the f_w.

We proceed by contradiction. Suppose that for each $k \in G$ and μ-a.e. $x \in \Lambda \cap k^{-1}(\Lambda)$ we have $Dk_x E^-(x) = E^-(kx)$. Then for each k, a.e. $x \in M$, and all nonzero $v \in E^-(x)$, $D(kfk^{-1})_x^m v$ converges to 0 exponentially. Notice that $kfk^{-1} = \exp(\mathrm{Ad}(k)(X))$. Therefore, for each $w \in W$, f_w also contracts each vector of E^- exponentially. It follows that the same holds for

$$f_0 := \prod_{w \in W} f_w = \exp\left(\sum_{w \in W} \mathrm{Ad}(k_w) X \right).$$

But this is a contradiction since the element $\sum_{w \in W} \mathrm{Ad}(k_w) X \in \mathfrak{a}$ is invariant under the action W, so it must be 0 by theorem 7.8.4, and f_0 is the identity. $\qquad\square$

We call the (generally only measurable) subbundle E^- the *contracting subbundle* of f. If f is a uniform partially hyperbolic diffeomorphism (i.e., $\Lambda = \Lambda_k$, for some $k \in \mathbb{N}$, in the notation of section 9.4), E^- is continuous on Λ.

By the previous lemma, if E^- (for some f as in the lemma) is C^r, then the group B defined earlier preserves a C^r H-structure, where H is a subgroup of some $GL(n, m, \mathbb{R})$ (as in example 5 of section 6.4), but the C^r algebraic hull of G must properly contain H since, otherwise, G would also leave E^- invariant, contrary to the conclusion of the lemma.

In order to apply the C^r rigidity theorem effectively, we would also like to know whether the algebraic hull H_0 for the subgroup G_0 (using the notation of theorem 10.4.1) contains a normal subgroup of the algebraic hull H for G. If F is the largest normal subgroup of H contained in H_0, we need to factor out F before applying theorem 10.4.1; in other words, we can still apply the theorem to the principal H/F-bundle P/F, and the price of doing so is that we lose any information that may be in some way associated with F. Therefore, the ideal situation corresponds to F being trivial.

The next theorem is a very special result, but it should help bring the conditions of the C^r rigidity theorem into focus. It will be clear that most of the ideas we are going to explore in the proof hold in much greater generality.

Theorem 10.5.3 Suppose that $SL(3, \mathbb{R})$ acts smoothly, by automorphisms, on a vector bundle $E \to M$ with fiber dimension 3, where M is a compact manifold, such that the action leaves invariant:

1. an ergodic Borel probability measure μ on M whose support is M;
2. a smooth, nonvanishing, alternating 3-form Ω on E.

Also suppose that for some f in the diagonal subgroup of $SL(3, \mathbb{R})$, the contracting and expanding subbundles $E_f^\pm \subset E$ are C^r and one of them is nontrivial. Then there exist everywhere linearly independent C^r sections X_1, X_2, X_3 of E and a smooth homomorphism $\rho : SL(3, \mathbb{R}) \to GL(3, \mathbb{R})$ such that

$$g_* X_i = \sum_{j=1}^{3} \rho(g)_{ij} X_j$$

for each $g \in SL(3, \mathbb{R})$. In particular, E must be a trivial bundle.

Notice that the theorem is not vacuous since its hypotheses are satisfied by the following example. Let L be any Lie group containing $SL(3, \mathbb{R})$ as a subgroup, and let Λ be a cocompact lattice in L. Then $SL(3, \mathbb{R})$ acts on the trivial bundle $E = G/\Lambda \times \mathbb{R}^3$ by left translations on $M = G/\Lambda$ and matrix multiplication on $\mathbb{R}^3$.

Exercise 10.5.4 Show that the previous example indeed satisfies the conditions of the theorem.

The remainder of the section is dedicated to the proof of the theorem. The discussion will be done for the case $r \geq 0$, where one key difficulty arises. Namely, we are claiming that the straightening section is defined over the entire manifold M, rather than only on an open dense set, and that it is C^r.

After possibly having to replace f by f^{-1}, we may assume that E_f^- has fiber dimension 1. In fact, the existence of the invariant form Ω implies that either E_f^- or E_f^+ has fiber dimension 1. Also notice that $E_f^+ = E_{f^{-1}}^-$.

Let H be the C^r algebraic hull of the action. Due to the presence of Ω, we must have $H \subset SL(3, \mathbb{R})$ (see example 2 in section 6.4). Applying the C^r rigidity theorem and the previous lemma we conclude that H contains a proper normal subgroup F such that H/F is isomorphic to a homomorphic image of $SL(3, \mathbb{R})$, where the homomorphism is the one given in the definition of a straightening section. But the only images of $SL(3, \mathbb{R})$ under a continuous homomorphism are $SL(3, \mathbb{R})$ itself and the trivial group. (Notice that $SL(3, \mathbb{R})$ has trivial center.) Therefore, $H = SL(3, \mathbb{R})$, F is trivial, and there exist a nontrivial homomorphism $\rho : SL(3, \mathbb{R}) \to GL(3, \mathbb{R})$ and C^r sections X_1, X_2, X_3 of $E|_U$, for some open dense $SL(3, \mathbb{R})$-invariant subset $U \subset M$, such that the conclusion of the theorem holds over U. The sections satisfy $\Omega(X_1 \wedge X_2 \wedge X_3) \equiv 1$.

It only remains to show that these sections extend to the entire manifold M and are everywhere C^r.

It will be convenient to distinguish $G = SL(3, \mathbb{R})$ and $H = SL(3, \mathbb{R})$, where the former is the group acting on E and the latter is the algebraic hull of the action. Let $p : P \to M$ denote the smooth H-reduction of the frame bundle of E determined by Ω. The C^r subbundle E_f^- can be expressed in terms of a C^r H-equivariant map φ from P into the projective space $P^2(\mathbb{R})$, as in example 5 of section 6.4. By the lemma, φ is invariant under the action of the subgroup $B \subset G$, which in the present case is conjugate to the subgroup of upper-triangular matrices in $SL(3, \mathbb{R})$.

We introduce a C^r section σ of $P|_U$, defined by the relation $X_i(x) = \sigma(x)e_i$ for each $x \in U$, where $\{e_i\}$ is the standard basis of $\mathbb{R}^3$. We also define a C^r map $\Phi : G/B \times P \to P^2(\mathbb{R})$ by $\Phi(gB, \xi) = \varphi(g^{-1}\xi)$.

Lemma 10.5.5 There is $w \in P^2(\mathbb{R})$ such that $\Phi(gB, \xi) = h_\xi \rho(g)w$ for all $g \in G$ and $\xi \in P|_U$, where h_ξ is the unique element of H such that $\xi = \sigma(p(\xi))h_\xi^{-1}$.

Proof. Define $\theta = \varphi \circ \sigma : U \to P^2(\mathbb{R})$. We claim that $\theta(bx) = \rho(b)\theta(x)$ for all $x \in U$ and $b \in B$. In fact,

$$
\begin{aligned}
\theta(bx) &= \varphi(\sigma(bx)) \\
&= \varphi(b\sigma(x)\rho(b)^{-1}) \\
&= \rho(b)\varphi(b\sigma(x)) \\
&= \rho(b)\varphi(\sigma(x)) \\
&= \rho(b)\theta(x).
\end{aligned}
$$

The same argument used in the proof of proposition 10.1.2 to show that a map denoted there by A is constant can be used here to conclude that θ is constant. The constant value will be denoted by $w \in P^2(\mathbb{R})$. Therefore, for each $\xi = \sigma(x)h_\xi^{-1} \in P|_U$,

$$
\begin{aligned}
\Phi(gB, \xi) &= \varphi\big(g^{-1}\sigma(x)h_\xi^{-1}\big) \\
&= \varphi\big(\sigma(g^{-1}x)\rho(g)^{-1}h_\xi^{-1}\big) \\
&= h_\xi\rho(g)\varphi(\sigma(g^{-1}x)) \\
&= h_\xi\rho(g)w.
\end{aligned}
$$
□

Lemma 10.5.6 For each $\xi \in P$, there exists a unique $h_\xi \in H$ such that $\Phi(gB, \xi) = h_\xi\rho(g)w$ for all $g \in G$, where w is the element of $P^2(\mathbb{R})$ obtained in the previous lemma.

Proof. We have, by the previous lemma, an open dense subset $U \subset M$ and for each $\xi \in P|_U$ an element $h_\xi \in H$ such that $\Phi(gB, \xi) = h_\xi\rho(g)w$. We need to show that the same is true for ξ outside $P|_U$.

Let ξ be an arbitrary element of P, and let $\xi_i \in P|_U$ be a sequence converging to ξ. Define maps η and η_i from G/B into $P^2(\mathbb{R})$ by

$$\eta(gB) := \Phi(gB, \xi) \text{ and } \eta_i(gB) := \Phi(gB, \xi_i) = h_{\xi_i}\rho(g)w.$$

Since Φ is C^r on P, η_i converges uniformly to η.

Proposition 6.7.5 implies that the sequence h_{ξ_i} is bounded in H. Let h be a limit point of this sequence. Then $\eta(gB) = h\rho(g)w$, as claimed.

It remains to show that h is unique. But if $h\rho(g)w = h'\rho(g)w$ for all $g \in G$, it would follow that $h^{-1}h'$ fixes pointwise each element of Hw. This means that $h^{-1}h'$ is contained in the intersection of the subgroups $h_1 H_w h_1^{-1}$, where h_1 ranges over all of H and H_w is the isotropy subgroup of w. In particular, $h^{-1}h'$ should belong to a proper normal subgroup of H. But H is a simple Lie group (with trivial center) so $h = h'$. □

It will be useful to improve on the preceding uniqueness argument somewhat. It was used in the previous proof that the group $\bigcap_{h\in H}hH_wh^{-1}$ is trivial, since it is a normal subgroup of $SL(3, \mathbb{R})$. In fact, by the descending chain condition, there is a finite number of $h_i \in H$, $i = 1, \ldots, k$, such that the intersection of the $h_i H h_i^{-1}$ is trivial. Let $g_i \in G$ be such that $\rho(g_i) = h_i$. Then the equations $\Phi(g_i B, \xi) = h_\xi\rho(g_i)w$ uniquely determine h_ξ for each $\xi \in P$.

Define a C^r map

$$\mathcal{G} : P \to P^2(\mathbb{R}) \times \cdots \times P^2(\mathbb{R})$$

by $\mathcal{G}(\xi)=(\Phi(g_1 B, \xi), \ldots, \Phi(g_k B, \xi))$. Then, by the lemma, $\mathcal{G}$ maps each element of P into the H-orbit of $\bar{w} = (\rho(g_1)w, \ldots, \rho(g_k)w)$. But we have chosen the g_i so that the isotropy subgroup $H_{\bar{w}}$ is trivial. Therefore, $\mathcal{G}$ is a C^r H-equivariant map into H. Therefore, we obtain a C^r section $\bar{\sigma}$ of P defined by $\mathcal{G}(\bar{\sigma}) = e$.

It is immediate that $\bar{\sigma}$ coincides with σ over U and that it is a straightening section for the G-action on the frame bundle of E. This concludes the proof of the theorem.

10.6 The Theorems of Zimmer and Margulis

From now on, we restrict the discussion to the measurable case only. The theorems in this section hold for more general fields than $\mathbb{R}$. The interested reader should consult [36] or [19].

Theorem 10.6.1 (Zimmer) Let G be a connected semisimple Lie group with real rank at least 2 and no compact simple factors. Suppose that G acts by automorphisms on a principal H-bundle P over a manifold M, preserving an ergodic Borel probability measure μ on M. Let L be a representative of the measurable algebraic hull of the G-action, and let Q be a measurable G-invariant reduction of P with group L. Let $\pi : L \to L_1$ be a homomorphism onto a connected noncompact simple Lie group L_1 with trivial center, and form the associated principal L_1-bundle $Q_1 = Q \times_L L_1$. Then there exists a measurable straightening ρ-section σ of Q_1, where ρ is a surjective rational homomorphism from G to L_1.

Proof. We will show that the theorem holds with a smooth ρ. For the proof that ρ is rational we refer to [36].

First notice that the measurable algebraic hull L_1' for the G-action on Q_1 is equal to L_1; otherwise we would get a measurable reduction of Q with $\pi^{-1}(L_1')$ as the structure group of the bundle. By Moore's ergodicity theorem, the $\mathbb{R}$-split one-parameter subgroups also act ergodically on M and no subgroup of L_1 contains a nontrivial normal subgroup of L_1, since L_1 is simple. Therefore, all that we need in order to apply the previous theorem to the present situation is to show that the measurable algebraic hull of some subgroup of G is a proper subgroup of G.

We first claim that there exists a proper algebraic subgroup B of L_1 such that L_1/B is compact. To see this, let $R : L_1 \to GL(d, \mathbb{R})$ be a faithful representation of L_1, where R is a morphism (say, the adjoint representation), and consider the algebraic action of L_1 on the projective space $P^{d-1}(\mathbb{R})$ induced from R. Any orbit of this action having smallest dimension will be closed (hence compact), due to corollary 4.9.2. On the other hand, if B denotes the isotropy group of a point in that closed orbit, it follows by corollary 4.9.3 and theorem 1.2.2 that L_1/B is compact. Since the action is real algebraic, B is a real algebraic subgroup. B is also a proper subgroup since, otherwise, L_1 would fix a point in $P^{d-1}(\mathbb{R})$, which means that the representation R would have a one-dimensional invariant subspace. But the kernel of this one-dimensional representation would be a normal subgroup of L_1 of codimension 1, and this is not possible since L_1 is semisimple.

Let $G_0 \cong \mathbb{Z}$ be the group generated by an element $\tau \neq e$ contained in an $\mathbb{R}$-split Cartan subgroup. Define the associated bundle $Q_1 \times_{L_1} L_1/B$ and apply proposition 6.7.3 to it. (In the proposition, $Q_1 \times_{L_1} L_1/B$ takes the role of P_V.) It follows that Q_1 admits a reduction with group L_2, where L_2 is an algebraic subgroup of L_1 that leaves invariant a Borel probability measure ν_0 on L_1/B. L_2 is a proper subgroup of L_1 since, otherwise, L_1 would fix ν_0, which is not possible by corollary 4.9.5. (By this corollary, L_1 would have fixed points on L_1/B, but no fixed points exist since B is a proper subgroup. Notice that the subgroup of L_1 generated by its one-parameter algebraic subgroups is a normal subgroup, so it is equal to L_1 since L_1 is simple.)

It follows that the algebraic hull of G_0 on Q_1 is a proper subgroup of the algebraic hull of G acting on the same bundle. This is the last condition that remained to be checked in order to apply theorem 10.3.1. This concludes the proof.   □

For the following result, recall that if L is a connected algebraic group, there is a solvable algebraic subgroup $R(L)$, called the radical of L, such that $L/R(L)$ is semisimple. (This is the Levi decomposition, theorem 4.10.1.) Denote by $\pi : L \to L_1$ the projection from L into any one of the noncompact simple factors of $L/R(L)$. (We assume that π quotients out the finite center of $L/R(L)$ so that L_1 has trivial center.) We will be referring to π in the next corollary.

Theorem 10.6.2 (Margulis) Let Γ be a lattice of a connected noncompact simple Lie group G of real rank at least 2. Let $R : \Gamma \to GL(m, \mathbb{R})$ be a linear representation of Γ, and denote by L the connected component of the Zariski closure of $R(\Gamma)$. Let $\pi : L \to L_1$ be as defined before and define the finite-index subgroup $\Gamma_0 = \Gamma \cap R^{-1}(L)$ of Γ. Then there is a homomorphism $\rho : G \to L_1$ such that $\rho|_{\Gamma_0} = \pi \circ R|_{\Gamma_0}$.

Proof. We write $M = G/\Gamma$. Let μ be the G-invariant Borel probability measure on M, which exists by virtue of Γ being a lattice. Let $P = G \times_\Gamma L_1$ be the principal L_1-bundle over M defined by the suspension construction of section 1.3 (see also exercise 6.1.4).

By exercise 6.5.3, the measurable algebraic hull of the G-action on P is L_1. Then the previous corollary implies the existence of a measurable section σ of P and a smooth homomorphism $\rho' : G \to L_1$ such that for μ-a.e. $x \in M$ and each $g \in G$, we have $g\sigma(x) = \sigma(gx)\rho'(g)$. Fix $x_0 = g_0\Gamma$ in this conull set and write $\sigma(x_0) = [g_0, l_0]$. Then x_0 is a fixed point for the action of $g_0\Gamma g_0$ on M and $g_0\gamma g_0^{-1}\sigma(x_0) = \sigma(x_0)\rho'(g_0\gamma g_0^{-1})$, for each $\gamma \in \Gamma$. Notice that

$$\sigma(x_0)\rho'\big(g_0\gamma g_0^{-1}\big) = \big[g_0, l_0\rho'\big(g_0\gamma g_0^{-1}\big)\big].$$

On the other hand, for each $\gamma \in \Gamma$,

$$g_0\gamma g_0^{-1}[g_0, l_0] = \big[g_0\gamma g_0^{-1}g_0, l_0\big] = [g_0\gamma, l_0] = [g_0, (\pi \circ P)(\gamma)l_0],$$

so $l_0\rho'(g_0\gamma g_0^{-1}) = (\pi \circ P)(\gamma)l_0$, for each $\gamma \in \Gamma$. Therefore, the conclusion of the corollary holds for the homomorphism

$$\rho(\cdot) = l_0\rho'(g_0)\rho'(\cdot)\rho'(g_0)^{-1}l_0^{-1}. \qquad \square$$

Theorem 10.6.3 (Mostow–Margulis) Let G and G' be connected noncompact simple Lie groups with trivial center, and suppose that the real rank of G is at least 2. Let $\Gamma \subset G$ and $\Gamma' \subset G'$ be lattices and consider an isomorphism $\pi : \Gamma \to \Gamma'$. Then π extends to a smooth isomorphism $\pi : G \to G'$.

Proof. We may identify G' with the connected component of e of a real algebraic group, by means of the adjoint representation. By the Borel density theorem (theorem 4.9.7), the Zariski closure of Γ' is G'. (Notice that the smallest algebraic subgroup of G' containing all the algebraic one-parameter subgroups of G' is a normal subgroup, whence coincides with G'.) Therefore, the present theorem is an immediate consequence of the previous one. $\qquad \square$

Exercise 10.6.4 Let Γ be a lattice of $SL(n, \mathbb{R})$, acting on a manifold M of dimension n. Suppose that the action is smooth and leaves invariant a volume form, that is, a nonvanishing alternating n-form. Suppose that the action is ergodic with respect to the probability measure defined by the volume form. Show that there exist measurable vector fields $X_1, \ldots, X_n$ on M, almost everywhere linearly independent, and a linear representation ρ of $SL(n, \mathbb{R})$, such that

$$\gamma_* X_i = \pm\sum_j \rho(\gamma)_{ij} X_j$$

for each i. (The unspecified sign means that the vector fields only define a section of the frame bundle "modulo" the center of $SL(n, \mathbb{R})$. Notice that this center is trivial if n is odd and is $\{\pm I\}$ if n is even.)

10.7 The Lyapunov Spectrum and Entropy

We now show that the Lyapunov spectrum of a smooth measure-preserving action of a connected semisimple Lie group G of real rank at least 2 on a compact manifold M of dimension n is determined by the linear representations of (a covering group of) G in dimension n. It will be convenient to use the following notation. Let $\rho : G \to GL(n, \mathbb{R})$ be a linear representation. For each $g \in G$, let λ_i, $i = 1, \ldots, l$, be the distinct eigenvalues of $\rho(g)$, having multiplicities d_i. The set of values $\ln |\lambda_i|$, with multiplicities given by the sum of the d_i for which the $|\lambda_i|$ are equal, will be referred to as the *Lyapunov spectrum* of $\rho(g)$. We also define the *entropy* of $\rho(g)$ as

$$h_\rho(g) := \sum d_i \ln|\lambda_i|,$$

where the sum runs over the indices i such that $|\lambda_i| > 1$.

Theorem 10.7.1 Let G be a connected semisimple Lie group with finite center and simple factors of real rank at least 2. Suppose that G acts by bundle automorphisms on a vector bundle E over a compact manifold M, preserving an ergodic Borel probability measure μ on M. Let n be the fiber dimension of E, and denote by $\tilde{G}$ the universal covering group of G. Then there exists a linear representation $\rho : \tilde{G} \to GL(n, \mathbb{R})$ such that for each $g \in G$, the Lyapunov spectrum associated to the $\mathbb{Z}$-action on E generated by g is equal to the Lyapunov spectrum of $\rho(g')$, where g' is any element of $\tilde{G}$ that projects onto g.

Proof. The first remark is that the Lyapunov exponents are not affected by passing to a finite ergodic extension of the action, so we can apply proposition 6.5.7. Let E_i be as in that proposition. Furthermore, proposition 9.3.7 says that it is enough to prove the theorem for the G-action on E_{i+1}/E_i, $i = 1, \ldots, m - 1$. The key property of the G-action on these quotients is that the algebraic hulls arc reductive and Zariski-connected. Therefore, there will be no loss of generality in assuming that the measurable algebraic hull H for the G-action on E is Zariski-connected and reductive. Recall theorem 4.10.1, which says that $H = (H, H)Z$, where the commutator (H, H) is a Zariski-connected semisimple group, Z is a Zariski-connected subgroup of the center of H, and the product is almost direct, that is, (H, H) and Z centralize each other and their intersection is finite.

Let P be a measurable G-invariant H-reduction of the frame bundle of E, and form the associated bundle $\bar{P} = P/(H, H)$. This is a principal $\bar{Z}$-bundle, where $\bar{Z} = Z/(Z \cap (H, H))$. It is immediate that $\bar{Z}$ is the algebraic hull of the G-action on $\bar{P}$. Therefore, $\bar{Z}$ is compact by theorems 8.6.6 and 8.6.4. The conclusion is that H is a reductive group with compact center.

Collecting the center of H and the compact simple factors of (H, H) into one single compact group S_0, and the noncompact simple factors into a semisimple group S, we can write $H = S_0 S$. This is an almost direct product. Let F be the finite central subgroup of H such that

$$H/F = \bar{S}_0 \times \bar{S}_1 \times \cdots \times \bar{S}_k,$$

a direct product, where $\bar{S}_0 = S_0/(S_0 \cap F)$ and the $\bar{S}_i$, $i \geq 1$, are the simple center-free factors of $\bar{S} = S/(S \cap F)$. Denote by $\pi : H \to \bar{S}$ the natural projection.

Applying theorem 10.6.1 to the G-action on $\bar{P} = P/S_0 = P \times_H \bar{S}$, we conclude that there exists a surjective homomorphism $\bar{\rho} : G \to \bar{S}$ and a measurable straightening $\bar{\rho}$-section $\bar{\sigma}$ of $\bar{P}$.

Denote by ρ the homomorphism from $\tilde{G}$ onto S that lifts $\bar{\rho}$. We claim that there exists a measurable section σ of P and, for each $g \in G$ and each $g' \in \tilde{G}$ projecting onto g, there is a measurable function $s_0 : M \to S_0$ such that

$$g\sigma(x) = \sigma(gx)\rho(g')s_0(x).$$

In fact, viewing P as a principal S_0-bundle over $\bar{P}$, choose any measurable section $\eta : \bar{P} \to P$. Then $\sigma := \eta \circ \bar{\sigma}$ is a measurable section of P and it is immediate to check that $g\sigma(x)$ and $\sigma(gx)\rho(g')$ lie in the same fiber of $P \to P/S$. Therefore, there exists a unique $s_0(x) \in S_0$ satisfying the foregoing equality. Since σ is measurable, s_0 is also.

The theorem is now a consequence of exercise 10.1.3.   □

Corollary 10.7.2 Let G be a connected simply connected simple Lie group with real rank at least 2, and let $g \in G$. Suppose that G acts smoothly on a compact n-manifold M preserving an ergodic Borel probability measure μ, and denote by $h_\mu(g)$ the measure-theoretic entropy. Then, either $h_\mu(g) = 0$ or $h_\mu(g) = h_\rho(g)$ for some n-dimensional linear representation ρ of G.

Under the conditions of theorem 10.7.1, it turns out, the measurable algebraic hull of the action is already reductive. This is shown in [39]. Taking this fact into account, the proof of 10.7.1 yields the following stronger conclusion.

Theorem 10.7.3 Under the same assumptions and notation as in theorem 10.7.1, the following holds after possibly having to pass to a finite ergodic extension. The measurable algebraic hull $H \subset GL(n, \mathbb{R})$ can be written as an almost direct product of Zariski-connected subgroups S and Z, where S is semisimple and all of its simple factors are noncompact, and Z is compact. Furthermore, E admits a measurable trivialization given by a measurable section σ of the frame bundle $\mathcal{F}(E)$ that is "almost straightening," in the following sense: $g\sigma(x) = \sigma(gx)\alpha(g, x)$, where $\alpha(g, x)Z = \rho(g)$ for a surjective homomorphism $\rho : G \to H/Z$, for each $g \in G$ and μ-a.e. $x \in M$.

Suppose that M is a smooth n-manifold and that μ is a Borel probability measure on M that assigns positive measure to nonempty open sets. Also suppose that G, as in theorem 10.7.1, acts smoothly on M preserving μ. Let m be the dimension of G and consider the natural action of G on $E = TM$. It can be shown as a corollary of theorem 6.6.3 that there is a μ-conull G-invariant open subset $U \subset M$ where the action is locally free. Therefore, $E|_U$ contains a smooth G-invariant subbundle F whose fibers are tangent to the G-orbits and have dimension m. The linearization of the tangent action, that is, the action of G on F, is described in the proof of theorem 6.6.4, where it is shown that the frame bundle of F has a (smooth) straightening section associated to the adjoint representation of G. By the previous theorem, the G-action on E/F is similarly described (by means of a measurable trivialization of E/F) by a linear representation $G \to H \subset GL(n - m, \mathbb{R})$, modulo a cocycle, into a compact normal subgroup of H.

10.8 Lattice Actions

The results of the previous two sections also hold for measure-preserving actions of a lattice group $\Gamma \subset G$. The basic idea is to consider the suspension of the Γ-action, as defined in section 1.3.

Before stating those results for lattices, we need to establish some preliminary results relating a Γ-action and its suspension. Let Γ be a lattice in a connected Lie group G, and let ν be the G-invariant probability measure on G/Γ. It will be helpful to introduce a principal Γ-bundle $Q = (G \times \Gamma)/\Gamma$, where Γ acts on the product via

$$(g, \gamma) \cdot \gamma_0 = \left(g\gamma_0, \gamma_0^{-1}\gamma\right).$$

G acts on Q by principal bundle automorphisms: $g_0 \cdot (g, \gamma) := (g_0 g, \gamma)$. The next lemma is essentially [36, 4.2.19].

Lemma 10.8.1 Let T be a Γ-space and let $\Phi : Q \to T$ be a Γ-equivariant, essentially G-invariant function. That is, for each $g \in G$ and $\gamma \in \Gamma$, and for ν-almost every $[g'] \in G/\Gamma$ and each $\gamma' \in \Gamma$, we have

$$\Phi(g[g', \gamma']\gamma) = \gamma^{-1}\Phi([g', \gamma']).$$

Then there exists $[g_0, \gamma_0] \in Q$ such that $a := \Phi([g_0, \gamma_0])$ is a fixed point for the Γ-action on T.

Proof. By Fubini's theorem, there exists $[g_0, \gamma_0]$ such that

$$\Phi(g[g_0, \gamma_0]) = \Phi([g_0, \gamma_0]) =: a$$

for a.e. $g \in G$.

For each $\gamma \in \Gamma$, and for a.e. $g \in G$, we have

$$
\begin{aligned}
a &= \Phi([g\gamma, \gamma_0]) \\
&= \Phi([g, \gamma\gamma_0]) \\
&= \Phi\big([g, \gamma_0] \cdot (\gamma_0^{-1}\gamma\gamma_0)\big) \\
&= \big(\gamma_0^{-1}\gamma\gamma_0\big)^{-1} \cdot \Phi([g, \gamma_0]) \\
&= \big(\gamma_0^{-1}\gamma\gamma_0\big)^{-1} \cdot a.
\end{aligned}
$$

Therefore, a is a fixed point for the Γ-action on T.   □

Suppose that Γ acts (say, smoothly) on a manifold M, preserving a Borel probability measure μ. Define $\check{M} := (G \times M)/\Gamma$, the space of orbits for the action of Γ on $G \times M$ given by

$$(g, x) \cdot \gamma := (g\gamma, \gamma^{-1}x)$$

for $g \in G$, $\gamma \in \Gamma$, and $x \in M$. In fact, $\check{M}$ is the associated bundle $(Q \times M)/\Gamma$ over G/Γ, with typical fiber M, as the reader can easily verify. For each $g \in G$, recall that there is a diffeomorphism between M and the fiber $M_{[g]} = \pi^{-1}([g])$, $[g] \in G/\Gamma$, given by

$$i_g : x \in M \mapsto [g, x] \in \check{M}.$$

On each fiber $M_{[g]}$, there is a probability measure $\mu_{[g]}$ given by

$$\mu_{[g]} := (i_g)_*\mu.$$

Notice that this is well defined since, given another representative $g\gamma$ of $[g]$, we have

$$(i_{g\gamma})_*\mu = (i_g \circ \gamma)_*\mu = (i_g)_*\mu,$$

due to the Γ-invariance of μ.

We now define a measure λ on $\check{M}$ as follows: Given any Borel set $A \subset \check{M}$, we have

$$\lambda(A) := \int_{G/\Gamma} \mu_{[g]}(A \cap M_{[g]}) \, d\nu([g]).$$

Exercise 10.8.2 Show that λ is a G-invariant probability measure. Moreover, show that μ is ergodic for the Γ-action if and only if λ is ergodic for the suspended G-action. We indicate here the proof of the implication: the Γ-action is ergodic $\Rightarrow$ the G-action is ergodic. Let A be a G-invariant Borel subset of $\check{M}$, and let χ_A denote its characteristic function. Since $\check{M} = (Q \times M)/\Gamma$, we can define a map that associates to each $q \in Q$ the real-valued function $\varphi_q \in L^\infty(M)$ such that $\varphi_q(x) = \chi_A([q, x])$.

Check that the map $q \mapsto \varphi_q$ is G-invariant. Applying the lemma, conclude that for almost every $q \in Q$, the set

$$A_q := \{x \in M \,\|\, [q, x] \in A\}$$

is Γ-invariant. Therefore, either $\mu(A_q)$ is 0 for ν-a.e. q, or it is 1 for ν-a.e. q. Conclude that the measure of A is either 0 or 1.

Let now $p : P \to M$ be a principal H-bundle. Suppose that Γ acts by bundle automorphisms. Define $\check{P} = (Q \times P)/\Gamma$, a principal H-bundle over the manifold $\check{M}$ defined earlier. The group G naturally acts on $\check{P}$ by principal bundle automorphisms.

Lemma 10.8.3 The notation introduced before is in force. Let H be a real algebraic group, and suppose that Γ acts by bundle automorphisms on a principal H-bundle P over a manifold M. Suppose, furthermore, that the action on M is ergodic with respect to an invariant Borel probability measure μ and that H is the measurable algebraic hull of the action. Then the measurable algebraic hull for the G-action on the bundle $\check{P}$ is also H.

Proof. Suppose that $L \subset H$ is the measurable algebraic hull of the G-action on $\check{P}$. Therefore, there is a G-invariant measurable L-reduction of $\check{P}$, which corresponds to a G-invariant measurable H-equivariant map

$$\Phi : \check{P} \to H/L.$$

To each $q \in Q$, where Q is the same as in proposition 10.8.1, define a map $\varphi_q : P \to H/L$ by

$$\varphi_q(\xi) := \Phi([q, \xi]).$$

Then $q \mapsto \varphi_q$ is a G-invariant Γ-equivariant map from Q into the space $\mathcal{F}(P, H/L)$ of measurable H-equivariant functions from P into H/L (two functions being identified if they agree a.e.).

We now apply the previous lemma (with $T = \mathcal{F}(P, H/L)$) to conclude that there exists a Γ-invariant element of $\mathcal{F}(P, H/L)$. But such an element defines a Γ-invariant L-reduction of P, so $H = L$. $\square$

We are now ready to state and prove the counterpart of theorem 10.6.1 for lattices.

Theorem 10.8.4 Let G be a connected simple Lie group with finite center and real rank at least 2, and let Γ be a lattice in G. Suppose that Γ acts by automorphisms on a principal H-bundle P over a manifold M, preserving an ergodic Borel probability measure μ on M. Let L be a representative of the measurable algebraic hull of the Γ-action, and let R be a measurable G-invariant reduction of P with group L. Let $\pi : L \to L_1$ be a homomorphism onto a connected noncompact simple Lie group L_1 with trivial center, and form the associated principal L_1-bundle $R_1 = R \times_L L_1$. Then there exists a measurable straightening ρ-section of R_1, where ρ is a surjective rational homomorphism from G to L_1.

Proof. We may assume without loss of generality that the algebraic hull H is already simple with trivial center, and that $R = P$. The idea of the proof is to pass to the suspension of the action, which is a G-action on $\check{P}$. By the previous lemmas, the G-action on $\check{M}$ is ergodic and the measurable algebraic hull is H. Therefore, we can apply theorem 10.6.1 to conclude that $\check{P}$ admits a straightening section for the G-action.

A measurable section $\check{\sigma}$ of $\check{P}$ is determined by an H-equivariant measurable map $\Phi : \check{P} \to H$, defined by

$$\Phi(\check{\sigma}h) = h^{-1}\Phi(\check{\sigma}) := h^{-1}.$$

If the section is straightening, with homomorphism ρ, it follows from the the the equation $g\check{\sigma}(y) = \check{\sigma}(gy)\rho(g)$ that

$$\Phi(g\eta) = \Phi(\eta)\rho(g)^{-1}$$

for each $g \in G$ and all η over a subset of full measure in $\check{M}$.

In order to prove the existence of a straightening section for P, we need to find an H-equivariant measurable map $\varphi : P \to H$ such that for each $\gamma \in \Gamma$ and all $\xi \in P$ over a subset of M of full measure, we have $\varphi(\gamma x) = \varphi(x)\rho'(\gamma)^{-1}$ for some homomorphism ρ' from Γ into H. We will show that φ exists and that $\rho'(\gamma) = h_0\rho(\gamma)h_0^{-1}$ for some $h_0 \in H$ and all $\gamma \in \Gamma$.

Let Φ be defined as before in terms of the straightening section for the G-action. We define a map $q \mapsto \varphi_q$ from the bundle Q (as in proposition 10.8.1) into the space $\mathcal{F}(P, H)$ of measurable H-equivariant maps from P to H by

$$\varphi_q(\xi) := \Phi([q, \xi]).$$

It is an easy calculation to check that φ_q is H-equivariant and that for each $g \in G$, $\gamma \in \Gamma$, and a.e. $q \in Q$ and $\xi \in P$, the following identities hold:

$$\varphi_{g \cdot q}(\xi) = \varphi_q(\xi)\rho(g)^{-1},$$
$$\varphi_{q\gamma}(\xi) = \varphi_q(\gamma\xi).$$

Denote by $u : Q \to G$ the map that associates to each $q = [g_1, \gamma_1] \in Q$ the element $g_1\gamma_1 \in G$. This is a well-defined map.

For each $q = [g_1, \gamma_1] \in Q$ and $\gamma \in \Gamma$, we can write

$$q\gamma = \left(g_1\gamma g_1^{-1}\right) \cdot q \cdot \left(\gamma_1^{-1}\gamma^{-1}\gamma_1\gamma\right).$$

Applying the preceding identities, we obtain

$$\varphi_q(\gamma\xi) = \varphi_q\left(\gamma_1^{-1}\gamma^{-1}\gamma_1\gamma\xi\right)\rho\left(g_1\gamma^{-1}g_1^{-1}\right)$$

for each $\gamma \in \Gamma$ and a.e. $q \in Q$, $\xi \in P$. Writing $\gamma_0 = \gamma_1^{-1}\gamma^{-1}\gamma_1$ and $\eta = \gamma\xi$ we can simplify the previous expression to

$$\varphi_q(\eta) = \varphi_q(\gamma_0\eta)h_0\rho(\gamma_0)h_0^{-1},$$

where $h_0 = \rho(u(q))$.

Finally, by using Fubini's theorem we can reverse the quantifiers and write: There exists $q \in Q$ such that for all $\gamma \in \Gamma$ and a.e. $\xi \in P$, we have

$$\varphi_q(\xi) = \varphi_q(\gamma\xi)h_0\rho(\gamma)h_0^{-1}.$$

The straightening section of P is now the unique measurable section σ such that $\varphi_q(\sigma(x)) = e$ for a.e. $x \in M$. $\qquad\square$

The conclusions of theorem 10.7.1 (concerning the Lyapunov spectrum) and corollary 10.7.2 (concerning the measure-theoretic entropy) also hold for lattices in a connected simple Lie group G with real rank at least 2. These results for lattices are easily derived by using the previous theorem and the proofs of the corresponding results for G.

Appendix A

Lattices in $SL(n, \mathbb{R})$

A.1 $SL(n, \mathbb{Z})$

If Γ is a discrete subgroup of a connected semisimple Lie group G, then by theorem 8.1.8 G/Γ admits a nonzero G-invariant measure, which is unique up to a scalar multiple. When this measure is finite, Γ is called a *lattice* of G.

Γ is said to be a *cocompact* (or *uniform*) lattice if G/Γ is compact. We prove in this section the following theorem.

Theorem A.1.1 $SL(n, \mathbb{Z})$ is a lattice subgroup of $SL(n, \mathbb{R})$.

In the next section, we show how one can construct cocompact lattices in $SL(n, \mathbb{R})$. For more general results concerning lattices in semisimple Lie groups, the reader should consult [4], [3], [6], [28], [36].

The following discussion and exercises will lead to a proof of the theorem. Let $G = SL(n, \mathbb{R})$, $K = SO(n)$, N be the group of upper-triangular $n \times n$ matrices with 1s on the diagonal, and A be the group of diagonal matrices with determinant 1 and positive entries. By the Iwasawa decomposition (theorem 7.6.4), the map

$$(k, a, n) \in K \times A \times N \mapsto kan \in KAN = G$$

is a diffeomorphism.

The upper-triangular group N is diffeomorphic to $\mathbb{R}^{n(n-1)/2}$. For example, if $n = 3$,

$$\theta : \begin{pmatrix} 1 & x & z \\ 0 & 1 & y \\ 0 & 0 & 1 \end{pmatrix} \mapsto (x, y, z).$$

Notice that

$$\theta^{-1}(a, b, c)\theta^{-1}(x, y, z) = \theta^{-1}(x + a, y + b, z + ay + c),$$

$$\theta^{-1}(x, y, z)\theta^{-1}(a, b, c) = \theta^{-1}(x + a, y + b, z + bx + c).$$

Exercise A.1.2 Define the map θ in dimension n. Denote by x_{ij} the $((i, j)$-entry) coordinate function on the space of $n \times n$ matrices. Show that the (image under θ^{-1} of the) Lebesgue measure

$$d\mu = \prod_{1 \leq i < j \leq n} dx_{ij}$$

is both left- and right-invariant. Therefore, N is also a unimodular group. (We remark that, given a continuous map $\theta : X \to Y$ between locally compact spaces and a measure μ on X, then one defines the *pushed-forward* measure $\theta_*\mu$ on Y according to the equation

$$\int_Y \varphi \, d(\theta_*\mu) = \int_X \varphi \circ \theta \, d\mu$$

for all compactly supported continuous functions φ on Y.)

We turn our attention to the abelian subgroup A. Recall that A is the group of $n \times n$ diagonal matrices with positive entries and determinant 1. As a manifold, it is diffeomorphic to $\mathbb{R}^{n-1}$, and we can define a diffeomorphism $\Psi : \mathbb{R}^{n-1} \to A$ as the inverse map of

$$\Psi^{-1}(\mathrm{diag}[a_1, \ldots, a_n]) = \left(\log \frac{a_1}{a_2}, \ldots, \log \frac{a_{n-1}}{a_n} \right).$$

A simple computation shows that

$$\Psi(x_1, \ldots, x_{n-1}) = \exp\left(\sum_{l=i}^{n-1} x_l - \sum_{l=1}^{n-1} l x_l \right).$$

Also observe that the Lebesgue measure on $\mathbb{R}^{n-1}$ (which is translation-invariant) pushes forward under Ψ to a bi-invariant measure on A. (The measure on A is therefore invariant under the transformation that multiplies the ith coordinate of A by a positive constant α_i such that $\prod_{i=1}^n \alpha_i = 1$.) If λ denotes the Lebesgue measure on $\mathbb{R}^{n-1}$, then a (bi-invariant) Haar measure μ_A on A is characterized by the property that for any integrable function ρ on A,

$$\int_A \rho(a) \, d\mu_A(a) = \int (\rho \circ \Psi)(x) \, d\lambda(x).$$

Exercise A.1.3 Given a positive real number t, define the set

$$A_t := \left\{ a = \mathrm{diag}[a_1, \ldots, a_n] \in A \,\middle|\, \frac{a_i}{a_{i+1}} \leq t \right\}.$$

Also define the function $\rho : A \to \mathbb{R}^+$ by $\rho(a) := \prod_{i=1}^{n-1} \left(\frac{a_i}{a_{i+1}} \right)^{r_i}$, where the r_i are positive integers. Show that

$$\int_{A_t} \rho(a) \, d\mu_A(a) = \prod_{i=1}^{n-1} \frac{t^{r_i}}{r_i}.$$

Denote by B the group of upper-triangular matrices of determinant 1 with positive entries on the diagonal. Notice that B is the semidirect product of A and N, where N is the normal subgroup: $B = A \ltimes N$. Denote by $\mu_N, \mu_A, \mu_B, \ldots$, the left Haar measures on the respective groups.

Exercise A.1.4 If μ_H is a left Haar measure on some locally compact group H and $i : H \to H$ is the inverse map, then $i_*\mu_H$ is a right Haar measure. Define the *adjoint map* $\mathrm{Ad}_g := L_g \circ R_{g^{-1}}$. Show that if $a \in A$,

$$(\mathrm{Ad}_a)_*\mu_N = \rho(a)\mu_N,$$

where $\rho(a) = \prod_{1 \leq i < j \leq n} \frac{a_{ii}}{a_{jj}}$. Conclude that a left Haar measure μ_B on B can be obtained as $i_* \nu_B$, where ν_B is a right Haar measure defined as

$$d\nu_B(an) = \rho(a)\, d\mu_A(a)\, d\mu_N(n).$$

Define $\eta : (k, b) \in K \times B \mapsto kb \in G = KB$ and consider the measure

$$\mu = \eta_*(\mu_K \times \nu_B)$$

on G. The measure μ is invariant under left multiplication by K and right multiplication by B. But so is the (bi-invariant) Haar measure μ_G, and since the action

$$((k, b), g) \in (K \times B) \times G \mapsto kgb^{-1} \in G$$

is transitive we conclude that μ and μ_G coincide, up to a multiplicative constant. Therefore, we can write the Haar measure on G as

$$d\mu_G(kan) = \rho(a)\, d\mu_K(k)\, d\mu_A(a)\, d\mu_N(n).$$

Given any positive real number t and compact subset $\omega \subset N$, we define a *Siegel set* $S_{t\omega}$ in $SL(n, \mathbb{R})$ by

$$S_{t\omega} := K A_t \omega,$$

where A_t is the set defined in exercise 1.7. When

$$\omega = N_u := \{n \in N \mid |n_{ij}| \leq u, 1 \leq i < j \leq n\},$$

set $S_{tu} := S_{tN_u}$.

It is immediate that any Siegel set is stable under left multiplication by elements in K.

Exercise A.1.5   Consider the right action of $SL(2, \mathbb{R})$ on the upper half-plane $\mathbb{H} = \{z = x + iy \in \mathbb{C} \mid y > 0\}$, defined as follows: For $z \in \mathbb{H}$ and $g = \left(\begin{smallmatrix} a & b \\ c & d \end{smallmatrix}\right)$ in $SL(2, \mathbb{R})$,

$$z \cdot g := g^{-1}z := -\frac{b - dz}{a - cz}.$$

Define the map

$$\Phi : g \in SL(2, \mathbb{R}) \mapsto i \cdot g \in \mathbb{H}.$$

Show that Φ defines a bijection between B and $\mathbb{H}$. Draw a picture of the set $\Phi(S_{\frac{2}{\sqrt{3}}, \frac{1}{2}})$. Verify that $\Phi(S_{\frac{2}{\sqrt{3}}, \frac{1}{2}})$ contains a fundamental domain for the action of $SL(2, \mathbb{Z})$ on $\mathbb{H}$. We observe that the set

$$\mathcal{F} := \left\{ z = x + iy \in \mathbb{C} \ \middle|\ -\frac{1}{2} \leq x \leq \frac{1}{2}, |z| \geq 1 \right\}$$

is the closure of a fundamental domain. This can be seen by using the (easily obtained) fact that $SL(2, \mathbb{Z})$ is generated by the two matrices

$$\begin{pmatrix} 1 & 1 \\ 0 & 1 \end{pmatrix} \quad \text{and} \quad \begin{pmatrix} 0 & 1 \\ -1 & 0 \end{pmatrix}.$$

The theorem will be proved once it is shown that, for some t and u, (i) $G = S_{tu}\Gamma$ and (ii) S_{tu} has finite measure.

Lemma A.1.6 If $t \geq \frac{2}{\sqrt{3}}$ and $u \geq \frac{1}{2}$, then $G = S_{tu}\Gamma$, where $G = SL(n, \mathbb{R})$ and $\Gamma = SL(n, \mathbb{Z})$.

Proof. The proof is by induction. If $n = 1$ there is nothing to prove since $G = \{I\}$. Assume the lemma is proved for all $l < n$. Define for each $g \in G$ the function

$$\Phi_g : x \in \mathbb{R}^n \mapsto \|gx\| \in [0, \infty),$$

where $\| \cdot \|$ is the Euclidean norm of $\mathbb{R}^n$. Φ_g is a proper map since

$$\|g^{-1}\|^{-1}\|x\| \leq \|gx\| \leq \|g\| \, \|x\|.$$

Therefore, Φ_g assumes a minimum value ν on the set $\mathbb{Z}^n \backslash \{0\}$. Let $e \in \mathbb{Z}^n \backslash \{0\}$ be such that $\|ge\| = \nu$.

Let P_i be the $n \times n$ matrix in G that permutes the standard basis vector e_1 with $(-1)^i e_i$, keeping the other basis vectors fixed; and let $M_{ij}(\epsilon)$ be the matrix in N with entry $m_{ij} = \epsilon \in \{\pm 1\}$ and all other nondiagonal entries 0. Then by applying a sequence of such matrices to e (and using the remark that if $e = me'$, for some $e' \in \mathbb{Z}^n$, then $m = \pm 1$, due to the minimality of $\Phi_g(e)$), we conclude that there exists $\gamma \in \Gamma$ such that $\gamma e_1 = \pm e$. Therefore, for every $e' \in \mathbb{Z}^n \backslash \{0\}$, $\|g\gamma e_1\| \leq \|ge'\|$. But, as left translation by γ^{-1} is a bijection of $\mathbb{Z}^n \backslash \{0\}$, we have that $g' := g\gamma$ satisfies

$$\|g'e_1\| \leq \|g'e'\|$$

for all $e' \in \mathbb{Z}^n \backslash \{0\}$. In order to prove the lemma, it suffices to show that for any $g \in G$ (and g' as before), there exists $\gamma' \in \Gamma$ such that $g'\gamma' \in S_{tu}$.

Decompose g' according to the Iwasawa decomposition $g' = k'a'n'$. It clearly suffices to show that there exists γ' such that $h\gamma' \in S_{tu}$, where $h := a'n'$. Observe that $\|he_1\| \leq \|he'\|$ for all $e' \in \mathbb{Z}^n \backslash \{0\}$, since k is an isometry.

We write

$$h = \begin{pmatrix} \alpha & x \\ 0 & \beta T \end{pmatrix},$$

where T is a $(n-1) \times (n-1)$ matrix with determinant 1 and $\beta^{n-1} = \alpha^{-1} > 0$. By the induction hypothesis, there exists $\gamma' = \begin{pmatrix} 1 & 0 \\ 0 & \gamma'' \end{pmatrix} \in \Gamma$ such that $h\gamma' = kan$, where

$$a = \text{diag}[\alpha, \beta a_2', \ldots, \beta a_n'], \quad \text{with } a_i'/a_{i+1}' \leq 2/\sqrt{3}.$$

Choose $e' = e_2 + me_1$, where $m \in \mathbb{Z}$ will be specified later. If $n \in N$, then there exists $y \in \mathbb{R}$ such that

$$n(e_2 + me_1) = e_2 + (m + y)e_1,$$

as a simple computation shows. Choose m such that $|m + y| \leq \frac{1}{2}$. Therefore

$$
\begin{aligned}
a_{11}^2 = \|ae_1\| = \|kane_1\|^2 &= \|h\gamma'e_1\|^2 \\
&= \|he_1\|^2 \\
&\leq \|h\gamma'e'\|^2 \\
&= \|kane'\| \\
&= \|a(e_2 + (m+y)e_1)\|^2 \\
&= \|a_{22}e_2 + (m+y)a_{11}e_1\|^2 \\
&= a_{22}^2 + (m+y)^2 a_{11}^2.
\end{aligned}
$$

Therefore

$$
\frac{3}{4} a_{11}^2 \leq a_{22}^2,
$$

from which it follows that $\frac{a_{11}}{a_{22}} \leq \frac{2}{\sqrt{3}}$. We have thus shown that

$$
G = K A_{\frac{2}{\sqrt{3}}} N \Gamma.
$$

The lemma is now a consequence of the next exercise.   □

Exercise A.1.7 Show (by induction) that $N = N_{\frac{1}{2}}(\Gamma \cap N)$. Since $N_{\frac{1}{2}}$ is compact, this shows that $\Gamma \cap N$ is a uniform lattice in N.

All that remains to show in order to prove the theorem is that a Siegel domain S_{tu} has finite measure. Define $\eta : (k, b) \in K \times B \mapsto kb \in G = KB$ and consider the measure $\mu = \eta_*(\mu_K \times \nu_B)$ on G (recall from exercise 1.7 the definition of the right Haar measure ν_B on B). The measure μ is invariant under left multiplication by K and right multiplication by B. But so is the (bi-invariant) Haar measure μ_G, and since the action

$$
((k, b), g) \in (K \times B) \times G \mapsto kgb^{-1} \in G
$$

is transitive we can conclude that μ and μ_G coincide, up to a multiplicative constant. Therefore, we can write the Haar measure on G as

$$
d\mu_G(kan) = \rho(a) \, d\mu_K(k) \, d\mu_A(a) \, d\mu_N(n),
$$

so that

$$
\mu_G(S_{tu}) = \mu_K(K)\mu_N(N_u) \int_{A_t} \rho(a) \, d\mu_A(a),
$$

which is finite, due to the previous exercises. This concludes the proof of the theorem.

Exercise A.1.8 Show that $SL(n, \mathbb{R})/SL(n, \mathbb{Z})$ is not compact for all $n \geq 2$. (The following sketch of a proof points to the general fact that the presence of unipotent elements in the lattice prevents it from being cocompact. An element u of $GL(n, \mathbb{C})$ is said to be *unipotent* if

$$
(u - I)^k = 0
$$

for some positive integer k. For example, $u = \left(\begin{smallmatrix} 1 & 1 \\ 0 & 1 \end{smallmatrix} \right) \in SL(2, \mathbb{R})$ is unipotent. Denote by τ the matrix in $SL(n, \mathbb{R})$ that has u in the upper left corner, 1s on the diagonal, and

0s in all the other places. Let a_t be the matrix in $SL(n, \mathbb{R})$ that has $\begin{pmatrix} t^{-1} & 0 \\ 0 & t \end{pmatrix}$ in the upper left corner, 1s as the other diagonal entries, and 0s as all the other entries. Then verify that $\lim_{t \to \infty} a_t \tau a_t^{-1} = e$. Show, by contradiction, that $\pi(a_t)$ cannot have a limit point. Notice that if a sequence $\pi(a_{t_i})$ converges to some $\pi(h)$, there would be a sequence $\theta_i \in \Gamma$ such that $a_{t_i} \theta_i \to z$. Use now the fact that $w_i := a_{t_i} \tau a_{t_i}^{-1}$ approaches e to conclude that $\theta_i^{-1} \tau \theta_i$ also approaches e. But that would mean that $\theta_i^{-1} \tau \theta_i = e$ for large i, which cannot be since τ is not the identity element.)

A.2 Cocompact Lattices

The group $SL(n, \mathbb{R})$, as well as the other connected semisimple Lie groups, also contains uniform lattices. In the rest of the section we show how one such lattice can be obtained. The construction is essentially arithmetical, and it will be important to view $SL(n, \mathbb{R})$ not simply as a Lie group, but as an algebraic linear group, with some underlying $\mathbb{Q}$-structure.

The following proposition, taken from [28], gives the basic criterion for a lattice to be cocompact.

Theorem A.2.1 Let G be a locally compact second countable group and Γ a lattice subgroup. Denote by $\pi : G \to G/\Gamma$ the natural projection, and let g_m be a sequence in G. Then $\pi(g_m)$ has no convergent subsequence if and only if there exists a sequence $\gamma_m \in \Gamma$ such that

$$\lim_{m \to \infty} g_m \gamma_m g_m^{-1} = e.$$

If $G(\mathbb{R})$ is the set of real points of a linear algebraic group G defined over $\mathbb{Q}$ and $\Gamma = G(\mathbb{Z})$, then for large m, γ_m must be unipotent. In particular, if Γ does not have unipotent elements, it is a cocompact lattice.

Proof. We first show that γ_m must be unipotent for large m. Denote by $P_m(T) \in \mathbb{C}[T]$ the characteristic polynomial of γ_m. Then

$$P_m(\lambda) = \det(\gamma_m - \lambda e) = \det\left(g_m \gamma_m g_m^{-1} - \lambda e\right)$$
$$\to \det((1 - \lambda)e)$$
$$= (1 - \lambda)^n.$$

But $P_m(T)$ has integer coefficients so, for large m, $P_m(\lambda) = (1 - \lambda)^n$ and γ_m is unipotent.

Suppose that g_m is a sequence such that $\pi(g_m)$ does not admit a convergent subsequence. Let B_m be an increasing family of compact subsets of G such that $\bigcup_{m=1}^{\infty} B_m = G$. Denote by μ the Haar measure on G as well as the G-invariant measure on G/Γ. Since G/Γ has finite measure, we have

$$\epsilon_m := \mu(G/\Gamma \setminus \pi(B_m)) \to 0.$$

Choose a fundamental system of compact neighborhoods of e such that

$$\mu(V_m) > \epsilon_m,$$

and define $W := V_m^{-1} V_m$, also a fundamental system of compact neighborhoods of e. For every positive integer l there exists n_l such that $\pi(g_m)$ belongs to the complement of the compact set $\pi(V_l^{-1} V_l B_l)$ for all $m \geq n_l$. We claim that for all such m, $\pi(V_l g_m) \cap \pi(V_l B_l)$ is empty. In fact, if that is not the case there would be an element $h \in G$ such that $h = v g_m \in V_l g_m$ and $h = v' b \gamma$ with $(v', b, \gamma) \in V_l \times B_l \times \Gamma$. Therefore $v^{-1} v' b \gamma = g_m$, so $\pi(g_m) \in \pi(V_l^{-1} V_l B_l)$, which is a contradiction.

Observe that

$$\mu(\pi(V_l g_m)) \leq \mu\left(\pi(V_l B_l)^c\right) \leq \mu\left(\pi(V_l)^c\right) = \epsilon_l.$$

On the other hand, $\mu(V_l g_m) > \epsilon_l$. Therefore, we can find $v, v' \in V_l$ such that $v g_m = v' g_m \gamma_m$. Thus, we have found for any $m \geq n_l$ an element $\gamma_m \in \Gamma$ such that $\gamma_m \neq e$ and $g_m \gamma_m g_m^{-1} = v^{-1} v' \in W_l$. As $l \to \infty$, $W_l \to e$.

Conversely, suppose that $g_m \in G$ is a sequence for which we can find $\gamma_m \in \Gamma$, $\gamma_m \neq e$, such that $g_m \gamma_m g_m^{-1} \to e$. For a contradiction, assume that the sequence $\pi(g_m)$ has a limit point $\pi(z)$ for some $z \in G$. Passing to a subsequence, if necessary, we can find $\theta_m \in \Gamma$ such that $g_m \theta_m$ converges to z. But then,

$$w_n := g_m \gamma_m g_m^{-1} = (g_m \theta_m)\left(\theta_m^{-1} \gamma_m \theta_m\right)\left(\theta_m^{-1} g_m^{-1}\right)$$

converges to e, so

$$\varphi_m := \theta_m^{-1} \gamma_m \theta_m$$

converges to $z^{-1} e z = e$. Therefore, as $\varphi_m \in \Gamma$, we must have $\gamma_m = e$ for large m, so $\gamma_m = e$ for large m. But this is a contradiction. $\square$

The remainder of the section is dedicated to constructing a lattice in the special linear group that is without unipotents, hence cocompact.

The Lie algebra of $SL(n, \mathbb{C})$ is the linear subspace of $M(n, \mathbb{C})$ consisting of traceless matrices:

$$\mathbf{g}(\mathbb{C}) = \mathfrak{sl}_n(\mathbb{C}) = \{A \in M(n, \mathbb{C}) \mid \mathrm{Tr}\, A = 0\}$$

(with Lie bracket $[A, B] = AB - BA$). Recall the *Cartan involution* $\theta(A) := -A^*$, where $*$ denotes the conjugate transpose of a matrix. We denote by A^t the transpose of A. Also recall that θ is an automorphism of the Lie algebra $\mathbf{g}(\mathbb{C})$, that is,

$$\theta([A, B]) = [\theta(A), \theta(B)]$$

for $A, B \in \mathbf{g}(\mathbb{C})$.

The next exercise reviews some material from the beginning of chapter 7.

Exercise A.2.2   Consider the Lie algebra $\mathbf{g} = \mathfrak{sl}(n, \mathbb{R})$ of $SL(n, \mathbb{R})$, which consists of $n \times n$ real matrices with zero trace. The Cartan involution θ defined earlier, acting on the linear space $\mathbf{g}$, has eigenspace decomposition

$$\mathbf{g} = \mathfrak{k} \oplus \mathfrak{p},$$

where $\mathfrak{k}$ is the eigenspace associated with eigenvalue 1 and $\mathfrak{p}$ is associated with eigenvalue -1. Check that $\mathfrak{k}$ is the Lie algebra of $SO(n)$ and $\mathfrak{p}$ is the space of symmetric matrices (not a subalgebra). Moreover,

$$[\mathfrak{k}, \mathfrak{k}] \subset \mathfrak{k},$$

$$[\mathfrak{k}, \mathfrak{p}] \subset \mathfrak{p},$$

$$[\mathfrak{p}, \mathfrak{p}] \subset \mathfrak{k}.$$

Show that

$$\mathfrak{g}_u := \mathfrak{k} \oplus \sqrt{-1}\mathfrak{p}$$

is the Lie algebra of the compact Lie group (the *special unitary group*)

$$SU(n) = \{A \in SL(n, \mathbb{C}) \mid A^* A = I\}.$$

Denote by E_{ij} the $n \times n$ matrix with 1 as the (ij) entry and 0s as all the other entries. For each pair $1 \leq i, j \leq n$ define

$$X_{ij} := E_{ij} - E_{ji}, \qquad Y_{ij} := E_{ij} + E_{ji}, \qquad H_{ij} := E_{ii} - E_{jj}.$$

Then the set

$$\{X_{ij}, Y_{ij}, H_{i,i+1} \mid 1 \leq i < j \leq n\}$$

constitutes a basis of $\mathfrak{g}(\mathbb{C})$ such that $\{X_{ij} \mid 1 \leq i < j \leq n\}$ is a basis of $\mathfrak{k}$, and $\{Y_{ij}, H_{i,i+1} \mid 1 \leq i < j \leq n\}$ is a basis of $\mathfrak{p}$. We rename and reindex the vectors so that

$$\{X_{ij} \mid 1 \leq i < j \leq n\} =: \{U_\alpha\}_{\alpha \in I},$$

$$\{Y_{ij}, H_{i,i+1} \mid 1 \leq i < j \leq n\} =: \{V_\alpha\}_{\alpha \in J}.$$

The bracket relations then become

$$[U_\alpha, U_\beta] = \sum_\gamma A_{\alpha\beta}^\gamma U_\gamma, \qquad [V_\alpha, V_\beta] = \sum_\gamma B_{\alpha\beta}^\gamma U_\gamma, \qquad [U_\alpha, V_\beta] = \sum_\gamma C_{\alpha\beta}^\gamma V_\gamma,$$

and notice also that

$$\theta(U_\alpha) = U_\alpha, \qquad \theta(V_\alpha) = -V_\alpha.$$

The main conclusion to draw from the preceding relations is that all the coefficients involved are rational.

Define $u := \sqrt{2}$ and $v := 2^{\frac{1}{4}}$ and consider the field $\mathbb{Q}(u)$. Denote by σ a field automorphism of $\mathbb{C}$ such that $\sigma(u) = -u$. Define $\mathbb{Q}(u)$-linear spaces $\mathfrak{m}$ and $\mathfrak{m}_u$ as follows:

$$\mathfrak{m} := \mathbb{Q}(u)\text{-linear span of } \{\tilde{U}_\alpha := U_\alpha, \ \tilde{V}_\alpha := vV_\alpha\},$$

$$\mathfrak{m}_u := \mathbb{Q}(u)\text{-linear span of } \{\tilde{U}_\alpha := U_\alpha, \ \tilde{V}_\alpha := \sqrt{-1}vV_\alpha\}.$$

It follows from the previous exercise that by extending the field from $\mathbb{Q}(\sqrt{2})$ to $\mathbb{R}$ we obtain,

$$\mathfrak{m} \otimes_{\mathbb{Q}(\sqrt{2})} \mathbb{R} = \mathfrak{sl}_n(\mathbb{R}) =: \mathfrak{g}$$

$$\mathfrak{m}_u \otimes_{\mathbb{Q}(\sqrt{2})} \mathbb{R} = \mathfrak{su}(n) =: \mathfrak{g}_u.$$

The Lie brackets for $\mathfrak{m}$ are

$$[\tilde{U}_\alpha, \tilde{U}_\beta] = \sum_\gamma A_{\alpha\beta}^\gamma \tilde{U}_\gamma, \qquad [\tilde{V}_\alpha, \tilde{V}_\beta] = \sum_\gamma u B_{\alpha\beta}^\gamma \tilde{U}_\gamma, \qquad [\tilde{U}_\alpha, \tilde{V}_\beta] = \sum_\gamma C_{\alpha\beta}^\gamma \tilde{V}_\gamma$$

and those for $\mathfrak{m}_u$ are

$$[\tilde{U}_\alpha, \tilde{U}_\beta] = \sum_\gamma A_{\alpha\beta}^\gamma \tilde{U}_\gamma, \qquad [\tilde{V}_\alpha, \tilde{V}_\beta] = \sum_\gamma -u B_{\alpha\beta}^\gamma \tilde{U}_\gamma, \qquad [\tilde{U}_\alpha, \tilde{V}_\beta] = \sum_\gamma C_{\alpha\beta}^\gamma \tilde{V}_\gamma.$$

Observe that the coefficients of the second set of equations are the image under σ of the coefficients of the first.

We now define $\mathfrak{h} := \mathfrak{sl}_n(\mathbb{R}) \oplus \mathfrak{su}(n)$ and choose for $\mathfrak{h}$ the basis whose elements are

$$\xi_\alpha := (\tilde{U}_\alpha, \bar{U}_\alpha), \qquad \check{\xi}_\alpha := (u\tilde{U}_\alpha, -u\bar{U}_\alpha),$$

$$\eta_\alpha := (\tilde{V}_\alpha, \bar{V}_\alpha), \qquad \check{\eta}_\alpha := (u\tilde{V}_\alpha, -u\bar{V}_\alpha).$$

A tedious computation gives the bracket relations for $\mathfrak{h}$ (the key point to note is that the coefficients are all rational):

$$[\xi_\alpha, \xi_\beta] = \sum_\gamma A^\gamma_{\alpha\beta} \xi_\gamma, \qquad [\check{\xi}_\alpha, \xi_\beta] = \sum_\gamma 2A^\gamma_{\alpha\beta} \check{\xi}_\gamma,$$

$$[\xi_\alpha, \check{\xi}_\beta] = \sum_\gamma A^\gamma_{\alpha\beta} \check{\xi}_\gamma, \qquad [\eta_\alpha, \eta_\beta] = \sum_\gamma B^\gamma_{\alpha\beta} \check{\xi}_\gamma,$$

$$[\check{\eta}_\alpha, \check{\eta}_\beta] = \sum_\gamma 2B^\gamma_{\alpha\beta} \check{\xi}_\gamma, \qquad [\eta_\alpha, \check{\eta}_\beta] = \sum_\gamma 2B^\gamma_{\alpha\beta} \xi_\gamma,$$

$$[\xi_\alpha, \eta_\beta] = \sum_\gamma C^\gamma_{\alpha\beta} \eta_\gamma, \qquad [\check{\xi}_\alpha, \check{\eta}_\beta] = \sum_\gamma 2C^\gamma_{\alpha\beta} \eta_\gamma,$$

$$[\xi_\alpha, \check{\eta}_\beta] = \sum_\gamma C^\gamma_{\alpha\beta} \check{\eta}_\gamma, \qquad [\check{\xi}_\alpha, \eta_\beta] = \sum_\gamma C^\gamma_{\alpha\beta} \check{\eta}_\gamma.$$

Denote by $\mathfrak{h}_\mathbb{Q}$ the $\mathbb{Q}$-linear span of the set $\{\xi_\alpha, \check{\xi}_\alpha, \eta_\alpha, \check{\eta}_\alpha\}$. We see from the foregoing multiplication table that $\mathfrak{h}_\mathbb{Q}$ is a $\mathbb{Q}$-Lie subalgebra of $\mathfrak{h}$ such that $\mathfrak{h}_\mathbb{Q} \otimes_\mathbb{Q} \mathbb{R} = \mathfrak{h}$. Identify the Lie algebra $\mathfrak{h}_\mathbb{C} = \mathfrak{h} \otimes_\mathbb{R} \mathbb{C}$ with $\mathbb{C}^N$, $N = 2(n^2 - 1)$, so that the basis $\{\xi_\alpha, \check{\xi}_\alpha, \eta_\alpha, \check{\eta}_\alpha\}$ is made to correspond to the standard basis $\{e_1, \ldots, e_N\}$.

Before moving further, we recall that for any given connected semisimple (real) Lie group H with finite center $Z(H)$, $H/Z(H)$ has a natural structure of a *linear algebraic real group*. Let $\mathfrak{h}$ be the Lie algebra of H and $\mathfrak{h}_\mathbb{C} := \mathfrak{h} \otimes_\mathbb{R} \mathbb{C}$ its complexification. Identify $\mathfrak{h}_\mathbb{C}$ with $\mathbb{C}^N$ ($N = \dim_\mathbb{R} \mathfrak{h}$) by means of a basis of $\mathfrak{h}$. Denote by $H \subset GL(N, \mathbb{C})$ the group of Lie algebra automorphisms of $\mathfrak{h}_\mathbb{C}$. H is a linear algebraic k-group, where k is any subfield of $\mathbb{R}$ that contains the structure constants of the Lie algebra (see section 7.2).

The group $H(\mathbb{R})$ has Lie algebra $\mathfrak{h}$. In fact, the Lie algebra of the group of automorphisms of $\mathfrak{h}$ is the Lie algebra of derivations of $\mathfrak{h}$, which for a semisimple $\mathfrak{h}$ is canonically isomorphic to $\mathfrak{h}$ itself (theorem 7.2.5).

Denote by $\mathrm{Ad} : H \to H$ the adjoint representation of H, and recall that we are identifying $GL(\mathfrak{h}_\mathbb{C})$ and $GL(N, \mathbb{C})$. Given $X \in \mathfrak{h}$ and $h \in H$, define $\mathrm{Ad}(h)X = hXh^{-1}$. $Z(H)$ is then in the kernel of the adjoint map and $H/Z(H)$ is isomorphic to the identity component of $H(\mathbb{R})$. For example, if $\mathfrak{h} = \mathfrak{sl}_n(\mathbb{R}) \oplus \mathfrak{su}(n)$,

$$H(\mathbb{R})^0 = PSL(n, \mathbb{R}) \times PSU(n),$$

where $PSL(n, \mathbb{R})$ (resp., $PSU(n)$) is defined as the quotient of $SL(n, \mathbb{R})$ (resp., $SU(n)$) by the center. Another general remark is that for any algebraic linear group H defined over $\mathbb{R}$, $H(\mathbb{R})$ has only finitely many connected components (proposition 4.7.1).

Let H be as before. Let $G \subset GL(n^2 - 1, \mathbb{C})$ be an algebraic subgroup defined over $\mathbb{Q}(\sqrt{2})$ consisting of automorphisms of $\mathfrak{sl}_n(\mathbb{C}) = \mathfrak{m} \otimes_{\mathbb{Q}(\sqrt{2})} \mathbb{C}$, where $\mathfrak{m} \otimes_{\mathbb{Q}(\sqrt{2})} \mathbb{C}$ is identified with $\mathbb{C}^{n^2 - 1}$ by means of a basis contained in $\mathfrak{m}$. Similarly, define

$G_u \subset GL(n^2 - 1, \mathbb{C})$ as a group of automorphisms of $\mathfrak{m}_u \otimes_{\mathbb{Q}(\sqrt{2})} \mathbb{C}$, also a $\mathbb{Q}(\sqrt{2})$-group. Then $G(\mathbb{R})^0 = PSL(n, \mathbb{R})$ and $G_u(\mathbb{R})^0 = PSU(n)$.

We now set $H_1 = H(\mathbb{R})^0$ and define $\Gamma' := H(\mathbb{Z}) \cap H_1$. Let Γ be the preimage under the map $SL(n, \mathbb{R}) \to PSL(n, \mathbb{R})$ of the projection of Γ' by the map $\pi_1 : H_1 \to PSL(n, \mathbb{R})$.

Theorem A.2.3 Γ is a cocompact lattice in $SL(n, \mathbb{R})$.

Proof. As the surjective homomorphism π_1 has compact kernel, any discrete cocompact subgroup of H_1 must project onto a discrete cocompact subgroup of the factor group. Moreover, the preimage under the finite covering map $SL(n, \mathbb{R}) \to PSL(n, \mathbb{R})$ of a cocompact lattice in $PSL(n, \mathbb{R})$ is a cocompact lattice in $SL(n, \mathbb{R})$. Therefore, it is sufficient to prove that Γ' is a cocompact lattice in H_1.

From theorem 8.1.9 due to Borel and Harish-Chandra, Γ' is a lattice in H_1. We claim that Γ' does not contain unipotent elements. To see that, observe that Γ' is the subset of H_1 that acts on $\mathbb{R}^N$ preserving the integer lattice $\mathbb{Z}^N$. In other words, the action of Γ' on $\mathfrak{h}$ stabilizes the set

$$L := \left\{ \xi = \sum_\alpha (m_\alpha \xi_\alpha + \check{m}_\alpha \check{\xi}_\alpha + n_\alpha \eta_\alpha + \check{n}_\alpha \check{\eta}_\alpha) \in \mathfrak{h} \,\middle|\, m_\alpha, \check{m}_\alpha, n_\alpha, \check{n}_\alpha \in \mathbb{Z} \right\}.$$

Note that L projects under $p_1 : \mathfrak{h} \to \mathfrak{sl}_n(\mathbb{R})$ and $p_2 : \mathfrak{h} \to \mathfrak{su}(n)$ onto subsets of $\mathfrak{m}$ and $\mathfrak{m}_u$:

$$p_1(L) = \left\{ \sum_\alpha [(m_\alpha + \check{m}_\alpha \sqrt{2})\tilde{U}_\alpha + (n_\alpha + \check{n}_\alpha \sqrt{2})\tilde{V}_\alpha] \,\middle|\, m_\alpha, \check{m}_\alpha, n_\alpha, \check{n}_\alpha \in \mathbb{Z} \right\},$$

$$p_2(L) = \left\{ \sum_\alpha [(m_\alpha - \check{m}_\alpha \sqrt{2})\bar{U}_\alpha + (n_\alpha - \check{n}_\alpha \sqrt{2})\bar{V}_\alpha] \,\middle|\, m_\alpha, \check{m}_\alpha, n_\alpha, \check{n}_\alpha \in \mathbb{Z} \right\}.$$

It is clear from the preceding description of $p_1(L)$ and $p_2(L)$ that the restrictions of p_1 and p_2 to L are injective. Therefore, $\pi_1 : H_1 \to PSL(n, \mathbb{R})$ and $\pi_2 : H_1 \to PSU(n)$ restricted to Γ' are injective maps. Moreover, if $\gamma \in \Gamma'$ is unipotent, the same is true for $\pi_2(\gamma)$. But $\pi_2(\gamma)$ belongs to a compact group, so it is unipotent only when it is the identity. (Observe that every element of a linear compact group is diagonalizable over $\mathbb{C}$.) □

References

[1] W. Ballmann. *Lectures on Spaces of Nonpositive Curvature*. DMV Seminar, vol. 25. Birkhäuser, Basel, 1995.

[2] P. Billingsley. *Probability and Measure*. 2nd ed. Wiley, New York, 1986.

[3] A. Borel. Compact Clifford–Klein forms of symmetric spaces. *Topology* 2, (1963):111–22.

[4] A. Borel. *Introduction aux Groupes Arithmétiques*. Hermann, Paris, 1969.

[5] A. Borel. *Linear Algebraic Groups*. Springer-Verlag, New York, 1991.

[6] A. Borel and Harish-Chandra. Arithmetic subgroups of algebraic groups. *Ann. Math.* 75 (1962):485–535.

[7] R. Bowen. *Equilibrium States and the Ergodic Theory of Anosov Diffeomorphism*. Springer Lecture Notes in Math., no. 470. Springer-Verlag, New York, 1975.

[8] L. Conlon. *Differentiable Manifolds: A First Course*. Birkhäuser, Boston, 1993.

[9] I. P. Cornfeld, S. V. Fomin, and Ya. G. Sinai. *Ergodic Theory*. Springer-Verlag, New York, 1982.

[10] R. Ellis and M. Nerurkar. Weakly almost periodic flows. *Trans. of the A.M.S.* 313, no. 1 (1989):103–19.

[11] R. Feres and F. Labourie. Topological superrigidity and Anosov actions of lattices. Preprint, 1997.

[12] N. Jacobson. *Lie Algebras*. Dover, New York, 1979.

[13] R. Horn and C. Johnson. *Matrix Analysis*. Cambridge University Press, New York, 1985.

[14] M. Greenberg and J. Harper. *Algebraic Topology: A First Course*. Mathematics Lecture Note Series. Benjamin Cummings, Menlo Park, CA, 1981.

[15] M. Gromov. Rigid transformation groups. In *Géométric différentielle*, ed. D. Bernard and Y. Choquet-Bruhat, Travaux en cours, vol. 33, 65–139. Hermann, Paris, 1988.

[16] A. Katok and B. Hasselblatt. *Introduction to the Modern Theory of Dynamical Systems*. Cambridge University Press, New York, 1995.

[17] A. Knapp. *Lie Groups Beyond an Introduction*. Birkhäuser, Boston, 1996.

[18] R. Mañé. *Ergodic Theory and Differentiable Dynamics*. Springer-Verlag, New York, 1987.

[19] G. A. Margulis. *Discrete Subgroups of Semisimple Lie Groups*. Springer-Verlag, New York, 1991.

[20] A. L. Onishchik and E. B. Vinberg. *Lie Groups and Algebraic Groups*. Springer-Verlag, New York, 1990.

[21] R. Palais. A global formulation of the Lie theory of transformation groups. *Memoirs of the American Math. Soc.*, no. 22 (1957).

[22] Ya. Pesin. Families of invariant manifolds corresponding to non-zero characteristic exponents. *Math. USSR Izvestija* 10, no. 6 (1976):1261–1305.

[23] Ya. Pesin. Characteristic Lyapunov exponents and smooth ergodic theory. *Russian Math. Surveys* 32, no. 4 (1977):55–114.

[24] K. Petersen. *Ergodic Theory*. Cambridge University Press, New York, 1983.

[25] R. Phelps. *Lectures on Choquet's Theorem*. Van Nostrand, New York, 1966.

[26] V. Platonov and A. Rapinchuk. *Algebraic Groups and Number Theory*. Academic Press, San Diego, 1994.

[27] M. Pollicott. *Lectures on Ergodic Theory and Pesin Theory on Compact Manifolds.* London Mathematical Society Lecture Note Series, no. 180. Cambridge University Press, New York, 1993.

[28] M. S. Raghunathan. *Discrete Subgroups of Lie Groups*. Springer-Verlag, New York, 1970.

[29] M. Rosenlicht. A remark on quotient spaces. *Anais da Academia Brasileira de Ciencias* 35, no. 4 (1963):487–9.

[30] I. R. Shafarevich. *Basic Algebraic Geometry*. Springer-Verlag, New York, 1977.

[31] V. S. Varadarajan. *Lie Groups, Lie Algebras, and Their Representation Theory.* Springer-Verlag, New York, 1984.

[32] V. S. Varadarajan. *Geometry of Quantum Theory*. Springer-Verlag, New York, 1985.

[33] P. Walters. *An Introduction to Ergodic Theory*. Springer-Verlag, New York, 1982.

[34] F. Warner, *Foundations of Differentiable Manifolds and Lie Groups*. Springer-Verlag, New York, 1983.

[35] H. Whitney. The elementary structure of real algebraic varieties. *Annals of Math.* 66 (1957):545–56.

[36] R. Zimmer. *Ergodic Theory and Semisimple Groups*. Monographs in Mathematics. Birkhäuser, Boston, 1984.

[37] R. Zimmer. Actions of Semisimple Groups and Discrete Subgroups. In *Proc. Int. Congr. of Mathematicians (1986: Berkeley, CA)*, ed. A. M. Gleason, 1247–58. AMS, Providence, RI, 1986.

[38] R. Zimmer. Ergodic theory and the automorphism group of a G-structure. In *Group Representations, Ergodic Theory, Operator Algebras, and Mathematical Physics*, ed. C. C. Moore, 247–78. Springer-Verlag, New York, 1987.

[39] R. Zimmer. On the algebraic hull of an automorphism group of a principle bundle. *Comment. Math. Helvetici* 65 (1990):375–87.

Index

geometric structure, 92
geometric structure of algebraic type, 92
group actions, 1
group generated by one-parameter subgroups, 73

H-pair, 200
H-translation, 200
Haar measure, 150
Hausdorff metric, 118
homogenization, 62
homomorphism of principal bundles, 91
hyperbolic set, 12

induced action, 8
infinitesimal action, 43
inner derivation, 123
invariant H-pair, 204
invariant measure, 17
invariant set, 5
irreducible algebraic set, 64
irreducible component, 64
irreducible G-space, 156
irreducible representation, 51
isometry, 97
isomorphic measure-preserving
    transformations, 30
isotropy subgroup, 3
Iwasawa decomposition, 136

Jordan decomposition, 74

Kazhdan property, 167
kernel of an action, 3
Killing form, 123
Kolmogorov–Sinai entropy, 29

lattice, 153
left-action, 1
Levi decomposition, 76
Lie algebra, 41
Lie group, 41
linear algebraic group, 69
linear representation, 50
local action, 43
local product structure, 14
local stable manifold, 13
local trivialization, 88
local unstable manifold, 13
locally closed set, 4
locally effective action, 3
locally free action, 3
Lyapunov exponents, 174

Markov chain, 23
maximal H-pair, 204
measurable map, 17

measurable sets, 17
measure, 17
measure-preserving action, 22
measure-preserving map, 17
measure-preserving transformation, 18
measure-theoretic entropy, 29
minimal action, 5
modular function, 150
Moore's ergodicity theorem, 156
morphism, 65
multiplicative ergodic theorem, 178

nilpotent group, 75
nilpotent Lie algebra, 130
Noetherian ring, 63
nonsingular point, 66
nonuniform (completely) hyperbolic system, 193
nonuniform partially hyperbolic system, 193
nonwandering point, 14

orbit, 1
orbit space, 2, 3
ordinary topology of a variety, 68

periodic point, 5
Pesin set, 193
plaque, 40
polar decomposition, 128
polynomial functions, 64
polynomial map, 65
principal bundle, 90
principal open set, 61
probability measures, 17
projective variety, 62
proper action, 3
property T, 167
pseudo-Riemannian metric, 98
pull-back of a bundle, 89

quasi-invariant measure, 17
quasi-projective variety, 62
quotient topology, 3

radical, 75
Radon–Nikodym derivative, 19
Radon–Nikodym theorem, 19
rational functions, 64
rational map, 65
rationally independent numbers, 6
real algebraic action, 59
real algebraic group, 59, 70
real algebraic one-parameter group, 73
real points, 60
real rank, 131
real variety, 59
recurrent point, 5